A User's Guide to Selecting Electronic Components

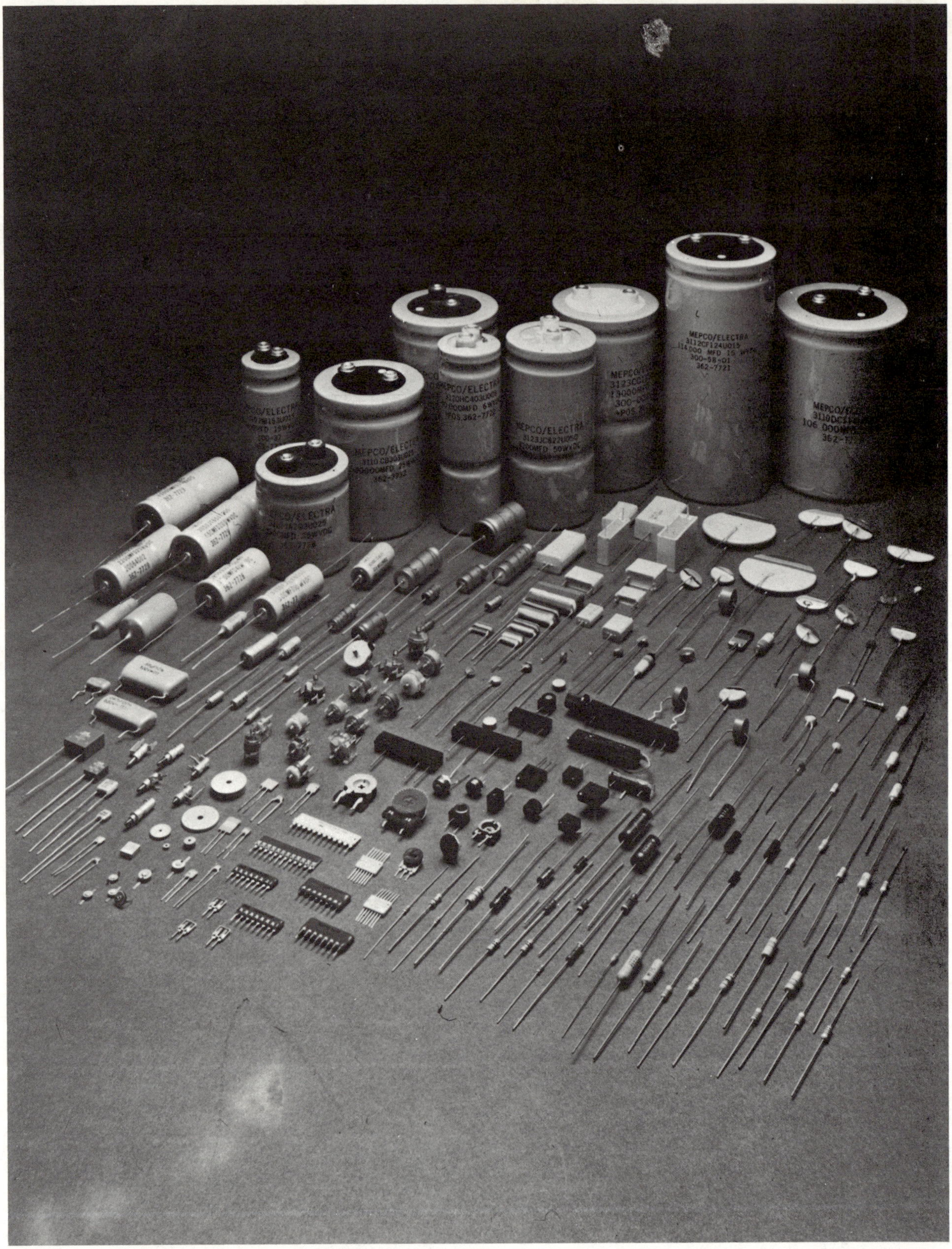

Mepco/Electra, Inc.

A User's Guide to Selecting Electronic Components

GERALD L. GINSBERG

Magnavox Government and Industrial Electronics Company General Atronics Corporation, a subsidiary

A WILEY-INTERSCIENCE PUBLICATION

JOHN WILEY & SONS, New York • Chichester • Brisbane • Toronto • Singapore

Library of Congress Cataloging in Publication Data:

Ginsberg, Gerald L
A user's guide to selecting electronic components.

"A Wiley-Interscience publication."
Bibliography: p.
Includes index.
1. Electronic apparatus and appliances–Handbooks, manuals, etc. I. Title.

TK7870.G533 621.3815'1 80-25197
ISBN 0-471-08308-9

Printed in the United States of America

10 9 8 7 6 5 4 3 2

To Judy

Preface

This book is designed to help the engineer select the components that he or she will specify. (Application information about specific types of component is given in the applicable section). With all components, the most important thing that a user must decide is which of the many types will best serve in the equipment being designed. Proper selection, in its broadest sense, is the first step in building reliable equipment.

To select properly, the user must know as much as possible about the types from which to choose. The user should know their advantages and disadvantages, their behavior under various environmental conditions, their construction, and their effect on circuits and the effect of circuits on them. The user should know what makes them fail. The user should also have an intimate working knowledge of all applicable specifications.

I hope that this book will help the engineer succeed in the quest for the most cost-effective component for each application.

GERALD L. GINSBERG

Philadelphia, Pennsylvania
November 1980

Acknowledgment

The author wishes to acknowledge the contributions of Esther Block, Susan Ginsberg, Carol Lawless, and Anne Yablon that have made this book possible.

Contents

A User's Guide to Selecting Electronic Components

PART ONE
Resistors

CHAPTER ONE
Resistors, General Information

CHAPTER TWO
Fixed-Value Resistors

CHAPTER THREE
Variable-Value Resistors

Table 1.1 Basic Resistor Types[1]

Resistor Type	Typical Construction	Application and Comments
Carbon composition	Powdered carbon held together with fillers and binders and molded into a small cylindrical shape with leads imbedded into the ends of the body.	General purpose—particularly if there is no need for any accuracy and no requirement to withstand the stress of temperature, humidity, etc. Low in cost and most readily available.
Carbon film (deposited carbon)	Pure carbon deposited on a ceramic or glass rod and sealed hermetically with a conformal coating of epoxy or special lacquer.	General purpose where a better level of performance is required for temp. and humidity conditions. Slightly lower in cost than carbon composition and generally used in new circuit design.
Metal-film and metal-oxide	Nichrome or Metal oxide deposited on a ceramic or glass rod, then conformal coated or transfer molded and sealed hermetically.	Used where very low temperature coefficient and accuracy are needed for critical circuitry. More costly than the previous types.
Wire-wound	Wire is wound on a berylium oxide core and silicone coated, epoxy molded or coated with vitreous enamel.	High-current (high-power) extreme accuracy circuit requirements. Take up much space and most expensive.

The purchased tolerance of the resistor to be used will allow it to differ from the nominal stated value, depending on the type of resistor specified. The temperature at which the resistor works, the voltage across it, and the environment in which it lives will affect the actual value at particular times.

For each size of resistor given type, quality, and so forth, there is an associated resistance range. In some cases, more than one range is possible, corresponding to the various temperature characteristics or resistance tolerances, or both, available for a given resistor size. The listed range in each case determines the limits within which standard resistance values can be obtained.

Within those limits, standard values can be selected in accordance with Table 1.2 and the specified resistance tolerance. For a particular tolerance, any resistance value in the appropriate column may be specified, as well as any decade multiple (or submultiple) of that value, within the stated range.

1.3 SELECTION CRITERIA

A. Select a resistor for each circuit application from the lists of standard types and values.

B. Be sure that the circuit being designed will work with any resistor whose resistance value is within the limits set by tolerance plus voltage coefficient plus temperature coefficient plus drift with time. Failure to take these precautions in designing equipment manufactured in quantity can produce some circuits that will not work under extreme conditions.

C. Various initial tolerances are available, depending on the type of resistor. It should be remembered that initial accuracies become meaningless if the inherent stability of the resistor does not support its initial accuracy.

D. Throughout its shelf life, as well as its operational life, any characteristic (resistance, inductance, power rating, dielectric strength, size, and so forth) of any part may change due to stresses caused by environmental changes of temperature, humidity, pressure, vibration, and other factors. Changes of characteristics caused by environmental stresses may be linear or nonlinear, reversible, or nonreversible (permanent), or combinations thereof.

E. For military applications, resistors should be selected from those listed in Figure 1.1 and Tables 1.3, 1.4, and 1.5.

1.4 RESISTOR TERMS AND DEFINITIONS

Attenuator A resistive network that reduces the amplitude of an electrical signal without appreciable phase shift or frequency distortion.

Bulk Resistivity Resistance measured between opposite faces of a cube of homogeneous material, typically 1 cm in all dimensions. The value is expressed in ohms per cubic centimeter.

Carbon Film Resistor A resistor formed by vacuum depositing of a thin carbon film on a ceramic form.

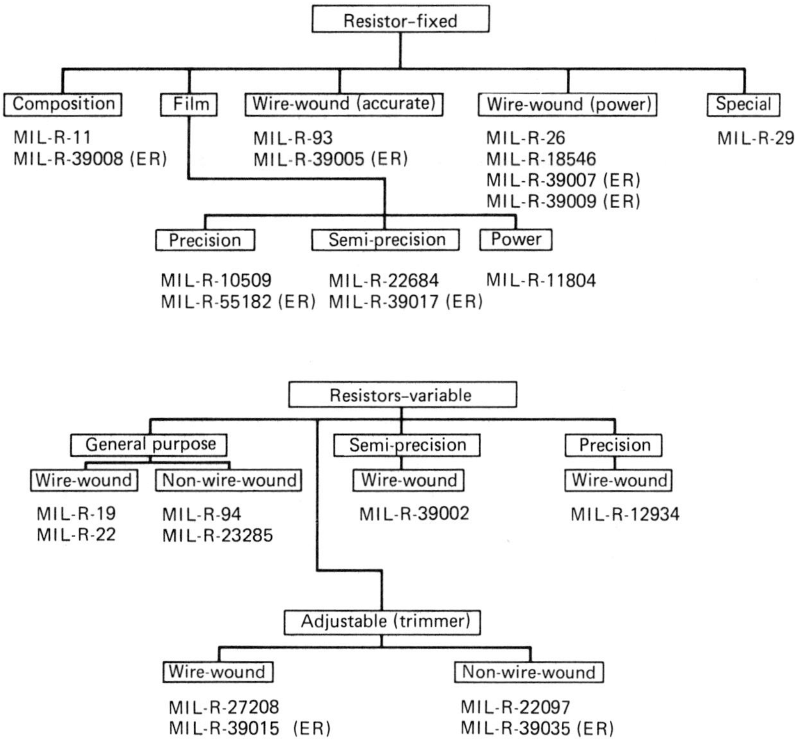

Figure 1.1 Military resistor specification categories.[2]

Table 1.2 Standard Resistance Values for the 10-to-100 Decade[3a]

RESISTANCE TOLERANCE (±%)

0.1% 0.25% 0.5%	1%	2% 5% 10%	0.1% 0.25% 0.5%	1%	2% 5% 10%	0.1% 0.25% 0.5%	1%	2% 5% 10%	0.1% 0.25% 0.5%	1%	2% 5% 10%	0.1% 0.25% 0.5%	1%	2% 5% 10%	0.1% 0.25% 0.5%	1%	2% 5% 10%
10.0	10.0	10	14.7	14.7	—	21.5	21.5	—	31.6	31.6	—	46.4	46.4	—	68.1	68.1	68
10.1	—	—	14.9	—	—	21.8	—	—	32.0	—	—	47.0	—	47	69.0	—	—
10.2	10.2	—	15.0	15.0	15	22.1	22.1	22	32.4	32.4	—	47.5	47.5	—	69.8	69.8	—
10.4	—	—	15.2	—	—	22.3	—	—	32.8	—	—	48.1	—	—	70.6	—	—
10.5	10.5	—	15.4	15.4	—	22.6	22.6	—	33.2	33.2	33	48.7	48.7	—	71.5	71.5	—
10.6	—	—	15.6	—	—	22.9	—	—	33.6	—	—	49.3	—	—	72.3	—	—
10.7	10.7	—	15.8	15.8	—	23.2	23.2	—	34.0	34.0	—	49.9	49.9	—	73.2	73.2	—
10.9	—	—	16.0	—	16	23.4	—	—	34.4	—	—	50.5	—	—	74.1	—	—
11.0	11.0	11	16.2	16.2	—	23.7	23.7	—	34.8	34.8	—	51.1	51.1	51	75.0	75.0	75
11.1	—	—	16.4	—	—	24.0	—	24	35.2	—	—	51.7	—	—	75.9	—	—
11.3	11.3	—	16.5	16.5	—	24.3	24.3	—	35.7	35.7	—	52.3	52.3	—	76.8	76.8	—
11.4	—	—	16.7	—	—	24.6	—	—	36.1	—	36	53.0	—	—	77.7	—	—
11.5	11.5	—	16.9	16.9	—	24.9	24.9	—	36.5	36.5	—	53.6	53.6	—	78.7	78.7	—
11.7	—	—	17.2	—	—	25.2	—	—	37.0	—	—	54.2	—	—	79.6	—	—
11.8	11.8	—	17.4	17.4	—	25.5	25.5	—	37.4	37.4	—	54.9	54.9	—	80.6	80.6	—
12.0	—	12	17.6	—	—	25.8	—	—	37.9	—	—	55.6	—	—	81.6	—	—
12.1	12.1	—	17.8	17.8	—	26.1	26.1	—	38.3	38.3	—	56.2	56.2	56	82.5	82.5	82
12.3	—	—	18.0	—	18	26.4	—	—	38.8	—	—	56.9	—	—	83.5	—	—
12.4	12.4	—	18.2	18.2	—	26.7	26.7	—	39.2	39.2	39	57.6	57.6	—	84.5	84.5	—
12.6	—	—	18.4	—	—	27.1	—	27	39.7	—	—	58.3	—	—	85.6	—	—
12.7	12.7	—	18.7	18.7	—	27.4	27.4	—	40.2	40.2	—	59.0	59.0	—	86.6	86.6	—
12.9	—	—	18.9	—	—	27.7	—	—	40.7	—	—	59.7	—	—	87.6	—	—
13.0	13.0	13	19.1	19.1	—	28.0	28.0	—	41.2	41.2	—	60.4	60.4	—	88.7	88.7	—
13.2	—	—	19.3	—	—	28.4	—	—	41.7	—	—	61.2	—	—	89.8	—	—
13.3	13.3	—	19.6	19.6	—	28.7	28.7	—	42.2	42.2	—	61.9	61.9	62	90.9	90.9	91
13.5	—	—	19.8	—	—	29.1	—	—	42.7	—	—	62.6	—	—	92.0	—	—
13.7	13.7	—	20.0	20.0	20	29.4	29.4	—	43.2	43.2	43	63.4	63.4	—	93.1	93.1	—
13.8	—	—	20.3	—	—	29.8	—	—	43.7	—	—	64.2	—	—	94.2	—	—
14.0	14.0	—	20.5	20.5	—	30.1	30.1	30	44.2	44.2	—	64.9	64.9	—	95.3	95.3	—
14.2	—	—	20.8	—	—	30.5	—	—	44.8	—	—	65.7	—	—	96.5	—	—
14.3	14.3	—	21.0	21.0	—	30.9	30.9	—	45.3	45.3	—	66.5	66.5	—	97.6	97.6	—
14.5	—	—	21.3	—	—	31.2	—	—	45.9	—	—	67.3	—	—	98.8	—	—

[a] Also usable in decade multiples and submultiples.

Table 1.3 Military Resistor Specifications[2]

SPECIFICATION	TITLE
MIL-R-11	Resistors, Fixed, Composition (Insulated), General Specification For.
MIL-R-19	Resistors, Variable, Wirewound (Low Operating Temperature), General Specification For.
MIL-R-22	Resistors, Variable (Wirewound, Power Type), General Specification For.
MIL-R-26	Resistors, Fixed, Wirewound (Power Type), General Specification For.
MIL-R-93	Resistors, Fixed, Wirewound (Accurate), General Specification For.
MIL-R-94	Resistors, Variable, Composition, General Specification For.
MIL-R-10509	Resistors, Fixed, Film (High Stability), General Specification For.
MIL-R-11804	Resistors, Fixed, Film (Power Type), General Specification For.
MIL-R-12934	Resistors, Variable, Wirewound, Precision, General Specification For.
MIL-R-18546	Resistors, Fixed, Wirewound (Power Type, Chassis Mounted), General Specification For.
MIL-R-22097	Resistors, Variable, Non-wirewound (Lead Screw Actuated), General Specification For.
MIL-R-22684	Resistors, Fixed, Film, Insulated, General Specification For.
MIL-R-27208	Resistors, Variable, Wirewound (Lead Screw Actuated), General Specification For.
MIL-R-39002	Resistors, Variable, Wirewound, Semi-Precision, General Specification For.
MIL-R-39005	Resistors, Fixed, Wirewound (Accurate), Established Reliability, General Specification For.
MIL-R-39007	Resistors, Fixed, Wirewound (Power Type), Established Reliability, General Specification For.
MIL-R-39008	Resistors, Fixed, Composition (Insulated), Established Reliability, General Specification For.
MIL-R-39009	Resistors, Fixed, Wirewound (Power Type, Chassis Mounted), Established Reliability, General Specification For.
MIL-R-39015	Resistors, Variable, Wirewound (Lead Screw Actuated), Established Reliability, General Specification For.
MIL-R-39017	Resistors, Fixed, Film (Insulated), Established Reliability, General Specification For.
MIL-R-39023	Resistors, Variable, Non-Wire-Wound, Precision, General Specification For.
MIL-R-39035	Resistors, Variable, Nonwire-Wound (Adjustment Type), Established Reliability, General Specification For.
MIL-R-55182	Resistors, Fixed, Film, Established Reliability, General Specification For.
MIL-R-83401	Resistor Networks, Fixed, Film, General Specification For.

Cermet A dielectric mixture used in making film-resistive elements. The first half of the term is derived from *cer*amic, the second from *met*al.

Chip Resistor Thick-film screened onto a ceramic or glass substrate; thin-film vacuum-deposited. Resistance is obtained by scribing, sandblasting, or otherwise adjusting the resistor element to tolerance.

Composition Resistor Resistive element consists of a carbon composition molded under extreme pressure, then enclosed in an insulating sleeve.

Critical Value of Resistance Value of resistance that will dissipate full-rated power at rated voltage.

Deposited Carbon Resistive element made of a thin film of crystalline carbon or a carbon alloy sputtered onto a ceramic rod.

Film Resistor A fixed resistor that consists of a thin layer of resistance material deposited on an insulated form with an overlaying insulating coat.

Inductance The property of a wound resistor that causes a voltage to be set up because of a change in current through its coils.

Insulation Resistance The resistance to a specified DC voltage impressed between the terminals and all other external conducting members.

Initial Tolerance The difference in percentage between the specified value and the actual value of a resistor.

Intermittent Duty Resistor A resistor capable of dissipating high overload currents for short periods of time without exceeding its maximum hot-spot temperature.

Ladder Network A repetitive arrangement of precision resistors in a ladderlike configuration.

Load Life The number of hours at which a device may dissipate rated power (P) under specified operating conditions while remaining within specified allowable degradations.

Load-Life Stability The relative change in resistance value in parts per million over a specified time period under normal load and temperature conditions.

Maximum Working Voltage The maximum voltage that may be applied across a resistor with consideration for the materials used, the allowable resistance deviation from the low-voltage value, and the shape of the resistor.

Metal-Film Resistor An electronic component in which the resistive element is an extremely thin layer of metal alloy, vacuum-deposited on a substrate.

Metal-Oxide Resistor A type of film resistor in which the resistance material deposited on the substrate is tin oxide, which provides good stability.

Noise An unwanted voltage variation generated within a resistor.

Noise Index The ratio of current noise voltage to average DC voltage produced by a passage of a specified DC current through a resistor.

Noninductive Resistor A wire-wound resistor type with a winding in which the magnetic field of one turn is canceled out by the magnetic field of an adjacent turn.

Ohm The unit of resistance symbolized by the Greek letter omega (Ω). It is defined as the resistance 0°C of a uniform column of mercury 106.300 cm long and weighing 14.4521 g.

Ohms Per Mil Foot The resistance in ohms of a wire with a diameter of 0.001 in. X 1 ft. long. (0.025 mm X .3 m)

Ohms Per Square A unit of measurement of resistance by topological shape. A resistor topology can be considered as consisting of continuous squares. The value of a resistor is equal to the number of squares times ohms per square.

Potential Drop The difference in potential between the two ends of a resistance with a current flowing through it.

Power Coefficient The maximum rise in the temperature of the hot spot per watt of internal power dissipation, expressed in degrees centigrade per watt.

Power Dissipation The dispersion of the heat generated within a device or component when a current flows through it.

Power Rating The maximum power that can be dissipated across the total resistive element under specified conditions while meeting specified performance requirements.

Pull-Down Resistor (1) A resistor connected across the output of advice or circuit to hold the output equal to or less than the zero input level of the following digital device. Also used to lower the output impedance of a device. (2) A resistor connected to a negative voltage or to ground.

Pull-Up Resistor A resistor connected to the positive supply voltage of a transistor circuit, as from the collector supply to the output collector.

Residual Resistance In a potentiometer or trimmer, the resistance between the center terminal and either end terminal with the shaft positioned against the stop at that terminal.

Resistance A property of conductors that, depending on their dimensions, material, and temperature, determines the current produced by a given difference of potential.

Table 1.4 Military Fixed Resistor Selection Guide[2]

Specification	Type	Styles Available in Standard	Power and Maximum Voltage Ratings	Resistance Tolerance (± %)	Ohmic Range[c]	Temperature Range (° C)[d]	Maximum Body Size
MIL-R-10509	Film (high stability)	RN75	1 W/500 V	0.1, 1.0	10–5.11 M	70–150 125–175	1.124 × 0.437
MIL-R-11804	Film (power type)	RD60	1 W/350 V	2	10–0.464 M	25–235	0.656 × 0.235
		RD65	2 W/500 V		10–1.47 M		1.000 × 0.359
		RD70	4 W/750 V		20.5–3.83 M		2.12 × 0.359
MIL-R-26	Wire-wound (power type)	RW29	11 W	5, 10	0.1–5.6 K	25–350	1.812 × 0.500
		RW31	14 W		0.1–6.8 K		1.562 × 0.594
		RW33	26 W		0.1–18 K		3.062 × 0.594
		RW35	55 W		0.1–39 K		4.062 × 0.906
		RW37	113 W		0.1–82 K		6.062 × 1.312
		RW38	159 W		0.1–0.15 M		8.062 × 1.312
		RW47	210 W		0.1–0.18 M		10.562 × 1.312
		RW56	14 W		0.1–8.2 K		2.094 × 0.562
MIL-R-18546	Wire-wound (power type, chassis mounted)	RE77	75 W[b]	1	0.05–28.7 K	25–275	3.594 × 1.781 × 2.843
		RE80	120 W[b]		0.1–34.8 K		4.594 × 2.219 × 3.031
MIL-R-39008	Composition (insulated), established reliability	RCR05	0.125 W/150 V	5, 10	2.7–22 M	70–130	0.160 × 0.066
		RCR07	0.25 W/250 V		2.7–22 M		0.281 × 0.098
		RCR20	0.5 W/350 V		2.7–22 M		0.416 × 0.161
		RCR32	1 W/500 V		2.7–22 M		0.593 × 0.240
		RCR42	2 W/500 V		10–22 M		0.728 × 0.336
MIL-R-55182	Film, established reliability	RNR50	0.05 W/200 V 0.1 W/200 V[a]	0.1, 0.5, 1 (RNR50–RNR75)	10–0.796 M	125–175 (RNR50–RNC90)	0.170 × 0.080
		RNR55	0.1 W/200 V 0.125 W/200 V[a]		10–1.21 M		0.281 × 0.140
		RNR60	0.125 W/250 V 0.25 W/300 V[a]		10–4.02 M		0.437 × 0.165
		RNR65	0.25 W/300 V 0.5 W/350 V[a]		12–6.26 M		0.656 × 0.250
		RNR70	0.5 W/350 V 0.75 W/500 V[a]		10–15 M		0.875 × 0.328
		RNR75	1 W/750 V 2 W/750 V		10–5 M		1.124 × 0.437
		RNC90	0.3 W/300 V[a] 0.6 W/300 V	1, 0.5, 0.1, 0.05, 0.01	40.2–50.5 K		0.320 × 0.345 × 0.120

MIL-R-39005	Wire-wound (accurate), established reliability	RBR52	0.5 W/600 V		0.1–0.806 M		1.020 X 0.390
		RBR53	0.33 W/300 V		0.1–0.499 M		0.770 X 0.390
		RBR54	0.25 W/300 V		0.1–0.249 M		0.770 X 0.265
		RBR55	0.15 W/200 V	0.01, 0.05	0.1–0.150 M	125–145	0.520 X 0.265
		RBR56	0.125 W/150 V	0.1, 1	0.1–0.1 M		0.364 X 0.265
		RBR57	0.75 W/600 V		0.1–1.37 M		1.020 X 0.515
		RBR71	0.125 W/150 V		0.1–0.1 M		0.343 X 0.281
		RBR75	0.125 W/150 V		0.1–71.5 K		0.315 X 0.265
MIL-R-39007	Wire-wound (power type), established reliability	RWR74	5 W		0.1–12.1 K		0.937 X 0.343
		RWR78	10 W		0.1–38.3 K		1.842 X 0.406
		RWR80	2 W		0.1–1.21 K		0.437 X 0.125
		RWR81	1 W	0.1, 0.5, 1	0.1–464	25–275	0.281 X 0.105
		RWR82	1.5 W		0.1–		0.328 X 0.094
		RWR84	7 W		0.1–12.1 K		0.937 X 0.343
		RWR89	3 W		0.1–3.32 K		0.622 X 0.218
MIL-R-39017	Film (insulated), established reliability	RLR05	0.125 W/200 V		10–0.287 M		0.170 X 0.074
		RLR07	0.25 W/250 V		10–2.49 M		0.281 X 0.098
		RLR20	0.5 W/350 V	1, 2	4.42–1.0 M	70–150	0.416 X 0.161
		RLR32	1 W/500 V		10–1.0 M		0.593 X 0.205
		RLR42	2 W/500 V		10–1.5 M		0.728 X 0.336
MIL-R-39009	Wire-wound (power type, chassis mounted), established reliability	RER40	5 W		1–1.65 K		0.662 X 0.677 X 0.351
		RER45	10 W[b]		1–2.80 K		0.812 X 0.843 X 0.437
		RER50	20 W		1–6.04 K		1.124 X 1.125 X 0.593
		RER55	30 W	1	1–19.6 K	25–275	2.000 X 1.187 X 0.656
		RER60	5 W		0.1–3.32 K		0.662 X 0.677 X 0.351
		RER65	10 W		0.1–5.62 K		0.812 X 0.843 X 0.437
		RER70	20 W		0.1–12.1 K		1.124 X 1.125 X 0.593
		RER75	30 W		0.1–39.2 K		2.000 X 1.187 X 0.656

[a] Power rating at 70°C (full load ambient operating temperature).
[b] Mounted on a metal chassis.
[c] M, megohms; K, kilohms.
[d] Full load ambient operating temperature and zero load temperature, respectively.

Table 1.5 Military Variable-Value Resistor Selection Guide[2]

Specification	Type	Styles Available in Standard	Power Rating (W)	Nominal Total Resistance	Temperature Range (°C)	Resistance Temperature Coefficient (ppm/°C)	Maximum Body Size
MIL-R-94	Composition (insulated)	RV4	2, 1	50–5 M			1.156 × 0.750
		RV6	0.5, 0.25	100–5 M	70–120		0.516 × 0.593
MIL-R-19	Wire-wound (low operating temperature)	RA20	2, 1.1	3–15 K			1.310 × 0.700
		RA30	4, 2.2	3–25 K	40–105		1.710 × 0.810
MIL-R-22	Wire-wound (power type)	RP05	5	10–5 K	25–340 (RP05–RP25)		0.525 × 0.687
		RP06	12.5	1–3.5 K			0.906 × 0.751
		RP10	25	2–5 K			1.680 × 1.410
		RP15	50	1–10 K			2.410 × 1.440
		RP20	75	2–10 K			2.810 × 1.780
		RP25	100	2–10 K			3.190 × 1.780
		RP30	150	2–10 K	25–390		4.060 × 2.030
MIL-R-12934	Wire-wound, precision	RR0900	1.25	100–10 K			0.880 × 0.812
		RR1000	2.0	100–50 K			0.880 × 1.625
		RR1100	1.5	100–20 K		±30	1.067 × 0.812
		RR1300	2.0	100–40 K	85–150	±100	1.442 × 1.062
		RR1400	3.0	200–0.2 M			1.442 × 2.250
		RR2000	4.0	100–60 K			2.005 × 1.312
		RR2100	5.0	200–0.25 M			2.005 × 2.250
		RR3000	6.0	200–0.1 M			3.005 × 1.312
MIL-R-39002	Wire-wound, semiprecision	RK09	1.5	10–50 K	85–135	±70 $R \geq 50\ \Omega$ ±200 $R < 50\ \Omega$	0.515 × 0.650

MIL-R-27208	Wire-wound (adjustment type)	RT26	0.25	10–2 K	85–150	±50	0.195 X 0.270 X 0.270
MIL-R-22097	Non-wire-wound (adjustment type)	RJ12 RJ26 RJ50	0.75 0.25 0.25	10–1 M 10–1 M 10–1 M	85–150	±100 ±250	1.282 X 0.200 X 0.330 0.270 X 0.195 X 0.270 0.270 X 0.250 X 0.270
MIL-R-23285	Non-wire-wound	RVC6	0.5	100–2.5 M	125–175	±250	0.516 X 0.593
MIL-R-39023	Non-wire-wound, precision	RQ090 RQ110 RQ150 RQ200 RQ300	1.0 1.25 1.50 2.00 3.00	100–1 M 100–1 M 100–1 M 100–1 M 100–1 M	70–125		0.880 X 0.810 1.067 X 0.810 1.442 X 1.06 2.005 X 1.31 3.005 X 1.31
MIL-R-39015	Wire-wound (lead-screw actuated), established reliability	RTR12 RTR22 RTR24	0.75 0.75 0.75	10–10 K 10–10 K 10–5 M	85–150	±50	1.260 X 0.200 X 0.330 0.510 X 0.230 X 0.510 0.390 X 0.245 X 0.390
MIL-R-39035	Non-wire-wound (adjustment type), established reliability	RJR12 RJR24 RJR26 RJR28 RJR32	0.75 0.5 0.25 0.3 0.5	10–1 M 100–1 M 100–1 M 100–2 M 10–2 M	85–150	±100 ±250	1.260 X 0.330 X 0.200 0.390 X 0.195 X 0.420 0.270 X 0.195 X 0.270 0.510 X 0.110 X 0.180 0.770 X 0.260 X 0.200

Resistance Drop The voltage drop that occurs across two points on a conductor when current flows through the resistance between those points.

Resistance Element A continuous, unbroken length of resistive material without joints, bonds, or welds, except at the junction of the element, and the electrical terminals connected to each end of the element or at an intermediate point, such as a center tap.

Resistance-Temperature Characteristic (Temperature Coefficient) The magnitude of change in resistance due to temperature, usually expressed in percent per degree centigrade or parts per million per degree centigrade (ppm/°C). If the changes are linear over the operating temperature range, the parameter is known as temperature coefficient.

Resistance Tolerance The deviation of the manufactured resistance value, expressed in percent, from the specified nominal resistance value at standard or stated environmental conditions.

Resistivity (1) A measure of the resistance of a material to electrical current either through its volume or on a surface. The unit of volume resistivity is the ohm-centimeter; the unit of surface resistivity is the ohm. (2) The ability to resist current; the reciprocal of conductivity.

Resistor A basic component in electrical and electronic circuits that produces heat while opposing the flow of electrical current.

Resistor Matching (Ratio Matching) Maximum deviation permissible from the ratio of nominal resistance values of two resistors measured at a given temperature. It is usually specified in percent or parts per million with one resistor selected as the reference resistor.

Self-Heating Coefficient The influence of internally generated heat on the resistance value. It is expressed in + percent, or parts-per-million resistance change per watt of dissipation. This coefficient is the product of the temperature coefficient times the power coefficient.

Stability The overall ability of a resistor to maintain its initial resistance value over extended periods of time when subjected to any combination of environmental conditions and electrical stresses.

Static Noise Also known as current noise. A measure of the voltage generated within a resistor when an external DC potential is applied.

Strain Relief Subjecting a wirewound resistor to hot and cold cycles for the purpose of equalizing internal stresses that resulted from work-hardening during winding. Sometimes called normalizing or stress equalizing.

Temperature Coefficient of Resistance (TCR) TCR is the change in resistance caused by a change in temperature. It has units of percent per degree centigrade or parts per million per degree centigrade.

Temperature Coefficient of Resistance Tracking TCR tracking is the difference in TCR of two or more resistors.

Thermistor A resistor with a high-temperature coefficient of resistance, exhibiting a definite, reliable, and repeatable response to temperature change. Negative TC types show an exponential decrease in resistance with an increase in temperature. Positive TC types show an increase in resistance with an increase in temperature.

Thick-Film Resistor Resistive element formed by laying a resistive ink or paste on a substrate through a pattern or screen. The material is then fired to harden it.

Thin-Film Resistor Resistive element formed in a vacuum through evaporation of a resistive material that is deposited on a substrate.

Time Stability The change in resistance per unit change in time over and above all other changes—the so-called "shelf stability" expressed in percent or ppm per 100 hr or per year.

Tracking A measure of the amount that two or more resistors will vary in resistance over a range of temperatures relative to one another.

Varistor (Voltage-Dependent Resistor) A special type of resistor whose resistance changes appreciably, nonlinearly, and consistently in response to voltage across the devices terminals.

Voltage Rating The maximum DC or peak AC voltage the resistor is designed to withstand across its terminals, which should not be exceeded under any combination of working conditions.

Wattage Rating The maximum power a resistor can dissipate under standard ambient temperature conditions and consideration for the load condition.

Wire-Wound Resistor A resistor in which the resistance element is a length of high-resistivity wire or ribbon, wound onto an insulating form.

CHAPTER TWO

Fixed-Value Resistors

2.1 FIXED-VALUE FILM RESISTORS[3–5]

Figure 2.1 depicts the typical construction of a metal-film resistor. The core is normally a solid body composed of an insulating material ranging from steatite to a high-alumina content ceramic. Each material has its advantages and disadvantages in terms of strength, surface properties, moisture absorption, heat, conductivity, and cost, and each must be selected accordingly. Normally, for higher-power density resistor designs, a high-alumina-content core is chosen for its superior heat-conduction characteristics.

A metal or crystallized-carbon film is deposited on the surface of the core by any number of possible methods. Among them are evaporation under low pressure, sputtering, electroless deposition, and spraying. The typical metal film consists of an alloy of nickel-chrome with various other metals added that determine both initial resistance and temperature coefficient.

After film deposition, each core has an initial end-to-end resistance (blank value), which for metal-film units may range from 0.1 to 3000 Ω or more and from 0.5 to 30,000 Ω for carbon-film units, depending on core dimensions. For higher-stability resistor requirements, the lowest blank value applicable to a finished value is normally preferred.

After metal deposition, a filming batch undergoes a heat-treatment process consisting of 10-30 hours at temperatures of 200–400°C. This treatment oxidizes the film surface changing the blank value slightly and raising the temperature coefficient of the units to the desired value.

Electrical contact with the resistive film is normally provided by metal caps press fitted onto the ends of the core and designed to have a sufficient friction fit to prevent pull-off in lead-strength tests.

The capped blank resistor is then helixed to value. A thin grinding stone or laser beam is employed to cut a helical insulating groove in the metal film to provide a relatively long spiraled resistive path from cap to cap.

Prior to helixing, the resistance of each blank is measured, enabling automatic setting of the helix pitch at a constant value for each resistor for maximum utilization of the film length. Servocontrolled equipment monitors the resistance value as the cut is made, stops the cut when the desired value is reached, and ejects the resistor.

Usually, the resistor leads are welded to the caps after helixing, although some manufacturers use a staked cap and lead assembly and helix their resistors with the leads attached.

The coating applied to a film resistor serves a number of purposes, including (1) protecting the film from outside influences such as moisture, abrasion, and contamination, (2) providing dielectric strength and insulation resistance, and (3) serving as a base for marking of color coding. Normally two or more coating layers are applied, usually of modified epoxy formulations, in a

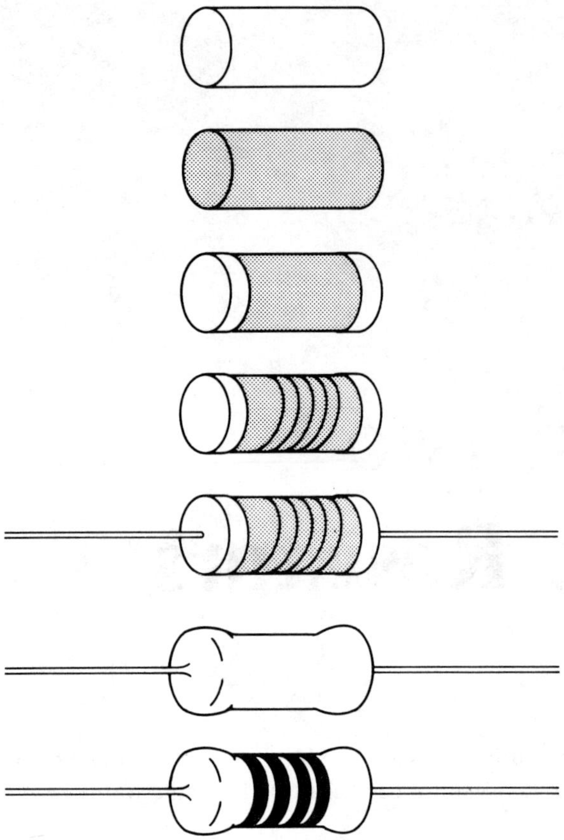

Figure 2.1 Fixed-value film resistor.[4]

roll-on operation, although some manufacturers produce molded-epoxy types.

2.1.1 Selection

Selecting a fixed-value film resistor begins with knowing the operating conditions imposed by the circuit. The voltage and power dissipation are established by the resistance value and the drift specifications, such as TCR, VCR, and stability dictated by system requirements. Also important is knowing whether the application is general-purpose or high-stability. When applying the power rating to the circuit consider the following:

A. Resistors can be operated in ambient temperatures higher than specified, if the power dissipation is less than the rated maximum.

B. Resistors can dissipate higher wattages than rated if both the ambient temperature and the voltage rating are not exceeded.

C. Resistors can be operated at higher power levels if forced air or coolants are used. Carbon films are full-load rated to operate at 25°C, but metal-alloy and oxide films are rated much higher. At 25°C these latter two types can operate at 200% of the rated load.

2.1.2 Types[6]

Metal-film resistors come in three styles: carbon-film, metal-alloy, and metal-oxide. It is important to know how to select the correct type for an application. (See Table 2.1.)

Most parameters for resistor operation in low-voltage circuits apply to resistor operation at high voltages. Such parameters as power rating, temperature coefficient of resistance (TCR), and stability do not change. At high voltages, however, it is necessary to be more aware of such parameters as voltage breakdown and voltage coefficient of resistance (VCR).

Selecting the proper resistor is important, but the circuit layout and operating temperatures must be considered as well. Part spacing, cooling, and insulating materials become more significant at high voltages.

Table 2.1 Film Resistor Selection Guide[5]

Carbon-Film	Metal-Alloy	Metal-Oxide
	Advantages	
Highest-power capability	Highest operating temperature	Best load stability
Highest-voltage capability	Good load stability	Best resistance tolerance
Highest resistance capability	Low VCR	Lowest TCR
Wide resistance range		High operating temperature
Most styles and terminations		Wide resistance range
	Disadvantages	
Low operating temperature	Lowest-power capability	Low-power capability
High TCR	Lowest-voltage capability	Low-voltage capability
Poorest load stability		

Each type of film resistor is suitable for specific applications: metal-oxide for high-stability requirements, carbon-film for general-purpose use, and metal-alloy for some high-stability and general-purpose requirements. Use high-stability types in voltage dividers or meter multipliers; use general-purpose types for noncritical functions, such as current limiting and bleeder resistors. (See Table 2.2.)

A. ***Carbon Film*** Carbon-film resistors are the workhorses of the high-voltage resistors. They offer the highest resistance values at the lowest breakdown voltages, as much as 10^{12} Ω at up to 125 kV. They come in $\frac{1}{4}$ to 100-W power ratings with an initial resistance tolerance as low as 1%. In addition, carbon-film resistors come in the widest variety of styles.

B. ***Metal Film*** Metal-film resistors are used for applications ranging from consumer appliances and commercial instrumentation, through industrial process control and computer periphers, to the most demanding scientific, aerospace, and military equipment and systems. Standard tolerances from ±0.1 to ±1% with time/temperature stabilities, voltage/power coefficients, and noise levels approaching those of wire-wound resistors, are possible.

Very low reactance makes metal-film resistors suitable for use from DC to microwave frequencies. They are available in a wide range of package styles, sizes, and constructions, ranging from conformally coated miniatures, through molded compacts, to hermetically sealed designs providing the ultimate in environmental immunity. The consumer/industrial types rival carbon-film designs in price/performance ratios.

1 ***Metal-Alloy Resistors*** Made from a microcrystalline film improved by forming a passivated film over it, this process gives metal-alloy resistors greater stability and higher operating temperatures than is true of carbon resistors. Nevertheless, metal-alloy resistor types are manufactured in lower power ratings than are carbon types, and they come in fewer styles.

2 ***Metal-Oxide Resistors*** The elite of the metal-film resistors, metal-oxide resistors have 50-ppm TCR, 30-kV breakdown voltage, high-resistance range, and outstanding stability. Sometimes called glaze films, they are cermets formed of metal powders and glass. They are similar to the metal-alloys resistors in that they are small parts with about the same power-handling capabilities. (See Table 2.3.)

Table 2.2 Film Resistor Types[6]

Parameter	Deposited Carbon Resistors	Ultraprecision Metal Film	Semiprecision and Precision Film Resistors	
Type	Precision carbon-deposited film resistors	Ultraprecision evaporated thin-film resistors	Precision metal film glaze metal	High-stability metal film
Description	Proprietary molded, conformally coated carbon film resistors with axial leads. Reliability backed by 20 years manufacturing experience.	Proprietary metal film formulations providing extreme high stability.	Glaze metal thick-film precision resistors, axial lead molded.	Evaporated thin film conformally coated and molded resistors. To MIL-R-10509 and MIL-R-55182.
Application	Telecommunications, applications requiring high-resistance values to 100 MΩ applications requiring the negative TCs of carbon film resistors.	Instrumentation, including DVMs, high-stability divider circuits, and analog computer functions. High-frequency applications where precision wirewounds cannot be used due to inductive characteristics. Also available in packaged or matched configurations.	Universal application, for all low-power resistor requirements, high-volume needs.	Instrumentation, analog, divider circuits and other applications that require low initial tolerances and high stability.
MIL/EIA type	RN	RNR-90	RL, RLR, RN	RN, RNR
Power rating	$\frac{1}{2}$-2 W	$\frac{1}{20}$-$\frac{1}{4}$ W	$\frac{1}{8}$-2 W	$\frac{1}{20}$-2 W
Tolerance	±1, 2, and 5%	±0.01–±1%	±1, 2, ±5%	±0.1–±1%

Table 2.3 Glaze-Metal Resistors[6]

Classification/ Configuration	Military Style	Military Range (Ω)	Tolerance (±%)	T.C. (±ppm/°C)	Power Rating	Voltage Rating (V)
General purpose and semiprecision	RC05	—	5	200	1/8 W@70°C	200
	RC07	—	5	200	1/4 W@70°C	250
	RL07	51–150 K	2, 5	200	1/4 W@70°C	250
	RL20	4.3–470 K	2, 5	200	1/2 W@70°C	350
	RL32	10–1 M	2, 5	200	1W@70°C	500
Low resistance	RL07	—	1, 2, 5	100	1/4 W@70°C	250
	RL20	—	1, 2, 5	100	1/2 W@70°C	350
Precision	RN55D	10–301 K	1	100	1/8 W@70°C	200
	RN55C	49.9–100 K	1	50	1/10 W@125°C	200
	RN60D	10–1 M	1	100	1/4 W@70°C	300
	RN60C	49.9–499 K	1	50	1/8 W@125°C	250
	RN65D	10–1 M	1	100	1/2 W@70°C	350
	RN65C	49.9–348 K	1	50	1/4 W@125°C	300
Established Reliability	RLR05/M	10–300 K	2, 5	200	1/8 W@70°C	200
	RLR07/R	10–2.49 M	1, 2	100	1/4 W@70°C	250
	RLR20/R	4.3–1 M	1, 2	100	1/2 W@70°C	350
	RLR32/P	10–1 M	1, 2	100	1W@70°C	500
	RNC55H	10–2 M	.5, 1	50	1/10 W@125°C	200
	RNC55K	10–2 M	.5, 1	100	1/10 W@125°C	200
	RNC60H	10–4.02 M	.5, 1	50	1/8 W@125°C	250
	RNC60K	10–4.02 M	.5, 1	100	1/8 W@125°C	250

Metal-oxide resistors are suitable for applications ranging from industrial/commercial photoelectric timing devices to nuclear and aerospace/military equipments. Resistance values up to 50 MΩ are possible in very small sizes.

Standard tolerances are from ±1 to ±5%, with temperature coefficients (TCs) as low as ±100 ppm/°C. They have excellent long-term stability and very low reactance and noise levels.

2.1.3 High Voltage[7] (See Figure 2.2)

When power-supply voltage is in the kilovolt range, unusual electrical, thermal, and mechanical effects can change resistance values so much that output voltage is unpredictable. Here, resistance values are critical and thick-film resistors are preferred over carbon types because they are more stable with the application of the high voltage. (See Table 2.4.) In addition, more care must be exercised in the mounting and connecting of high-voltage resistors.

2.1.4 Military[8]

Fixed-value film military-specification resistor types RN, RNR, RLR, and RD are characterized by the following:

A. Low tolerance; high stability; low environmental changes; low-temperature coefficient; spacing and weight saving; and low noise.

B. Nominal minimum resistance tolerance available is ±0.1% for fixed-film resistors.

C. Maximum practical full-power operating temperature should not exceed 125°C for metal-film RN and RNR types. Type RLR resistors conform to the 70°C rating. Type RD resistors are continuously derated from 25 to 275°C.

D. Operation at RF (above 100 MHz) may produce inductive effects on spiral-cut type fixed-film resistors.

E. Part numbering in accordance with Table 2.5.

F. Identification color codes in accordance with Table 2.6.

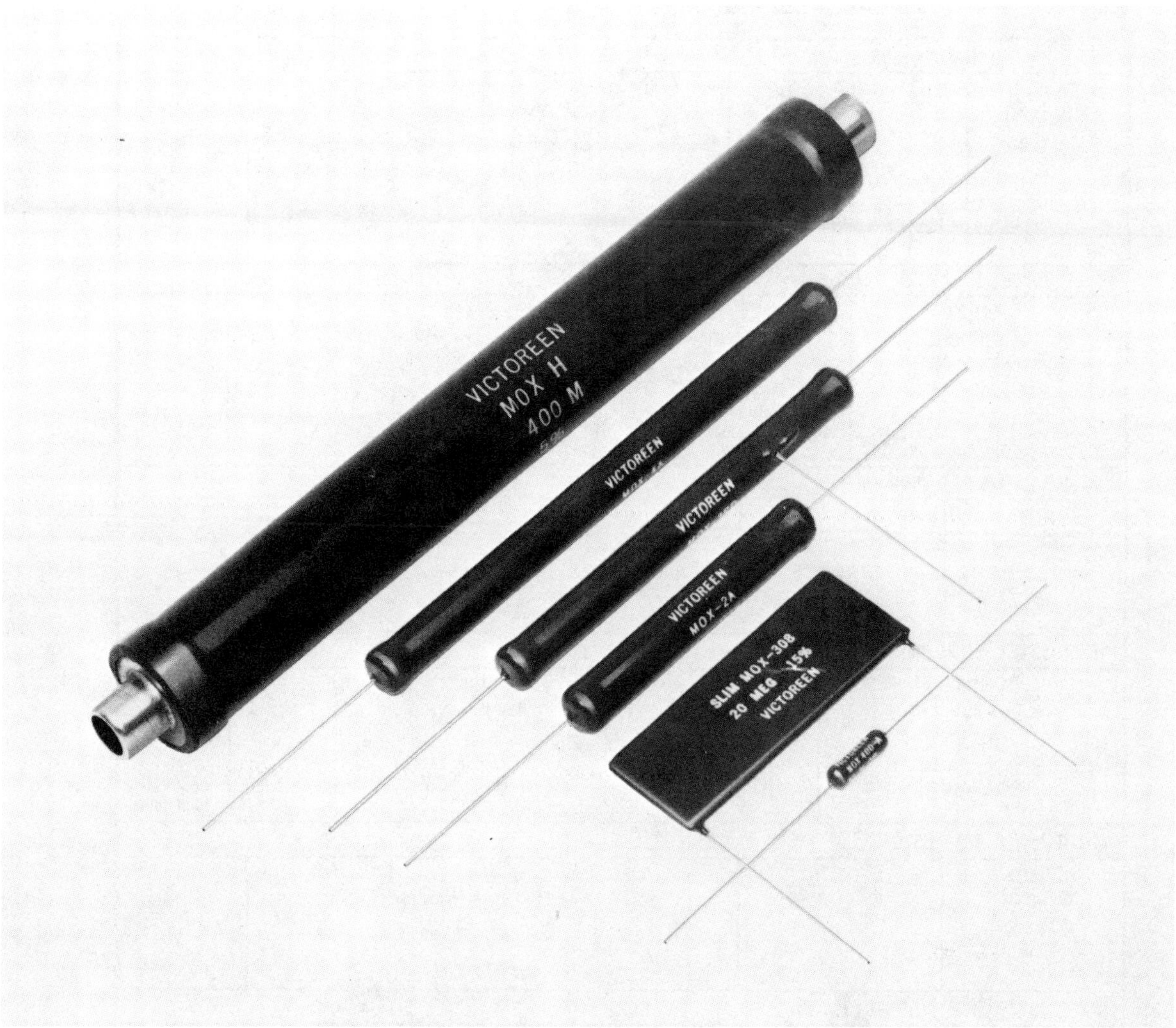

Figure 2.2 High-voltage film resistors. (Courtesy of Victoreen Inc.)

Table 2.4 High-Voltage Film Resistors[5]

Type	Resistance Range (Ω)	Tolerance (%)	Power Range (W)	Voltage Range (kV)	Temperature Coefficient Range (ppm/°C)	Voltage Coefficient Range (ppm/V)	Operating Temperature $(°C)_{max}$	Time Stability (1000 hr) (%ΔR)
Carbon film	10 K–1 M	1–20	¼–100	1–125	−500 to −1000	0.1–5	120	2–5
	1 M–1 G	1–20	¼–100	1–125	−1000 to −4000	0.1–15	120	2–10
	1 G–100 G	2–20	¼–100	1–125	−2000 to −7000	0.2–50	120	5–10
Metal alloy	1 K–100 K	1–10	½–5	3.5–15	−200 to −350	0.1–1	225–350	1
	100 K–100 M	1–10	½–5	3.5–15	−350 to −2500	0.1–5	225–350	1–2
	100 M–1 G	2–10	½–5	3.5–15	−2500 to −4000	0.5–10	225–350	2–5
Metal oxide	1 K–1 M	½–5	¼–6	1.5–30	±50 to ±200	0.1–3	225	0.1–1
	1 M–100 M	½–5	¼–6	1.5–30	±50 to ±200	0.5–5	225	0.5–2
	100 M–10 G	½–10	¼–6	1.5–30	±50 to ±600	1–50	225	1–5

Table 2.5 Military Fixed-Value Resistor Numbering System (Reproduced from Ref. 8.)

MIL-R-11 (RC) RC07GF153K

RC07 Style/Power	GF Characteristic	153 Resistance Value	K Tolerance
(See Tables 1 and 2)	"G" = 70° C max. ambient temperature for full load operation. "F" = temperature coefficent which varies (with resistance) from ± 625 ppm/° C to ± 3100 ppm/° C.	First two digits are significant, 3rd digit "number of zeros." 153 = 15,000 Ω	(See Table 4) K = ± 10%

★MIL-R-39008 (RCR) RCR07G153JS

RCR07 Style/Power	G Characteristic	153 Resistance Value	J Tolerance	S Failure Rate
(See Tables 1 and 2)	"G" indicates a max. ambient temperature of 70° C for full load operation, and a TC which varies (with resistance) from ± 625 ppm/° C to ± 1900 ppm/° C.	First two digits are significant, 3rd digit "number of zeros." 153 = 15,000 Ω	(See Table 4) J = ± 5%	(See Table 5) S = .001%/ 1000 hours

MIL-R-10509 (RN) RN55D1003F

RN55 Style/Power	D Characteristic	1003 Resistance Value	F Tolerance
(See Tables 1 and 2)	(See Table 3) D = ± 100 ppm/° C	First three digits are significant, 4th digit "number of zeros." 1003 = 100,000 Ω	(See Table 4) F = ± 1%

★MIL-R-55182 (RNR) RNR55H1003FS

RNR55 Style, Terminal and Power	H Characteristic	1003 Resistance Value	F Tolerance	S Failure Rate
(See Tables 1 and 2) RNR = Solderable Leads RNC = Solderable/Weldable Leads RNN = Nickel Leads	(See Table 3) H = ± 50 ppm/° C	First three digits are significant, 4th digit "number of zeros." 1003 = 100,000 Ω	(See Table 4) F = ± 1%	(See Table 5) S = .001%/ 1000 hours

MIL-R-22684 (RL) RL07S153J

RL07 Style/Power	S Terminal (Lead)	153 Resistance Value	J Tolerance
(See Tables 1 and 2)	"S" = Solderable	First two digits are significant, 3rd digit "number of zeros." 153 = 15,000 Ω	(See Table 4) J = ± 5%

★MIL-R-39017 (RLR) RLR07C1502GR

RLR07 Style/Power	C Terminal (Lead)	1502 Resistance Value	G Tolerance	R Failure Rate
(See Tables 1 and 2)	"C" = Solderable/Weldable	First three digits are significant, 4th digit "number of zeros." 1502 = 15,000 Ω	(See Table 4) G = ± 2%	(See Table 5) R = .01%/ 1000 hours

★ **NOTE:** The Established Reliability specification (i.e., MIL-R-39008, MIL-R-39017, and MIL-R-55182) supersede MIL-R-11, MIL-R-22684, and MIL-R-10509 respectively, for all *new* design. Resistors qualified to the three Established Reliability specifications may be substituted, without limitation, wherever the older MIL devices are specified.

TABLE 1 – Resistor Style

RC	Fixed Composition Resistor (MIL-R-11)
RCR	Fixed Composition Resistor, Established Reliability (MIL-R-39008)
RL	Fixed Film Resistor (MIL-R-22684)
RLR	Fixed Film Resistor, Established Reliability (MIL-R-39017)
RN	Fixed Resistor, High Stability (MIL-R-10509)
RNR	Fixed Film Resistor, Established Reliability (MIL-R-55182)

TABLE 2 – Resistor Power

Nominal Body Length" x Dia."	Size	Power (@ 70° Unless Otherwise Stated)
RC - RCR - RL - RLR		
.145 x .062	05	1/8
.250 x .090	07	1/4
.375 x .138	20	1/2
.562 x .225	32	1
.688 x .138	42	2
RN (Characteristic C) - **RNR**		
.150 x .065	50	1/20 @ 125° C
.250 x .109	55	1/10 @ 125° C
.375 x .125	60	1/8 @ 125° C
.625 x .188	65	1/4 @ 125° C
.750 x .250	70	1/2 @ 125° C
1.062 x .375	75	1 @ 125° C
RN (Characteristic D)		
.250 x .109	55	1/8
.375 x .125	60	1/4
.625 x .188	65	1/2
.750 x .250	70	3/4

TABLE 3 – Characteristics

RN	
B	± 500 ppm/°C
C	± 50 ppm/°C (T2)
D	± 100 ppm/°C (T0, T1)
E	± 25 ppm/°C (T9)
F	± 50 ppm/°C
RNR/ RNC	
H	± 50 ppm/°C (T2)
J	± 25 ppm/°C (T9)
K	± 100 ppm/°C (T0, T1)

NOTE: There is no temperature coefficient designation for the RL numbering system, all units are ± 100 ppm/°C.

TABLE 4 – Tolerance

K = ± 10%
J = ± 5%
G = ± 2%
F = ± 1%
D = ± 0.5%
C = ± 0.25%
B = ± 0.10%

TABLE 5 – Failure Rate 1000 Hours (60% Confidence)

M = 1.0%
P = 0.1%
R = .01%
S = .001%

Table 2.6 Military Color Code for Film-Type Resistors

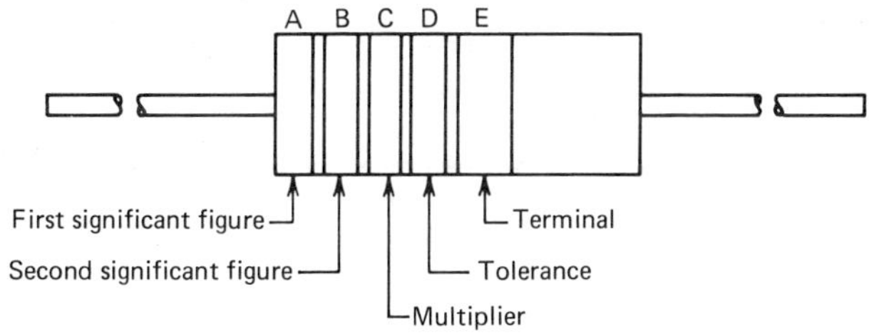

Color	1st digit A	2nd digit B	Multiplier C	Tolerance D	Terminal E
Black	0	0	1	–	–
Brown	1	1	10	–	–
Red	2	2	100	–	–
Orange	3	3	1,000	–	–
Yellow	4	4	10,000	–	–
Green	5	5	100,000	–	–
Blue	6	6	1,000,000	–	–
Violet	7	7	10,000,000	–	–
Gray	8	8	100,000,000	–	–
White	9	9	–	–	Solderable
Gold	–	–	0.1	±5%	–
Silver	–	–	0.01	–	–

G. The resistance temperature characteristic is fairly low (±500 ppm/°C and +200 ppm/°C) for thick-film types (RD and RLR) and very low (±25 ppm/°C) for metal-film types (RN and RNR).

2.2 FIXED-VALUE WIRE-WOUND PRECISION RESISTORS[9–11]

The unique facility with which the wire-wound resistor can provide resistances ranging from the very low to the very high, carry small currents, dissipate many watts of power, be designed in many shapes and sizes, and meet a great variety of special requirements long ago established it as the "workhorse" of the resistor world. Almost all resistor applications requiring the dissipation of more than 1 or 2 W of power, either DC or commercial-frequency AC, and going up through the audio frequency range, are well handled by wire-wounds.

Wire-wound resistors may be classified in many ways: by shape, type of coating, operating temperature, type of termination, fixed or adjustable value, method of mounting, environmental capabilities, inductive or non-inductive winding, and many more classes or subclasses that cut across each other.

In the fabrication of a wire-wound resistor the resistance alloy wire is space-wound at a predetermined pitch with its ends welded or brazed to the terminals. An immense range of resistance values is obtained by selection of the resistance alloy, the wire size, and the number of turns. The most important wire alloys are listed in Table 2.7.

The pitch, or distance from center to center of adjacent turns, can be a distance slightly more than the wire diameter up to not more than five times the wire diameter in order to achieve the desired total ohms.

A protective coating embeds and completely covers the turns of wire in order to (1) mechanically protect the wire, (2) keep the turns fixed in position to avoid shorting, (3) insulate the turns against voltage breakdown across the coil, (4) prevent oxidation of the wire, (5) protect against corrosion from humidity or salt water, and (6) help carry away and distribute the heat from the wire.

Axial-leaded resistors, which now constitute the major portion of the column of wire-wound resistors produced, differ from the lug- or tab-type primarily in substituting caps over the ends of the core, replacing the bands. Wire leads welded to the caps serve to connect and to support the resistor.

Wire-wound resistors can provide lower temperature coefficients (1 ppm/°C), attain greater stability and accuracy, and dissipate more power (> 2000 W) than is possible with composition and film resistors. Even though film resistors are improving and may eventually supersede wire-wounds in both cost and performance, wire-wounds still have the edge at the extreme ratings.

Table 2.7 A Representative Group of Resistance Alloys (Reproduced from Ref. 9.)

ASTM Alloy Class*	Alloy Composition (Approximate)	Ohms per CMF	Trade Names	Mean Temperature Coeff. of Res. ppm/°C	Temperature Range for TC °C	Resistance Range for Which Used	†Average Resistance Change at Full Load
1a	Nickel base, non-magnetic Ni 75%, Cr 20% plus Al, Cu, Fe, etc.	800	Evanohm Karma Moleculoy Nikrothal L	0 ± 20	−65 to +250	Very high, medium and up, for low temp. coeff.	Under ±1% to ±2%
1b		800		0 ± 10	−65 to +150		
2a	Iron base, magnetic Fe 73%, Cr 22.5%, Al 4.5% (plus Co in one alloy)	800	Alloy 815-R Kanthal DR Mesaloy	0 ± 20	−65 to +200	Alternate sometimes for Class 1	Under ±1% to ±2%
2b		800		0 ± 10	0 to +150		
3a	Nickel-Chromium 80%—20%	650	Chromel A Nichrome V Nikrothal B Protoloy A Tophet A	+80 ± 20	−65 to +250	High and medium	+4 to +6%
3b		675		+60 ± 20			
4	Nickel—Chromium—Iron 60%—16%—24%	675	Chromel C Electroloy Nichrome Nikrothal 6 Tophet C	+140 ± 30	−65 to +200	High and medium	+5 to +8%
5a	Copper-Nickel 55%—45%	300	Advance Copel Cupron Cuprothal 294 Neutroloy	0 ± 20	−65 to +150	Low and low to medium for low temp. coeff.	Under ±1% to ±2%
5b				0 ± 40			
6	Manganin 13% Mn, 87% Cu	290	Manganin	0 ± 15	+15 to +35	Low and low to medium for low TC near 25°C	Under ±1% to ±2%†
7	Copper-Nickel 77%—23%	180	180 Alloy Cuprothal 180 Midohm	+180 ± 30	−65 to +150	Very low	+5% to +8%
9	Copper-Nickel 90%—10%	90	90 Alloy 95 Alloy Cuprothal 90	+450 ± 50	−65 to +150	Very low	+5% to +10%

*American Society for Testing Materials, Tentative Specification B267-68. †—For resistor with 300°C hot spot rise from 25°C ambient except 54°C rise for Manganin.

2.2.1 Precision

Resistors, whatever the type, with better than ∓1% tolerances are classified generally as precision resistors. The typical precision wire-wound, however, is rated with a ∓0.1% absolute-value initial tolerance. Moreover, most other precision resistor types are limited to less than 1-W ratings.

Precision wire-wound resistors come in a large variety of body sizes and shapes with axial, radial, or printed-circuit-mounting leads. Resistance wire (usually a nickel-chromium alloy) in various diameters, together with size limitations, determines specific resistance ranges. Typical production units range from 0.1 to 200KΩ, but the low-resistance end of the range generally must be larger than 0.1 Ω for very high precision units, because of lead-resistance errors.

Normal initial tolerances for high-production units can be controlled to ±0.002%. However, ±0.005% is a more common "best" initial tolerance and costs less. Unfortunately, a precision wire-wound resistor's tolerance error band increases with age, handling, and storage. Absolute resistance values will drift with age and temperature use, but these changes are small, typically 0.003%/ year.

2.2.2 Wire

The series of metals listed in addition to a few others are the special alloys made and drawn in wire form to produce carefully controlled resistance per unit length, temperature coefficient of resistance, tensile strength, specified ductility, and ability to operate for long periods at elevated temperatures.

The perfection of alloys combining high specific resistance with low TC (thermal coefficient of resistance), has made possible small precision-power resistors. Some of the 800-Ω/cmf (circular mil foot) low-TC alloys are limited to a maximum temperature of 275°C, if they are to have minimum change of resistance after full-load use.

The actual TC of the finished resistor depends on more than the wire and is influenced by such parameters as the core diameter, winding tension, thermal treatment after winding, and the nature of the core and coating materials.

For maximum reliability, many specifications place limitations on the use of the very fine wire (sizes as small as 0.0005 in. in diameter are in use). Military specification MIL-R-26 limits the wire size to a nominal of 0.00175 in. in diameter. Some specifications for small, axial-lead resistors permit wire of 0.001 in. or even 0.0008 in. in diameter. Another limitation for reliability by specification requires lower operating temperatures.

2.2.3 Coatings

Vitreous enamel is the standard high-temperature coating. This completely inorganic material is a complex mixture of minerals that is generally applied as an aqueous suspension of fine powder, dried, and then fired at temperatures around 760°C to fuse it into a smooth glassy covering, well bonded to the core. There are variations in composition among the vitreous enamels used that affect the firing temperature, match of coefficient of expansion with the core, capability of withstanding thermal cycling without cracking, maximum operating temperature, insulation resistance, and other characteristics.

Vitreous enamel coatings are customarily fairly thin and conformal. However, molded axial-lead resistors with a thick, uniform, cylindrical coating are available with high insulation values and other improved characteristics.

Silicone coatings generally consist of mixtures of silicone resins with fine inorganic powders (such as silica), which produce opacity, body, increased hardness, controlled thermal expansion, and improved heat resistance. These coatings are generally conformal, but molded silicone-ceramic coatings on axial-lead resistors are also produced.

2.3 FIXED-VALUE WIRE-WOUND POWER RESISTORS[9–11]

Power wire-wound resistors can safely dissipate large amounts of energy and can handle body temperatures up to 350°C. They are the workhorses of the resistor world and they come in a great variety of styles. Some are silicone coated and others are vitreous coated. Many have metal cooling fins and mounting ears.

Power ratings are limited by the maximum permissible hot-spot temperatures that can develop in the resistive element or core. Wire-wound units as large as 2000 W with surface temperatures close to 400°C are fairly common. Of course, it is best to keep them as cool as possible. Accordingly, a ceramic core provides good conduction for heat removal.

Wire-wound resistors used in electronic circuitry vary from tiny 1-W units (approximately $\frac{9}{16}$ in. long × $\frac{1}{8}$ in. in diameter) to 1500-W units (20 in. long × $2\frac{1}{2}$ in. in diameter). Resistances vary from less than 1 Ω to hundreds of thousands of ohms and megohms.

Wire-wound resistors are generally catalogued as shown in Table 2.8, with most types available in three different kinds of coating, namely vitreous enamel (the generally accepted industry standard), silicone (recently developed coverings), and cermets (limited-application group). Figure 2.3 shows examples of these various types.[12] Figure 2.4 is a cutaway cross section of a typical tab-type or lug-type resistor. A ceramic tube, generally steatite, but also supplied in other vitreous and refractory bodies, is equipped with two metal bands. These bands are typically made of a nickel–iron low-expansion alloy chosen so that once the ends are drawn tight and

Table 2.8 Wire-wound Power Resistor Types (Reproduced from Ref. 9.)

Type	Wattage Range	Resistance Range (ohms)	Winding Possibilities	Other Features
Axial Lead	1 to 14	1-200k	Inductive or Non-inductive	Fixed
Other Small Wire-Lead Resistors	3 to 20	0.5-100k	Inductive or Non-inductive	Fixed
Tab Terminal Tubular*	5 to 1000	0.1-1.7 meg.	Inductive or Non-inductive	Fixed, Tapped or Adjustable
Tab Terminal Tubular with Exposed Corrugated Ribbon Winding	10 to 1500	0.1-110	Inductive	Fixed, Tapped or Adjustable
Ferrule Equipped	5 to 220	0.1-100k	Inductive or Non-inductive	Fixed, Tapped or Adjustable
Edison Screw Base	10 to 215	1-185k	Inductive or Non-inductive	Fixed
Live Bracket	5 to 220	0.1-1 meg.	Inductive or Non-inductive	Fixed, Tapped or Adjustable
Strip Type Miniature & Regular	10 to 95	0.1-150k	Inductive or Non-inductive	Fixed, Tapped or Adjustable
Metal Clad	5 to 250	0.5-60k	Inductive or Non-inductive	Ratings based on heat-sink mountings
Bare (Heavy Current) Units: Edgewound or Round Wire	220 to 1200	0.4-42.1	Inductive	Fixed or Movable Taps

*Tab styles provide for soldering, bolted connections, and quick-connect shapes for push-on connectors, and also wire-wrap types.

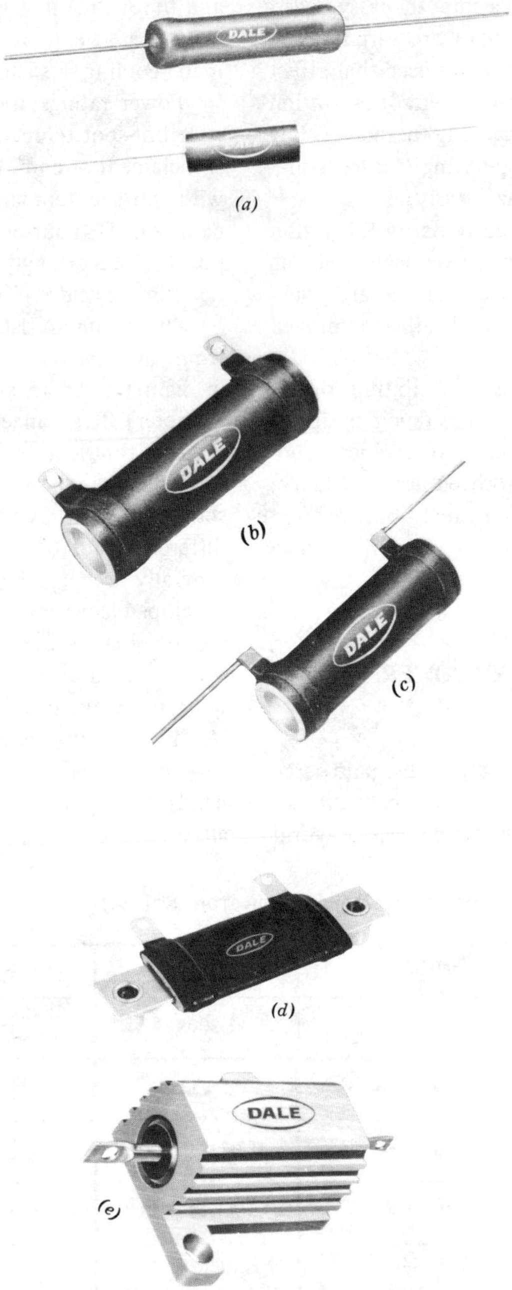

Figure 2.3 Wire-wound power resistors. (*a*) Axial Lead, (*b*) tubular (lug type), (*c*) tubular (lead type), (*d*) flat, (*e*) aluminum housed. (Courtesy of Dale Electronics, Inc.)

welded, the band will remain in place during heating and cooling cycles.

A vitreous or silicone coating over the wire element transforms the surface of a wire-wound resistor into a better, more uniform heat radiator. Leads often serve to conduct heat from the core.

The core must be durable, low cost, and able to withstand very high temperatures, and must have a high dielectric strength. Furthermore, it must be inert and stable and should be extrudable into many shapes. A ceramic core can meet all these requirements.

Table 2.9 lists the ceramic materials commonly used in power wire-wound resistors. As heat conductivity increases, the physical size of a resistor can be decreased

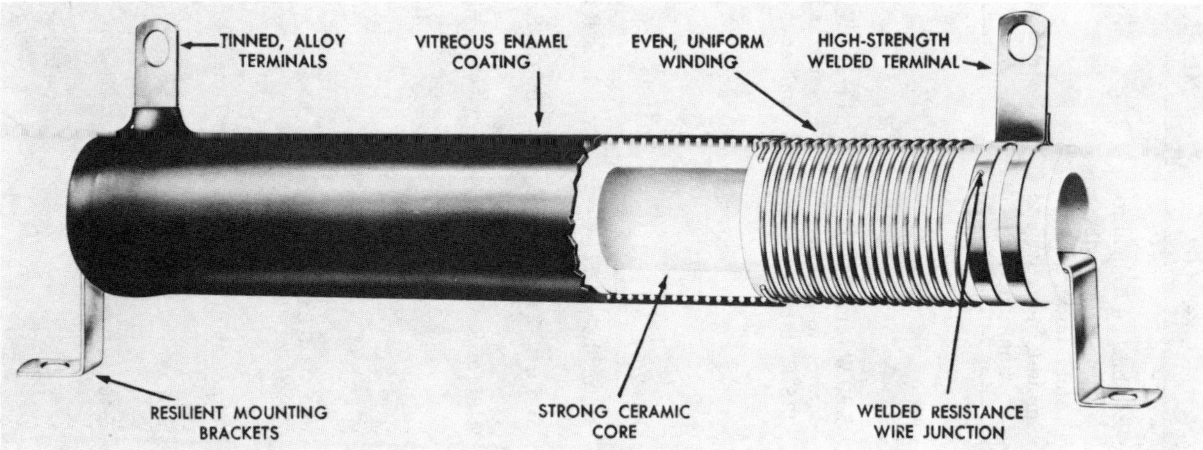

Figure 2.4 Cutaway view of vitreous enamel lug-type fixed-value power resistor (Courtesy of Ohmite Mfg. Co.).

for a given power rating. But, of course, the heat must have somewhere to dissipate when it leaves the core, hence adequate heat sinking must be provided.

Power resistors are usually wound in a single layer in order to take advantage of the heat-sinking capability of the core. But a single layer limits the maximum resistance attainable, although values as high as 275 Ω can be obtained. In addition, military specifications (see Tables 2.10 and 2.11) establish minimum wire size, which further restricts achieving high resistance values. Also, the requirement for noninductively wound units cuts the practically attainable upper limit by about one-half that of ordinary inductive types.

At the low end of the resistance range, at about 0.1 Ω, military specifications also impose limitations, that is, "no more than 20% of the overall body length shall remain uncovered by the winding." Commercial part values down to 5-mΩ can be achieved.

2.3.1 Watt Rating

Such factors as the enclosure, high ambient temperatures, and the presence of nearby materials or components of lower temperature rating affect the wattage that a resistor should be required to dissipate in a particular application. To allow for such considerations, it is general engineering practice to operate resistors at a fraction of nominal rating. On the other hand, resistors can safely handle pulses of power much higher than their rating, depending on the duration of the pulse and the amount of off time.

Table 2.9 Core Material[11]

Material	Cost Factor	Core Conductivity, BTU-ft/ft^2-h-°F
Steatite	low	1.5
Aluminum oxide	medium	8.0
Beryllium oxide	high	64.0

Figure 2.5 shows various important application conditions as well as the factors to be used to keep the resistor's maximum temperature rise within desired limits. The actual watts multiplied by the products of the factors gives the free-air watt size required. The following are the major conditions to be considered.

A. ***Ambient Temperature*** Above-standard ambients restrict the permissible temperature rise for a given hot-spot temperature, thereby reducing the permissible wattage.

B. ***Enclosure*** The enclosure limits the flow of heat from the resistor. An accurate estimate is almost impossible, except by comparison with similar known measured conditions. Calculation results in only approximate accuracy unless many factors are precisely known.

C. ***Grouping*** The general derating effect due to the close proximity of other heat sources can be significant.

D. ***Altitude*** The increase in ambient altitude reduces the cooling ability of the ambient air as it becomes rarefied.

E. ***Pulse Operation*** The temperature rise is a function of the avearage power, the thermal capacity of the resistor, and other physical properties.

F. ***Cooling Air*** Forced circulation of air (or fluids) increases the resistors rating by removing the heat more rapidly.

Table 2.10 Military Wirewound Resistors[11]

Parameter	MIL-R-26 RW Series Characteristic "V" General Purpose	MIL-R-26 RW Series Characteristic "U" Precision	MIL-R-18546 RE Series (metal-body chassis mount)	MIL-R-39007 Established Reliability Type RWR Series. Axial lead, precision, inductive and noninductive	MIL-R-39009 Established Reliability Type RER Series. Metal-body, chassis mount, inductive and noninductive	MIL-R-19365 RX Series Adjustable
Tolerances (± %)	5	0.1 0.5 1.0	1	0.1, 0.5, 1.0	1	5
Wattage ratings @ 25°C ambient	3-240 W derated to zero 350°C ambient	1-10 W derated to zero 275°C ambient	Characteristic G: 5-120 W derated to zero at 275°C ambient Characteristic N: non-inductive version of "G"	1-10 W derated to zero at 275°C ambient	5-30 W (mounted) derated to zero at 275°C ambient	11-210 W derated to zero at 350°C ambient
Maximum Tempco (±ppm/°C)	U characteristic: 30 for R > 10 Ω 50 for 1 < R < 10 Ω 90 for R < 1 Ω V characteristic: 260 for R > 20 Ω 400 for R < 20 Ω		30 for R > 2 kΩ 50 for R < 2 kΩ	30 for R > 10 Ω 50 for 1 < R < 10 Ω 90 for R < 1 Ω	30 for R > 20 Ω 50 for 1 < R < 19.6 Ω 100 for R < 1 Ω	260 for R > ρ 400 for R < ρ where ρ = 0.25 Ω/in.² of winding area
Maximum ΔR with life (25 C, rated power)	±(3% + .05 Ω) for RW 20-69. ±(0.5% + .05 Ω) for RW 70-81) Time: 2000 hours		±(1% + 0.05 Ω) Time: 1000 hours	±(0.5% + 0.05 Ω) Time: 2000 hours	±(1% + 0.05 Ω) Time: 2000 hours	±5% Time: 1000 hours
Maximum ΔR with shorttime over-load	±(0.2% + 0.5 Ω) for RW 70, 81. ±(2% + 0.05 Ω) for others. 10-times rated power for 5 sec		±(0.5% + 0.05 Ω) 5-times rated power for 5 sec	±(0.2% + 0.05 Ω) 5-times rated power for 5 sec (RWR71, 80-82, 89). 10-times rated power for 5 sec (others)	±(0.3% + 0.05 Ω) 5-times rated power for 5 sec	±2% 10-times rated power for 5 sec

Table 2.11 Wirewound Power Resistors[12]

Application	Applicable MIL Spec and Types	Wattage Rating	Resistance Range	Core Sizes	Terminals	Mountings	Tolerance
For limited space requiring high power-to-size ratio. Vertical or horizontal stacking	MIL-R-26 RW-20 thru RW-24	30-95 W	0.1–150K Ω	Length 1 1/4 to 6″	Lugs	Spring push-in	±5% (10% below 1 Ω)
General purpose wirewound resistors with a broad power range	MIL-R-26 RW-29, 30, 31, 32, 33, 35, 36, 37, 38, 47	5-225 W	0.1–1.3M Ω	O.D. 1/4 to 1-1/8″ Length 1-10 1/2″	Lugs	Push-in bracket or thru-bolt	±5% (10% below 1 Ω)
For resistance or voltage adjustment	MIL-R-19365C RX-29, 32, 33, 35, 36, 37, 38, 47	12-225 W	1–100K Ω	O.D. 5/16 to 1-1/8″ Length 1 1/2 to 10 1/2″	Lugs	Push-in bracket or thru-bolt	±5%
For voltage divider networks	MIL-R-26 RW-20, 21, 22, 23, 24, 33, 35, 36, 37, 38, 47	11-225 W	0.1–1.1M Ω	O.D. 5/16 to 1-1/8″ Length 1 1/2 to 10 1/2″	Lugs	Push-in bracket or thru-bolt	±10% each section (±10% total)
Axial leads. For applications requiring high performance at low cost	MIL-R-26 RW-67, 68, 69	2.5-13 W	0.1–100K Ω	Body Dia 0.188 to 0.375″ Body Length 0.500 to 1.781″ Leads 1.5 to 2″	Leads	Terminals	±5% (10% below 1 Ω)

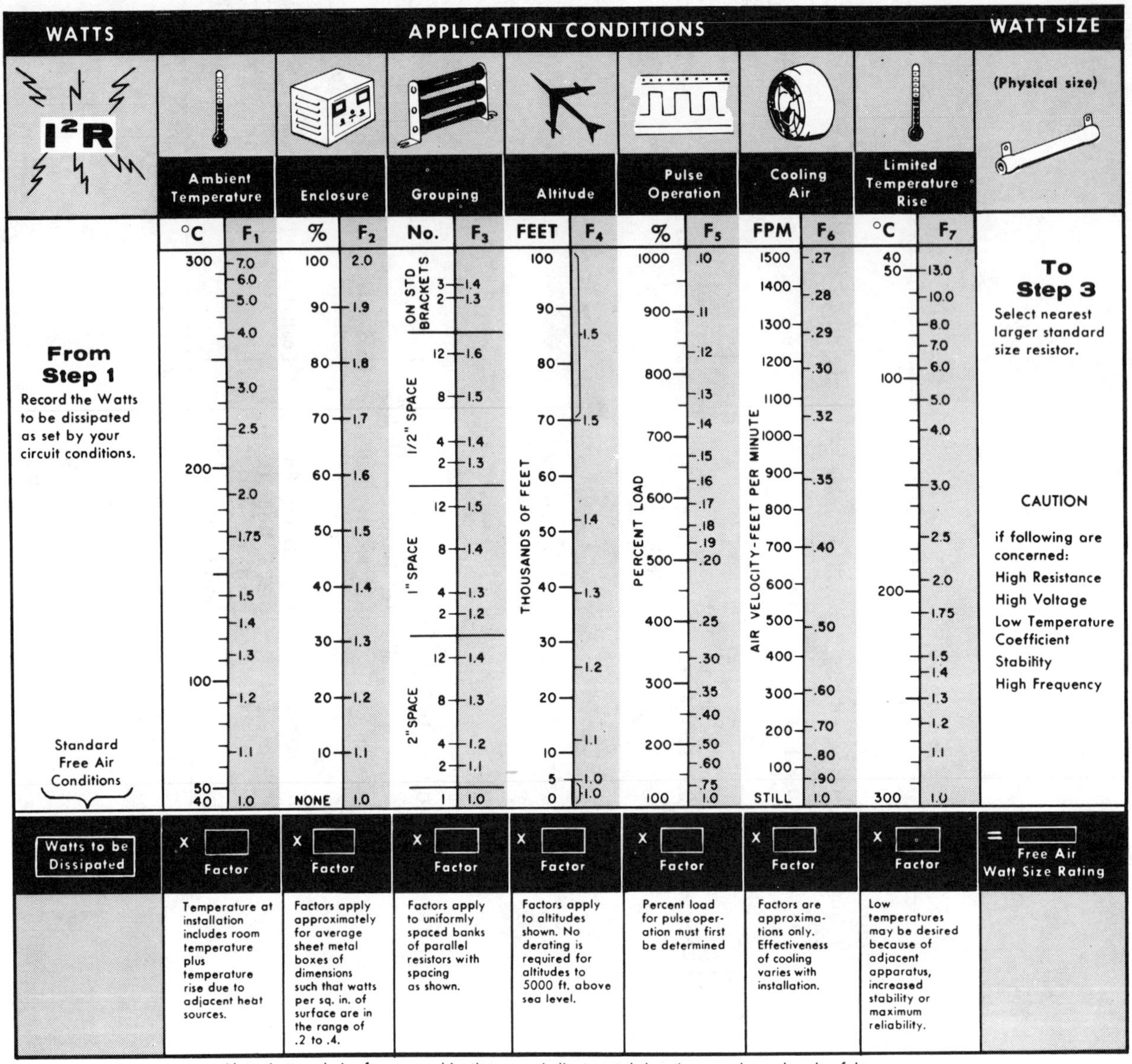

Note that analysis of some combinations may indicate total derating may be reduced safely.

Figure 2.5 Selecting a power resistor. (Reproduced from Ref. 10.)

G. ***Limited Temperature Rise*** Temperature rise less than maximum may be desired to protect adjacent components, to reduce change in resistance with load, and to increase the resistor's reliability.

H. ***Other*** Very high resistances, high voltages, radiofrequency or high-frequency pulse applications, and so forth, generally require special consideration.

2.4 FIXED-VALUE COMPOSITION RESISTORS[13]

Fixed-value composition resistors are among the most reliable, widely used components available for general industrial and military applications where low-cost, nonprecision resistors are required.

2.4.1 Construction

Axial-lead molded carbon composition resistors (Figure 2.6) are made by combining carbon powder with nonconductive fillers and binders. The resistance element and outer insulating shell are molded of similar materials into solid homogeneous structures that resist catastrophic failure and erratic resistance changes in severe environments.

Resistor lead design adds additional reliability to the unit. The lead end configurations allow for maximum

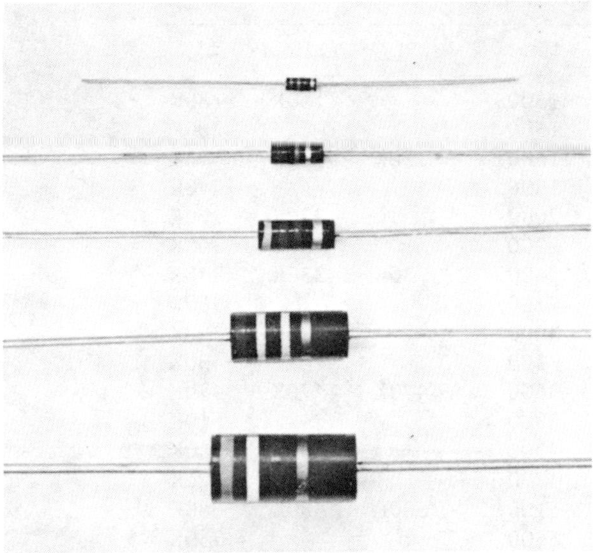

Figure 2.6 Axial-lead molded carbon composition resistors. (Courtesy of Stackpole Carbon Co.)

contact with the resistance material and assurance against pullout. The molded-in leads form a solid integral unit with the body.

2.4.2 Resistance

Standard carbon composition resistor values vary from 10 Ω to 22.0 MΩ with tolerances of 5, 10, and 20%, as shown in Table 2.12.

2.4.3 General Characteristics

The general characteristics of established-reliability composition resistors are as follows:

A. Nominal minimum resistance tolerance available is ±5%. Combined effects of climate and operation on unsealed types may raise this tolerance to ±15% from the low value due to aging, pressure, temperature, humidity, voltage gradient, and so forth.

B. High-voltage gradients will produce resistance change during operation.

C. High "Johnson" noise levels at resistance above one megohm preclude use in critical circuits of higher sensitivity.

D. RF will produce end-to-end shunted capacitive effects because of short resistor bodies and small internal distances between both ends.

E. Operation of VHF or higher frequency reduces effective resistance due to losses in the dielectric (the so-called "Boella" effect).

F. Exposure to humidity may have two effects on the resistance value: surface moisture may result in leakage paths that will lower the resistance values, or absorption of moisture into the element may increase the resistance. This phenomenon is more noticeable in higher ranges, since it depends on the resistance value.

G. The resistance-temperature characteristic is the highest for general-purpose resistor styles covered by military specifications.

2.4.4 Wattage

Standard values for fixed composition resistor wattage are $\frac{1}{8}$, $\frac{1}{4}$, $\frac{1}{2}$, 1, and 2 W. Figure 2.7 can be used to determine the voltage that may be applied to various-size resistors without exceeding the rated wattage. Figure 2.7 can also help determine what size resistor must be used if the voltage and resistance value of a circuit are fixed.

2.4.5 Color Coding

Color coding for composition resistors is shown in Figure 2.8.

2.5 RESISTOR NETWORKS[14-17]

Here is a review of how resistor networks stack up against discrete resistors:

A. "On-board" cost of a resistor network is often lower than that of discrete resistors.

B. Provide an added advantage of close tracking, generally in the area of 50-100 ppm or better. To achieve this performance using discrete resistors would require the specifying of matched sets at a much higher price.

C. Simplify parts inventory and stocking, in that the four or five different resistance values in a network can be stocked under one part number.

D. Are packaged in configurations compatible with semiconductors devices and integrated circuits, that is, dual-in-line packages (DIPs).

E. Save printed wiring board space and simplify board conductor layouts.

F. Offer tighter resistance tolerances than are generally available in discrete resistors of comparable price.

G. Are compatible with programmed automatic component insertion equipment.

2.5.1 Types

Precision resistor networks come in three basic constructions: discrete, thick film, and thin film, but they share many crossover points in their characteristics.

Table 2.12 Standard Carbon Composition Resistor Values

±20%	±10%	±5%	±20%	±10%	±5%	±20%	±10%	±5%
10	10	10			1300		180K	180K
		11	1500	1500	1500			200K
	12	12			1600	220K	220K	220K
		13		1800	1800			240K
15	15	15			2000		270K	270K
		16	2200	2200	2200			300K
	18	18			2400	330K	330K	330K
		20		2700	2700			360K
22	22	22			3000		390K	390K
		24	3300	3300	3300			430K
	27	27			3600	470K	470K	470K
		30		3900	3900			510K
33	33	33			4300		560K	560K
		36	4700	4700	4700			620K
	39	39			5100	680K	680K	680K
		43		5600	5600			750K
47	47	47			6200		820K	820K
		51	6800	6800	6800			910K
	56	56			7500	1.0M	1.0M	1.0M
		62		8200	8200			1.1M
68	68	68			9100		1.2M	1.2M
		75	10.0K	10.0K	10.0K			1.3M
	82	82			11.0K	1.5M	1.5M	1.5M
		91		12.0K	12.0K			1.6M
100	100	100			13.0K		1.8M	1.8M
		110	15.0K	15.0K	15.0K			2.0M
	120	120			16.0K	2.2M	2.2M	2.2M
		130		18.0K	18.0K			2.4M
150	150	150			20.0K		2.7M	2.7M
		160	22.0K	22.0K	22.0K			3.0M
	180	180			24.0K	3.3M	3.3M	3.3M
		200		27.0K	27.0K			3.6M
220	220	220			30.0K		3.9M	3.9M
		240	33.0K	33.0K	33.0K			4.3M
	270	270			36.0K	4.7M	4.7M	4.7M
		300		39.0K	39.0K			5.1M
330	330	330			43.0K		5.6M	5.6M
		360	47.0K	47.0K	47.0K			6.2M
	390	390			51.0K	6.8M	6.8M	6.8M
		430		56.0K	56.0K			7.5M
470	470	470			62.0K		8.2M	8.2M
		510	68.0K	68.0K	68.0K			9.1M
	560	560			75.0K	10.0M	10.0M	10.0M
		620		82.0K	82.0K			11.0M
680	680	680			91.0K		12.0M	12.0M
		750	100K	100K	100K			13.0M
	820	820			110K	15.0M	15.0M	15.0M
		910		120K	120K			16.0M
1000	1000	1000			130K		18.0M	18.0M
		1100	150K	150K	150K			10.0M
	1200	1200			160K	22.0M	22.0M	22.0M

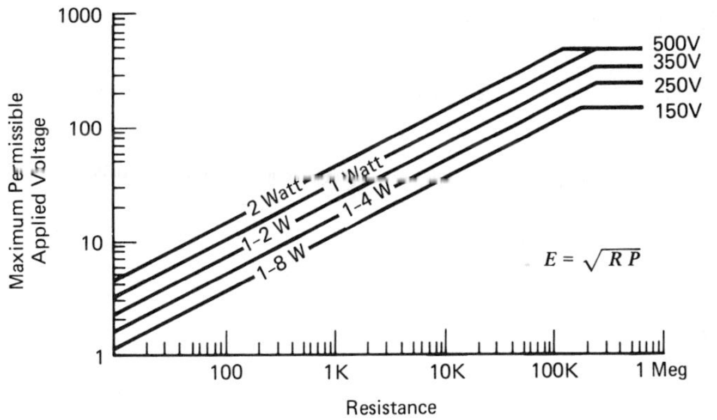

Figure 2.7 Maximum ratings of composition resistors.[13]

Discrete resistors are often combined to form packaged networks. Networks take precision one step further into technology. Networks made from discrete components have basically the same precision properties as do their constituent parts.

Thick-film and thin-film networks are individually designed to parameter limits to provide the desired precision properties, which depend on the type of material used in the film. A quick comparison of the types of networks shows that discrete networks are bulky when compared with the film networks, but they can operate with up to 12 and 14 bits in digital applications. They usually give the best initial accuracy and stability. Deposited-coefficient types, however, provide the best temperature coefficient tracking. But these networks are usually limited to about eight bits.

In general, thick-film resistor networks are used when cost is the main consideration. Thin-film networks are used in applications for which performance needs outweigh cost considerations.

A. ***Discrete Resistor Networks*** Discrete resistor networks have special performance characteristics that make them better suited than thick-film or thin-film in certain applications. Discrete resistor networks can be made to handle higher-power levels than is possible with the other two network types.

Changes in resistance values and tolerances can be made much more easily in discrete networks. In applications for which ultraprecision is the primary consideration, discrete resistors are often the only means to achieve it.

B. ***Thick-Film Resistor Networks*** Thick-film resistor networks are made by screen printing a cermet material on a substrate. Generally, thick-film networks have a higher temperature coefficient of resistance (TCR) and TCR tracking than do thin-film networks, and they cannot be adjusted as tightly as thin-film networks. However, the TCR of thick-film networks is still better than

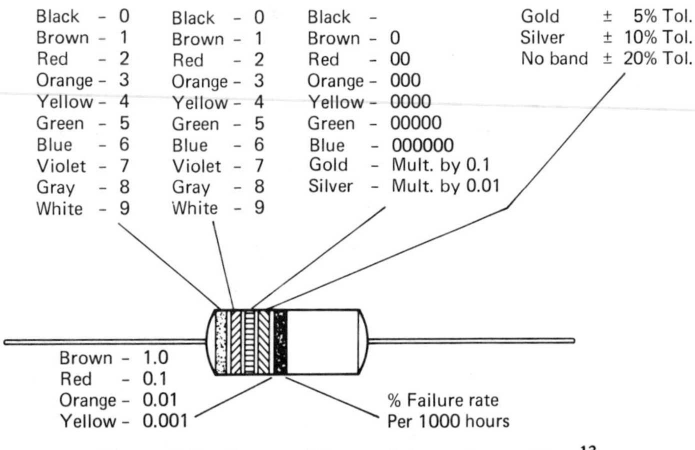

Figure 2.8 Composition resistor color coding.[13]

that of descrete-carbon film and carbon-composition resistors.

Thick-film networks suit broad general-purpose applications, ones in which characteristics such as tracking capability, tolerance, and stability need to be from fair to good. These constitute by far most applications and usually are the ones having the highest volume requirements.

C. ***Thin-Film Resistor Networks*** Thin-film resistor networks are made by vapor depositing a metal film on a substrate. Because all the resistors are on the same substrate, their characteristics closely match one another. The main advantages of thin-film networks are their low TCR, tight TCR tracking, and their ability to be trimmed to tight tolerance.

Thin-film networks are used where precision is the major requirement. Probably the most significant feature of thin-film networks is tracking—that is, the ability of individual resistors to react nearly identically to temperature changes and to age virtually identically.

Thin-film resistor networks are easily specified. The following are key parameters:

1 ***Storage Temperature Range,*** that is, the temperature range that the devices may encounter without power being applied in environments encountered during storage and shipment.
2 ***Operating Temperature Range***, that is, the temperature range the network will encounter in a powered-up state, and over which parameter specifications are to be maintained.
3 ***Value and Tolerance*** of the individual resistors in the network required over the operating temperature range, specified in ohms and percent.
4 ***Matching or Tracking*** between resistors in the network and the percent accuracy of matching required over the operating temperature range.
5 ***Maximum Power Dissipations*** of the individual resistors in the network.
6 ***Maximum Voltage*** allowed across any resistor in the network.
7 ***Frequency Response*** required.

Table 2.13 compares the thick-film and thin-film resistor technologies with respect to tolerance, TCR, TCR tracking, and matching. The values given are meant to be representative. As custom resistor networks are often specified, a wide range of performance value is possible.

2.5.2 Applications

There are many designs for which resistor networks have been applied. Generally used where a printed wiring board layout permits multiple resistors to be combined

Table 2.13 Thick-Film and Thin-Film Performance Characteristics[14]

Characteristic	Thick Film	Thin Film
Tempco	250 ppm typical; 100 ppm easily attainable	50 ppm typical; 10 ppm easily attainable
Tolerance	10–100 Ω: ±2 Ω typical; ±1 Ω easily attainable	±0.5%–1% typical
	100–200 k: ±2% typical; ±1% easily attainable	±0.05%–0.1% easily attainable
	200 k–1 M: ±5% typical; ±2% attainable > 1 M–5–10%	
Matching	1% matching with two resistors and 2% matching with three or more resistors easily attainable	5 ppm typical; 2 ppm easily attainable
Tracking	100 ppm typical; 50 ppm easily attainable	

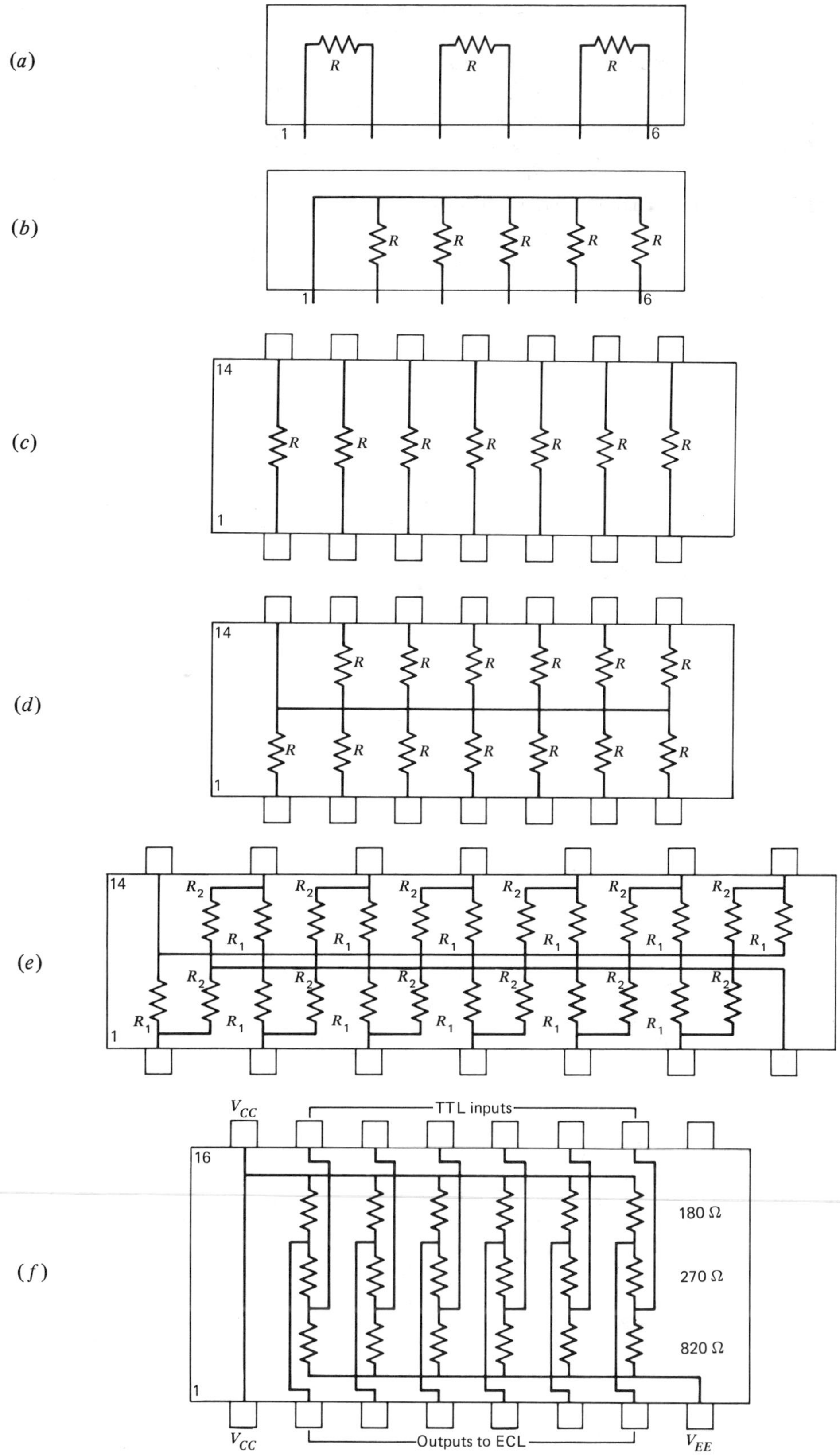

Figure 2.9 Typical resistor networks.[17] (*a*) Six-pin SIP with three resistors, (*b*) six-pin SIP with five resistors, (*c*) 14-pin DIP with seven resistors, (*d*) 14-pin DIP with 13 resistors, (*e*) 14-pin DIP with pulse-squaring network, (*f*) 16-pin DIP with 18 resistors.

into one part, they have been used in circuits requiring pull-up resistor arrays for unused TTL gates, parallel high-speed circuitry, wired OR configurations, pull-down applications, TTL-MOS interfacing, digital-pulse squaring, transmission line termination, current limiting, logic level translation, TTL-to-ECL translators, fixed-voltage attenuation with impedance matching, and so on. These applications occur frequently and have enabled the resistor network manufacturer to develop several useful network configurations, as shown in Figure 2.9.

Type RZ resistor networks for military applications are characterized as follows:

A. Nominal minimum resistance tolerance of ±1%.

B. Continuously derated from 70 to 125°C.

C. Operation at RF (above 100 MHz) may produce capacitive effects.

D. The resistance temperature characteristic is fairly low (±300 and ±100 ppm/°C).

CHAPTER THREE

Variable-Value Resistors

3.1 GENERAL INFORMATION

Variable resistors[18,19] are variable voltage dividers. Specifically, they are electromechanical transducers that depend on the relative position of a moving contact (wiper) and a resistance element for operation. They can be divided into three groups, as shown in Table 3.1: potentiometers (pots), trimmers, and rheostats.

Whether a pot or a rheostat, there is a choice of single- and multiturnshaft capabilities. Multiturn types provide better resolution and setability but, unfortunately, cost more than an equivalent single-turn device. Multiturn potentiometers and trimmers also have more complex mechanical drive mechanisms. The cost of a multiturn device can range from 2 to 10 times that of a comparable single-turn unit. Also, some wiper drive mechanisms are more reliable than others.

3.1.1 Size

Whether the variable-value resistor case is rectangular or round, there are limitations on the power-handling capability. The larger the case, the larger the resistance element, hence the higher the power dissipation capability.

Cermet elements offer improved power ratings over carbon and, in some cases, over wire-wound elements. Wire-wound potentiometers and rheostats usually have the highest power ratings.

3.1.2 Element Material

Resistance elements can be made from many materials, most commonly wire, carbon film, metal film, cermet, conductive plastic, and bulk metal. Each has different electrical and mechanical properties that best fit widely divergent applications, as shown in Table 3.2. Wire-wound elements have a limited useful frequency range from DC to about 5 kHz, while cermet and some film elements have limited use if the required resistance is below 50 Ω.

Different chemical compositions are used for resistance elements, with the effect that elements may have different sizes and shapes, even if they are rated for the same resistance and power dissipation. In general, the larger the resistance element, the better the resolution and power dissipation. A mix of capabilities is provided by hybrid potentiometers, as in those units whose wiper path of a wire-wound element is coated with a conductive film, such as plastic or metal.

Wire-wound elements generate the least noise, while carbon-composition elements generate the most. The noise is random, stemming from several sources, such as

or conductive-plastic potentiometer depends on how important temperature resistance and power are to the specific application. If the stability of a few parts per million per degree is necessary, wire-wound potentiometers are better. If the circuit has components whose temperature coefficient (TC) will overcome the potentiometer's stability, a non-wire-wound potentiometer may be the better choice.

3.2.2 Configurations

Potentiometers fall into three classes, according to their mechanical configurations, as follows (see Table 3.3):

A. ***Single-Turn Devices*** Rotate 360° or less. These potentiometers are available in diameters of up to 3 in. with linear and nonlinear outputs. They can be ganged for single-shaft operation of several units.

B. ***Multiturn Devices*** Rotate more than 360°, with three- and 10-turn units common. Potentiometers in this class are also available in diameters of up to 3 in. and are more accurate than comparable single-turn units, as shown in Figure 3.1.

C. ***Other Devices*** Include those that operate by sliding the wiper linearly and sector pots, which rotate less than 360°. Because potentiometers are continuously varied, long-term mechanical life is important, as up to 50 million revolutions are necessary for some applications.

Table 3.3 Advantages of Potentiometer Types[18]

Single-Turn Wire-Wound
More suitable for nonlinear outputs
Lower operating-torque requirements
Take up less space
Greater number of ganged units possible
Withstand more severe shock and vibration
Generally less expensive
Multiturn Wire-Wound
Greater linearity accuracy
Better electrical resolution
Capable of dissipating more power, size for size
Non-Wire-Wound
Better high-frequency a-c performance
Higher available resistance values
Cermet-element pots capable of higher power dissipation
Superior unloaded rotational life

3.2.3 Electrical Characteristics (see Table 3.4)

When specifying a precision potentiometer, bear in mind that potentiometer design is the result of a series of tradeoffs that have proved generally useful and widely acceptable. The making of such tradeoffs should be based on answers to the following electrical considerations:

A. Linear or nonlinear? If a nonlinear, has the function been adequately specified?

B. What type of linearity or conformity (absolute, zero-base, or independent) and accuracy? Absolute conformity is usually required for direct readout potentiometers. Independent is sufficient when trimmers are used. Is it a closed-loop function? If so, express conformity tolerance as zero-to-peak or peak-to-peak performance.

C. What resistance value and tolerance? Between which terminals resistance is measured? Should you specify open-loop resistance? Does it include or exclude end trimmers? Can the potentiometer be shunted to get low resistance while maintaining good resolution? Can it be shunted by a very high resistance to meet a close resistance tolerance? Has the effect of wiper shorting been considered?

D. When specifying the resistance of sine/cosine potentiometers, has the required resistance/quadrant been specified (i.e., the effective resistance per quadrant after the loop is closed)?

E. What type (angular or voltage) and value resolutions are necessary?

F. What noise is tolerable in the system? Can certain frequencies be excluded in measuring noise, or can sharp spikes of short duration be permitted (or filtered out)? Is the standard noise test pertinent, or do you need a different one? At what speed should it be run?

G. Have you considered the effect wiper load is going to have on output accuracy? (Most loads can be compensated for by the vendor and thus achieve any desired accuracy.)

H. Are special taps needed? Are they to be used as voltage reference points (voltage tap) or as input points (current tap)? How are they to be located voltage-wise relative to the index points? Relative to mechanical stops? Relative to end taps (standard)?

I. What is the electrical angle? How is it related to the required function? How is it related to the electrical angles of functions in other cups (phasing)? Is overtravel necessary? How is it related to the mechanical angle?

J. Is end resistance critical? Should it be measured by resistance or voltage? (A resistance measurement includes

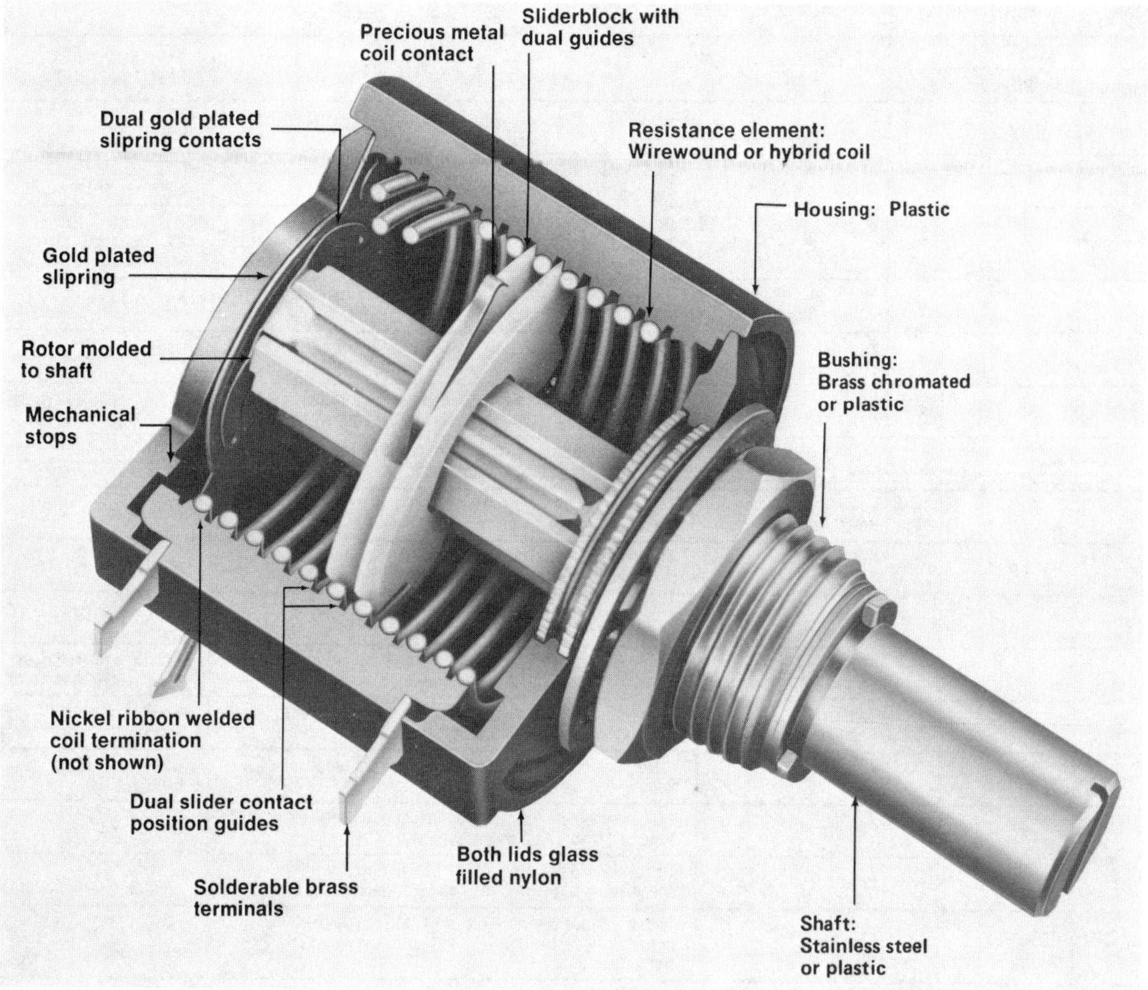

Figure 3.1 Multiturn precision potentiometer. (Courtesy of Beckman Instruments, Inc.)

wiper circuit resistance, which a voltage measurement does not.) Should the measurement be at the point of minimum resistance or voltage (standard) or at the stop (special)?

K. What voltage is applied? How much current will flow through the contact? Is this a rheostat application? What power is dissipated?

L. What maximum dielectric strength is desirable? At sea level or other altitude?

M. Is temperature coefficient of resistance important? Must it be matched to fixed resistors? Over what temperature range?

N. Are AC characteristics important? Quadrature voltage? Capacitive reactance? Other? At what frequency? At what wiper setting? Measured between what points?

O. How much backlash is tolerable?

3.2.4 Mechanical Considerations

Additional tradeoffs for mechanical considerations include the following:

A. Weldable or solderable leads? What size wires?

B. What type of mounting: servo, bushings, locking bushing, or three-hole? (A no-turn lug is usually advisable on a bushing mount.) Will the bushing length specified satisfy your maximum panel thickness with all hardware in place?

C. What kind of shaft: splined, flatted, or slotted? Must special diameter, length, or shoulder be considered? What runout is permissible relative to mounting surface? (Potentiometer manufacturers do not recommend machining operations on the shaft after assembly.)

Table 3.4 Potentiometer Electrical Parameters

APPLIES TO: Wirewound	Non-Wirewound	Both	PARAMETER NAME	BRIEF DEFINITION	TYPICAL VALUES OR SPECIFICATIONS
		•	Total Resistance, TR	The resistance between the end terminals	10Ω nom. to 5 MegΩ nom.
		•	Absolute Minimum Resistance, MR	The lowest value of resistance obtainable between the wiper and either end terminal	0.5Ω max. 1.0% (of TR) max.
		•	End Resistance, ER	The value of resistance obtained between the wiper and end terminal with the wiper positioned against the adjacent end stop	2% or 1Ω to 20Ω
		•	Minimum Voltage Ratio	The voltage ratio obtained with the wiper positioned to the Absolute Minimum Resistance	0.1% to 3.0%
		•	End Voltage Ratio	The voltage ratio obtained with the wiper positioned to the End Resistance	0 to 3.0%
		•	Contact Resistance, CR	The resistance of the wiper circuit from the wiper terminal to its ultimate contact with the element	.1Ω to 2.0% (of TR)
	•		Contact Resistance Variation, CRV	The instantaneous change in Contact Resistance that may occur as a result of wiper movement	±1% (of TR) ±3Ω
•			Equivalent Noise Resistance, ENR	Theoretical, lumped parameter resistance reflecting the magnitude of signal loss due to noise	100Ω max.
	•		Output Smoothness, OS	The instantaneous variation in output voltage from the ideal output voltage with wiper in motion	±0.1% (of TR)
		•	Adjustability of Resistance, A	The accuracy to which the wiper can be positioned to an arbitrarily selected resistance value	±0.1% (of TR)
		•	Adjustability of Output Voltage Ratio	The accuracy to which the wiper can be positioned to an arbitrarily selected voltage ratio	±0.05% (of Full Scale voltage ratio)
		•	Temperature Coefficient, TC	The change in total resistance that may occur due to a change in ambient operating temperature	w/w: ±50ppm/°C Non w/w: ±100ppm/°C
•			Resolution	A measure of the incremental changes in output voltage (or resistance) with wiper travel	0.01% (of TR) to 1.0% (of TR)
		•	Conformity	The allowable deviation of the actual output from a specified theoretical output function	0.1% to 2.0%
		•	Linearity	The allowable deviation of the actual output from a specified, straight line, theoretical output function	±.05% to ±1.0% (of Input Voltage)
		•	Insulation Resistance, IR	The resistance presented to a d.c. voltage applied between the terminals and all other external, conducting parts	1000 Meg Ω and higher

D. What materials are required for housing, shaft, terminals, and so forth?

E. Is there a shaft load or a speed requirement that necessitates ball bearings? Is a sleeve bearing satisfactory? Is there an axial load on the shaft?

F. What lateral runout, pilot diameter runout, shaft end play, or radial play is allowed?

G. Are stops required? What strength must they have? Does the application require a static or a dynamic load? (Avoid using potentiometer stops for anything but emergency use. Provide system stops.) What mechanical angle and tolerance are necessary?

H. If there is more than one cup, must they be phasable in the field?

I. Are any special terminals or terminal locations required?

J. Is very high or very low operating torque required? At room temperature or over a range of temperatures?

K. Is special marking required? Fungus treatment?

3.2.5 Operating Considerations (see Table 3.5)

Operation should also include the following considerations:

A. How many shaft revolutions will be required? At what rate per minute? To what value may any of the original electrical or mechanical requirements be permitted to degrade? Is the rotation continuous (over the bridge)? In one direction?

B. What is the operating temperature range? How much can the original requirements change over this temperature range?

Table 3.5 Potentiometer Application Guide[22]

To Avoid	Do
For all types of potentiometers	
Inoperative or damaged potentiometers	Mount potentiometers only on flat surfaces.
	Tighten mounting screws or nuts with reasonable torque.
	Be gentle with terminals or leads (excessive force can have a delayed reaction).
	Use solvents sparingly. Before immersing any pot in solvent, find out how much exposure it was designed to withstand.
Pot failure or degradation resulting from the effects of heat (either self-generated or externally produced)	Operate pots within their power and current ratings.
	Avoid subjecting the wiper circuit to currents in excess of its rating (which often is less than that permissible through the pot's resistive element).
	When using a pot as a rheostate, connect the unused end to the wiper. This helps with heat dissipation and reduces the wiper's burden when at or near minimum resistance.
	Never use a common VOM (multimeter) to measure end resistance in the wiper circuit. Use a digital ohmmeter instead.
	When the circuit permits, connect a fixed resistor in series with a pot to limit the current through it to a safe value.
	Design the associated circuits so that they can't cause excessive current flow through the element or wiper circuit of a pot.
Damage to associated circuit components caused by pot settings	When designing circuits containing pots, make sure that no pot setting will subject other components to damaging conditions. Look for the hidden domino effect.
For rectangular or square, screw-mounted trimmers	
Inoperative or damaged potentiometers	Never "economize" by attempting to use a trimmer's body as a cable clamp.
For pots with pin terminals	
Open circuits or noisy connections	Bend or form pins with care. Avoid overbending or excessive re-forming or twisting.
For pots with solder terminals	
Mechanical and electrical degradation	Apply only sufficient heat to make a good solder joint. If wave soldering, keep travel time to the minimum that will give good joints. Excessive time allows heat buildup.
Noisy or intermittent contacts	Be sparing with flux. Excessive fluxing can degrade the wiper's contact action.
For panel-mounted pots with threaded bushings	
Damaged housing, increased rotational torque, loss of linearity and shaft out of square	Provide a matching hole in the panel for the pot's antirotational pin (or cut it off).
Stripped bushing threads or loose assemblies	Use threaded bushings with sufficient length to permit full thread engagement by the nut. Never try to economize by making a pot do double duty as a fastener for other items.
Damaged mechanical stop	Use a reasonable size of knob so the mechanical stop is not overstressed. "Big knob on small pot" invites failure.
A variety of mechanical failures	Provide support for any panel or chassis that has a protruding pot shaft. Never let such an assembly lie on its face with the pot shaft supporting the weight.
	Don't use a pot as a convenient handle for picking up a chassis, panel or circuit board. The stress produced can easily result in mechanical and electrical damage.

C. Are there other special operating conditions, such as humidity, vibration, shock, acceleration, or altitude?

3.3 TRIMMERS[23,24]

There is a growing selection of trimming potentiometer, or "trimmer" types, sizes, electrical characteristics, and environmental capabilities. This wide field of choice can be used to advantage by reviewing basic considerations, determining the necessary requirements and specifications, and then selecting the best trimmer at the right price to do the job.

3.3.1 Style and Size

Commonly available trimmers come in the following styles and sizes:

A. Lead-screw-actuated rectangular style with side adjustment in sizes from $\frac{3}{4}$ in. to $1\frac{1}{4}$ in. long by $\frac{5}{32}$ in. to $\frac{5}{16}$ in. wide.

B. Lead-screw-actuated square style with top or side adjustment in sizes from $\frac{3}{4}$ in.down to $\frac{1}{4}$ in.

C. Single-turn round or cubic style with top or side adjustment in sizes from $\frac{1}{2}$ in. down to $\frac{1}{4}$ in. as shown in Figure 3.2.

Figure 3.2 Round trimmers. (Courtesy of Mepco/Electra Inc.)

If the trimmer is to be used on a printed-wiring board edge and is to be adjusted after the board is installed in a rack, the square and rectangular trimmers are usually considered. Both can be adjusted from the side with little trouble. If the trimmer is to be mounted in the center of a printed-wiring board, both round and square top-adjustment trimmers are available. If the trimmer is to be panel mounted, a rectangular or round trimmer may be desirable, as they are usually available with a threaded bushing for direct panel mounting.

When packaging calls for stacking trimmers, rivetless models are the most desirable. When trimmers stack on the same mounting screws, they should be derated to approximately 75% of their normal power rating.

3.3.2 Resistive Element

The four basic types of trimmer resistive elements are wire-wound, metal-film, cermet, and carbon (see Table 3.6). The choice between wire-wound and non-wire-wound elements usually depends on whether infinite resolution of the non-wire-wound element is more desirable than the better temperature coefficient (tempco) of the wire-wound element.

Wire-wound trimmers have virtually zero contact resistance and very low electrical noise. The resistance tolerance, percent error at full scale, can be as low as 3%.

Metal-film resistance elements compare favorably in tempco and contact resistance with wire-wound units, while offering infinite resolution, which is not possible with wire-wound devices. Metal-film devices perform best in medium- and low-resistance applications with an average tolerance of ±10%.

Cermet resistance elements possess a tempco of 100 ppm/°C, compared with a maximum of about 50 ppm/°C for wire-wound and metal-film elements. In addition, cermet trimmers have an inherent contact resistance of about 2% the total resistance. Tolerances are usually ±10%.

Carbon resistance elements are similar in performance to cermet devices, except that they are limited in lower operating temperature range and have a tempco of several hundred parts per million per degree centigrade. They are available in medium- and high-resistance ranges with tolerances of about ±20%. Carbon trimmers have the lowest cost, which makes them popular for many applications.

When an application calls for the qualities of both wire-wound and non-wire-wound elements, an excellent choice is often a metal-film trimmer. They are free of the high-temperature coefficients of other non-wire-wounds and are the only infinite-resolution trimmers that offer virtually zero contact resistance at the trimmed position.

Table 3.6 Trimmer Resistor Elements

Characteristics	Wire-Wound	Non-Wire-Wound Trimmers			
		Cermet	Hot-Mold Carbon	Carbon Film	Thin-Metal Film
Setting ability	Poor to good	Excellent	Excellent	Excellent	Excellent
Resistance range	Low to medium	Low through high	Medium to high	Medium to high	Low to medium
Power rating	Medium	High	Low	Low	Medium
Temperature coefficient	Lowest	Low to medium	High	High	Low
Environmental stability					
High temperature	Good to excellent	Excellent	Poor	Fair	Excellent
Load life	Good to excellent	Excellent	Fair	Fair	Excellent
Humidity	Good to excellent	Excellent	Poor	Poor	Excellent
Rotational life	Good	Excellent	Good	Good	Fair
Rheostat usage	Good	Fair	Poor	Poor	Fair
AC usage	Fair	Excellent	Excellent	Excellent	Excellent

A. ***Setting Ability*** The ability to adjust to fine settings is one of the most important requirements of a trimmer. The wire-wound trimmers are limited in this ability by resolution, or number of wire turns in the element, with each turn appearing as a definite step during fine adjustment.

The non-wire-wound trimmers are superior because their resolution, or the degree to which output voltage responds to adjustment, is virtually infinite. The lead-screw-actuated multiturn trimmers are somewhat easier to adjust accurately than are the single-turn units.

B. ***Resistance Range*** A wide resistance range is necessary to meet all applications. The cermet trimmers offer the widest range, from a few ohms to several megohms. The wire-wound trimmers are limited to a maximum of 100 K Ω, while the other non-wire-wound types exclude many high and low values.

C. ***Resolution*** Metal-film, cermet, and carbon trimmers have theoretically infinite resolution. The resolution of wire-wound elements can vary widely, even among trimmers of the same resistance.

When a circuit calls for the qualities of both wire-wound and non-wire-wound elements, an excellent choice is a metal-film trimmer. They are free of the high-temperature coefficients of other non-wire-wounds and are the only infinite-resolution trimmers offering virtually zero contact resistance at trimmed position.

3.3.3 Mechanical Considerations

Low-mass wiper designs play an important part in a trimmer's ability to hold a setting when subjected to shock, acceleration, and vibration. The wiper should be of one-piece construction to keep its mass to a minimum. Multipiece wiper construction increases mass and decreases reliability.

The wiper should transmit the wiper voltage to a single-piece commutating bar and external terminal. At the ends of its travel, the wiper should act as a clutch mechanism to protect the trimmer from damage if the adjustment screw is turned past the minimum or maximum setting positions. Wipers without a damage-proof clutch action may break due to over travel or rigorous environmental conditions, thereby exposing the system to open-circuit conditions or power surges.

External terminations are another critical area where mechanical characteristics must be considered. If the trimmer is to be safe from open circuits due to termination burnout or separation, the resistance element should be firmly attached, usually silver brazed, directly to the external terminals.

A. ***Operating Temperatures*** Although operation above 125°C is rare, resistance elements for wire-wound trimmers can be made using special alloys of precious metals for operation at elevated temperatures. When high-temperature applications are anticipated the mandrel on which the resistance wire is wound should be ceramic, as ceramic has greater dimensional stability than do conventional insulated copper-wire mandrels.

B. ***Humidity Proofing*** The use of immersion in solvents for flux removal from soldered printed-wiring boards necessitates the use of trimmers with antimoisture seals. Fortunately, virtually all styles and sizes of trimmers are available in sealed versions.

C. ***Environmental Stability*** It is important that both the total resistance and voltage setting of a trimmer remain unchanged during temperature extremes, long-term load life, and humidity cycling. The cermet trimmers show superior resistance stability. All the non-wire-wound trimmers show good voltage-setting stability due to essentially infinite resolution. Good setting stability during shock and vibration depends primarily on sound mechanical design. However, the non-wire-wound trimmers have an advantage because of their resolution.

3.3.4 Electrical Considerations

A. ***AC Usage*** The non-wire-wound trimmers perform satisfactorily in AC applications into the megahertz range, while wire-wound trimmers are limited to the low kilohertz range due to capacitance from wire to core.

B. ***Noise*** Electrical noise in wire-wound trimmers is random and usually increases during exposure to operational environments due to oxidation and wear debris. In contrast, noise in non-wire-wound trimmers is repeatable and remains essentially unchanged during exposure to operational and environmental life.

C. ***Failure*** Improved trimmer reliability in all applications is of growing importance. The cermet and carbon trimmers are less susceptible to catastrophic failure than are the wire-wound or thin-film types.

3.4 VARIABLE RESISTOR TERMS AND DEFINITIONS[20]

Absolute Minimum Resistance (Residual Resistance) The resistance between the wiper and the end terminal, with the wiper positioned mechanically to obtain the minimum resistance at the end being measured.

Adjustability The precision with which the output of a device can be set to a desired value.

Adjustment Shaft The mechanical input member of a trimming potentiometer which, when actuated, causes the wiper to traverse the resistance element, resulting in a change in output voltage or resistance.

Adjustment Travel (Electrical) The total travel of the adjustment shaft between minimum and maximum output voltages.

Attenuator A device used to reduce power, voltage, or current in controllable and known amounts.

Base (see Figure 3.3) A stamped or molded part used as a structural member to unite the terminals and the resistor element.

Bushing (see Figure 3.3) A mounting method–having a threaded part that serves as contact surface for the unit and the chassis.

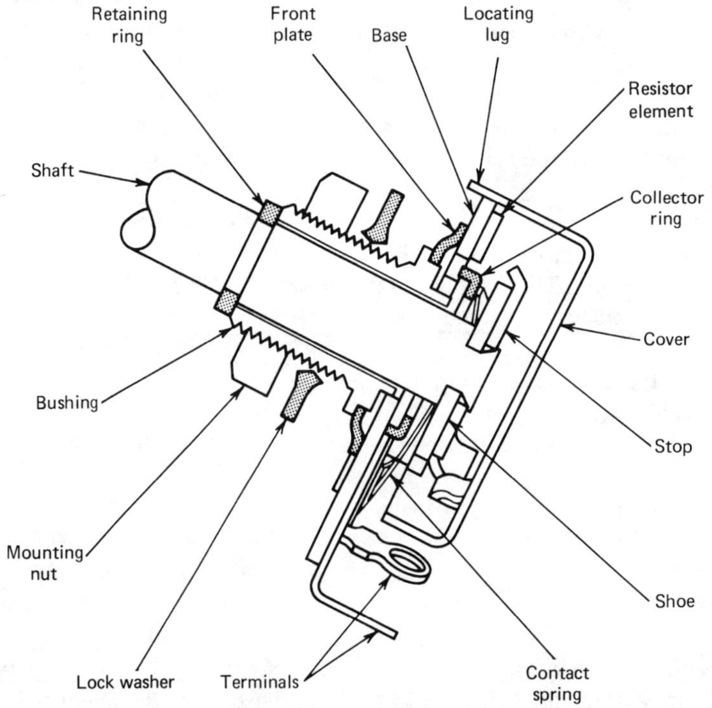

Figure 3.3 Basic variable-value resistor elements.[20]

Collector (see Figure 3.3) A portion of the center terminal—the "track" that permits electrical continuity during rotation.

Contact (see Figure 3.3) Single wiping: One point of contact with the resistive element. Double wiping: Two points of contact with the resistive element.

Contact Resistance In a potentiometer or trimmer, a variation in the electrical output not present in the electrical input, observed when the contact is moved over the element.

Contact Resistance Variation (CRV) The apparent resistance seen between the wiper and the resistance element when the wiper is energized with a specified current and moved over the adjustment travel in either direction at a constant speed. The output variations are measured over a specified frequency bandwidth, exclusive of the effects due to roll-on or roll-off of the terminations and is expressed in ohms or percent of total nominal resistance.

Contact Spring A metallic member that provides electrical continuity between the resistor element and the center terminal.

Cover A shield that protects all internal parts from mechanical damage.

Continuity Continuity is the maintenance of continuous electrical contact between the wiper and both end terminals of the resistive element.

Continuity Travel The total travel of the adjustment shaft over which electrical continuity is maintained between the slider and the resistance element.

DC Offset The DC component of contact resistance. It is relatively constant with a linear resistance slope and acts as a base level from which the AC component fluctuates.

Deadband (Dead Zone) The portion of the element that is shorted by a tap. When the wiper traverses this area, there is no change in output.

Dielectric Strength The ability of a potentiometer to withstand an impressed AC potential between the terminals and the exposed metallic surfaces without a breakdown (flash) for carbon composition elements or without exceeding a specified leakage current for wire-wound elements.

Dielectric Withstanding Voltage The ability to withstand the application of a specified potential of a given characteristic, between the terminals and all other external conducting members such as shaft, housing and mounting hardware without exceeding a specified leakage current value.

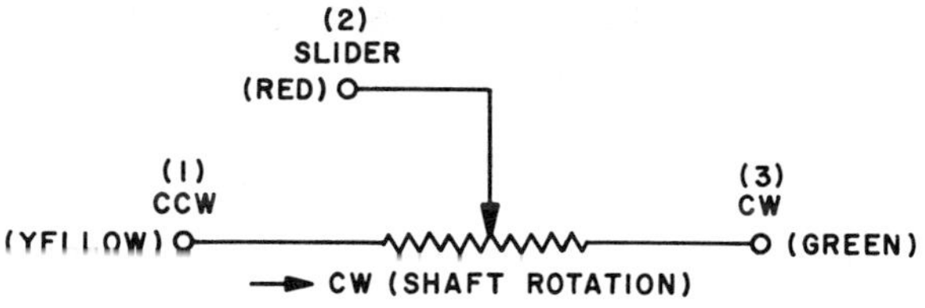

Figure 3.4 Direction of travel.[20]

Direction of Travel Clockwise (CW) or counterclockwise (CCW) rotation of the adjustment shaft when viewing the adjustment shaft end of the potentiometer. The designation of terminals shown in Figure 3.4 corresponds to the direction of slider travel.

Electrical Rotation In a potentiometer, the total rotation of the shaft over which electrical continuity between the contact and an end terminal is maintained. Electrical rotation includes the minimum flats.

Electrical Travel (Effective Rotation) The travel of the wiper on the element, during which a unidirectional change in output voltage ratio is experienced.

Element A resistive material to which is applied a voltage that is to be divided. It is normally a coil of wire on a card or mandrel, or a conductive film deposited on an insulator base material.

End Resistance (Re) In a potentiometer or trimmer, the resistance measured between the wiper terminal and an end terminal when the wiper is positioned at the corresponding end of mechanical travel. Also called absolute minimum resistance for continuous rotation trimmers.

Equivalent Noise Resistance (ENR) Wire-wound only. Any spurious variation in the electrical output not present in the input, defined quantitatively in terms of an equivalent parasitic, transient resistance in ohms, appearing between the contact and the resistive element when the shaft is rotated or translated.

Front Plate (see Figure 3.3) A basic assembly member that locates and aligns all external parts properly.

Hop-off In a potentiometer or trimmer, the sudden jump in resistance as a contact is rotated over the junction of two resistance slopes or over the junction of the minimum flat and the resistance slope.

Immersion Sealed The ability of the unit to withstand submersion in acceptable cleaning solutions used in normal soldering processes without performance degradation under specific environmental conditions.

Impedance The opposition to the AC current, equivalent to resistance in DC circuitry.

Insertion Loss The effect of an attenuator set at minimum resistance on the load current of a circuit. It is the ratio of the current with the attenuator in the circuit to the current if that attenuator were not present, generally expressed as a power loss in decibels.

Insulation Resistance The resistance to a specified DC voltage impressed between the terminals and all other external conducting members, such as shaft, housing, and mounting hardware.

Linearity It is the maximum deviation from the curve formed by plotting potentiometer electrical output versus mechanical travel, and the best straight line which may be drawn through this curve.

Load Life The number of hours during which a trimming potentiometer will dissipate rated power under specified operating conditions while remaining within specified allowable degradation.

Load Resistance In a potentiometer or trimmer, an external resistance as seen by the output voltage (connected between the wiper terminal and the designated reference point).

Locating Lug An extension on the front of a potentiometer that extends through the mounting panel to prevent relative motion.

Matching The selection of two elements at specified attenuation points or in specified attenuation ranges so that the ratio of the outputs of the elements at these points, calculated as a power ratio, does not exceed a specified limit.

Mechanical Adjustment The number of turns of shaft rotation necessary for complete mechanical wiper travel.

Mechanical Life The maximum number of complete cycles through which a potentiometer may be actuated without electrical or mechanical failure.

Mechanical Rotation The total number of degrees through which a potentiometer shaft can be rotated.

Mechanical Travel Clutching Action: The total travel of the adjustment shaft between the points where the clutch actuation begins. Continuous Rotation: The total travel where the slider continues to move as the adjustment shaft continues to be actuated. Solid Stops: The total travel of the adjustment shaft between stops.

Minimum Flats In a potentiometer or trimmer, the area at each end of the element that has no change of electrical output with rotation of the shaft.

Multiturn Adjustment In a potentiometer or trimmer, requiring more than 360° mechanical adjustment to cause the wiper to traverse the total resistance element.

Noise Any electrical signal or disturbance that interferes with the desired electrical output. It may be caused by faulty termination, foreign particles, or oxidation of the element.

Non-Wire-Wound Trimming Potentiometer A trimming potentiometer characterized by the continuous nature of the surface area of the resistance element to be contacted. Contact is maintained over a continuous, unbroken path. The resistance is achieved by using material compositions other than wire such as carbon, conductive plastic metal film, and cermet.

Operating Temperature Range The maximum and minimum ambient temperatures at which a potentiometer may be operated without permanent damage or failure.

Output Resistance The resistance measured between the slider terminal and the designated reference point. Unless otherwise specified, the designated reference point is the counterclockwise (CCW) terminal.

Output Voltage (*e*) The voltage between the slider terminal and the designated reference point. Unless otherwise specified, the designated reference point is the counterclockwise (CCW) terminal.

Output Voltage Adjustment Ratio (*e*/*E*) The ratio of the output voltage to the designed input reference voltage. Unless otherwise specified, the reference voltage is the total applied voltage.

Potentiometer An electromechanical device having a terminal connected to each end of the resistive element and a third terminal connected to the wiper contact. The electrical input is divided as the contact moves over the element, thereby making it possible to change the resistance mechanically.

Power Rating The maximum power that a trimming potentiometer can dissipate across the total resistive element under specified conditions while meeting specified performance requirements.

Precision Potentiometer A mechanical electrical transducer dependent on the relative position of a moving contact (wiper) and a resistance element for its operation. It delivers with a high degree of accuracy a voltage output that is some specified function of the applied voltage and shaft position.

Push Strength The axial force on the shaft toward the potentiometer that can be applied without causing mechanical or electrical damage.

Residual Resistance The resistance between the center terminal and either end terminal with the shaft positioned against the stop at that terminal.

Resistance Element A continuous, unbroken length of resistive material without joints, bonds, or welds, except at the junction of the element, and the electrical terminals connected to each end of the element or at an intermediate point such as a center tap.

Resistance-Temperature Characteristic (RTC) The difference between the total resistance values measured at a reference temperature of 25°C and the specified test temperature expressed as a percentage of the total resistance.

$$\text{RTC} = \frac{R_2 - R_1}{R_1} \times 100$$

where R_1 = resistance at reference temperature (25°C) in ohms

R_2 = resistance at the test temperature in ohms

Resistance Variation Uniformity in the rate of change of the electrical output. A variation is generally indicated as a backup on an ohmmeter.

Resolution In a wire-wound potentiometer, trimmer, or adjustable resistor, the average increment of output determined by the number of turns of the resistance wire. Carbon-composition elements have infinite resolution.

Retaining Ring (see Figure 3.3) Found only on units that have bushing-type construction. This ring is designed to retain the shaft within the unit.

Rheostat A variable resistor with two electrical terminals—one attached to one end of the resistance element, and the other attached to the movable contact or wiper.

Rotational Life In a potentiometer or trimmer, the number of cycles obtainable under specific operating conditions while remaining within specified allowable degradation. A cycle is one complete traversal of the wiper over the resistive element in both directions.

Rotational Torque The average torque required to turn a potentiometer shaft through its entire rotation.

Setting Stability In a potentiometer or trimmer, the amount of change in the output voltage without readjustment, expressed as a percentage of the total applied voltage.

Shaft (see Figure 3.3) A cylindrical part to which the contact is attached and is rotated to control the resistance.

Shaft Wobble Total movement of the shaft in any one plane, normally measured 1 in. from mounting surface.

Shoe (see Figure 3.3) An insulated member that transmits shaft movement to the contact spring.

Single-Turn Adjustment (ST) In a potentiometer, requiring 360° or less mechanical input to cause the wiper to traverse the total resistance element.

Slider The member in contact with the resistive element that allows the output to be varied with mechanical input adjustment.

Stacking The mounting of one trimming potentiometer adjacent to, or on top of, another utilizing the same mounting hardware.

Starting (Operating) Torque The maximum moment in the clockwise and counterclockwise directions required to initiate shaft adjustment anywhere in the mechanical travel.

Stop (see Figure 3.3) A part connected to the shaft that limits total mechanical rotation. This is often an integral part with the shoe.

Stop Clutch (SC) A device that allows the slider to idle at the ends of the resistance element, without damage, as the adjustment shaft continues to be actuated in the same direction.

Stop Solid (SS) A positive limit to mechanical or electrical adjustment, or both.

Stop Strength The torque that can be applied to a potentiometer in either the full clockwise or counterclockwise position without causing electrical or mechanical damage.

Stop Torque The maximum static moment that can be applied to adjustment shaft at each mechanical stop for a specified period of time without loss of continuity or mechanical damage, which would affect operational characteristics.

Tap A fixed electrical connection to a specified position on the element.

Temperature Coefficient of Resistance (TC) (Tempco) The unit change in resistance per degree Celsius change from a reference temperature, expressed in parts per million per degree centigrade.

Terminals External fittings for making electrical connections. These may be insulated stranded leads, solder lugs, printed circuit pins, turrets, solid wires, or any other type of electrical connector.

Terminal Strength The ability of the terminals to withstand specified mechanical stresses without sustaining damage that would affect utility of the terminals or operation of the trimming potentiometer.

Termination The connection between the end of the resistance element and the terminal.

Torque A twisting moment measured in ounce-inches, necessary to start the rotation of the shaft.

Total Resistance The DC resistance measured between the input terminals with the slider positioned at either end of the mechanical travel, or for continuous rotation trimmers, in the dead space.

Trimming Potentiometer An electrical mechanical device with three terminals—two connected to the ends of a resistive element and one connected to a movable conductive contact that slides over the element, thereby allowing the input voltage to be divided as a funtion of the mechanical input. It can function as either a voltage divider or rheostat.

Usable Range That portion of the element available for use after subtracting the end resistance. It is usually expressed as a percentage of the total resistance.

Voltage Coefficient A measure of the change of resistance with applied voltage. This term is applicable to carbon potentiometers only. The voltage coefficient is expressed in percent per volt and is computed from the change in total resistance noted when applying full-rated voltage and one-tenth-rated voltage for short duration to avoid resistance changes due to self-heating.

Voltage Ratio (Percent) A figure indicating the position of the wiper relative to the terminations of the element expressed in percentage of total applied voltage. It is calculated by dividing the output voltage by the total applied voltage times 100.

Wiper (Slider, Brush, or Movable Contact) A movable conductive part that divides the voltage according to its position on the resistance element and transmits it to an external electrical terminal.

Wire-Wound Trimming Potentiometer A trimming potentiometer characterized by a resistance element made up of turns of wire on which the slider contacts only a small portion of each turn.

PART TWO

Capacitors

CHAPTER FOUR
Capacitors, General Information

CHAPTER FIVE
Fixed-Value Electrolytic Capacitors

CHAPTER SIX
Fixed-Value Electrostatic Capacitors

CHAPTER SEVEN
Variable-Value Capacitors

CHAPTER FOUR

Capacitors, General Information

A capacitor[1,25,26] consists basically of two metal plates separated by an insulating material (dielectric), such as air, paper, ceramic, aluminum oxide, tantalum oxide, Teflon, and other insulating materials that have controlled impurities. The size of the capacitor is dependent on the area of the metal plates, the distance between them, and the dielectric used (see Figure 4.1).

The farad (F) is the unit of capacitance. However, the farad is inconveniently large for practical use, and submultiples of the farad are normally used in specifying the size of the capacitor:

A. 1 microfarad = 1 millionth of a farad (10^{-6} F).
B. 1 micromicrofarad = 1 picofarad (pF) = 1 millionth of a microfarad (10^{-12} F)

Capacitors connected in parallel effectively increase the area of the metal plates; the individual capacitors are simply added together to obtain the total capacitance of the parallel group. Conversely, capacitors connected in series will reduce the total capacitance of that circuit according to the formula

$$\frac{1}{C_t} = \frac{1}{C_1} + \frac{1}{C_2} + \frac{1}{C_3}$$

At very low frequencies and also at very high frequencies there is a loss that sets a limit to the practical use of a capacitor with any given dielectric. Figure 4.2 shows the approximate usable frequency range for capacitors with various dielectric materials.

4.1 TYPES

Capacitors are generally of two types: fixed-value and variable-value, with fixed-value capacitors being either electrolytic or electrostatic. Electrolytic capacitors are polarized devices with high capacitance ratings normally used for bypass and coupling applications. Electrostatic capacitors are of types that are usually differentiated by the dielectric material/medium, including plastic, paper, mica, glass, ceramic, air, and vaccum.

4.2 ELECTRICAL PROPERTIES[27,28]

In order to better understand the electrical properties of capacitors, some knowledge of dielectric materials is required. In addition, other highly significant properties must be understood, especially when capacitors are used in AC applications.

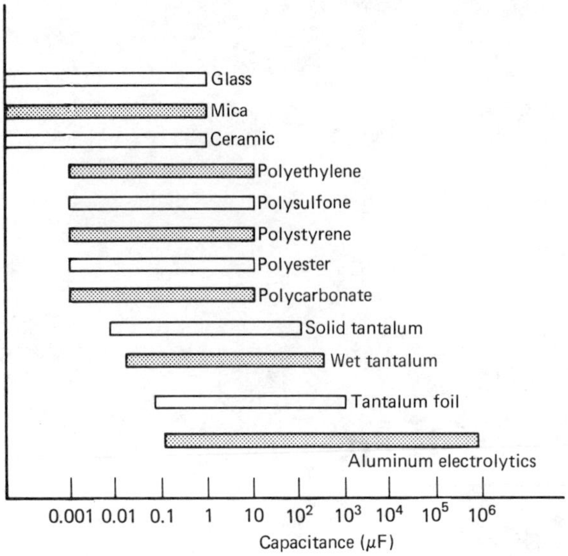

Figure 4.1 Capacitance vs capacitor type.[25]

4.2.1 Dielectric Constant

The dielectric constant (K) achieved in any insulator material occurs as the result of interactions between an applied electric field and localized charge centers within the dielectric. Charges are not transferred, but are displaced by the field, that is, positive charges shift toward the negative electrode and negative charges become displaced toward the positive electrode. In dielectrics with a high dielectric constant, easily polarized charges shift, creating a polarization effect that opposes the electric field applied, thereby drawing more charge onto the electrodes. Dielectric constants of materials are compared with the dielectric constant of a vacuum, as shown in Table 4.1.

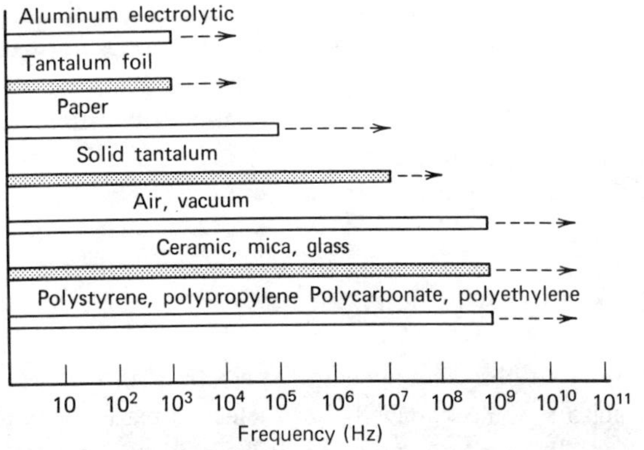

Figure 4.2 Capacitor frequency vs capacitor type.[25]

Table 4.1 Dielectric Constants @ +25°C[28]

Dielectric Material	K (Dielectric Constant)
Vacuum	1.0 (exact)
Air	1.0001
Teflon	2.0
Polystyrene	2.5
Polycarbonate	2.7
Mylar	3.0
Polysulfone	3.06
Kapton	3.2
Polyethylene	3.3
Kraft paper (impregnated)	2.0–6.0
Mica	6.8
Aluminum oxide	7.0
Tantalum oxide	11.0
Ceramics	35.0–6000+

4.2.2 Dissipation Factor/Power Factor

When voltage is applied to a capacitor, a charging current will flow. This current will delay the buildup of the full applied voltage, generating a time displacement of the alternating voltage behind the current.

In a "no-loss" (perfect) capacitor the phase-difference voltage and current would be exactly 90°. In a real dielectric, however, this ideal phase angle is inhibited for two reasons, namely,

A. No dielectric is a 100% perfect insulator, hence a small degree of leakage occurs.

B. The polarization mechanisms of the material, which must take place in order to store electrical charge, are not 100% efficient. The capacitor does not discharge 100% of the energy absorbed in the charging cycle. Furthermore, these mechanisms, notably space charge, dipolar, ionic, and electronic polarization are highly dependent on frequency and dielectric composition.

4.2.3 Quality Factor (Q)

In simple terms, the quality factor Q is the reciprocal of the loss factor, that is,

$$Q = \frac{1}{\delta}$$

A near-perfect capacitor or a high-Q capacitor, therefore, has a Q approaching infinity. The Q factor is dependent on frequency, that is, the polarization mechanism operating at any given frequency.

4.2.4 Temperature Coefficient (TC)

The temperature coefficient TC of a dielectric defines the deviance in capacitance that occurs over a given temperature range.

4.2.5 DC Voltage Coefficient (VC)

Dielectrics based on ferroelectric formulations will experience a decrease in dielectric constant or capacitance with increased voltage (DC) bias.

The variance occurs due to the restriction that voltage stress causes on the freedom of some of the polarization mechanisms. Increased DC bias (more volts per mil of dielectric) will inhibit portions of the polarization cycle that oppose the DC bias.

The voltage coefficient becomes more severe with higher-*K* dielectrics, which have ferroelectric domain orientation (shifting of ionic dipoles) as the major polarizing mechanism.

4.2.6 Insulation Resistance (IR)

The insulation resistance is a measure of the ability of the charged capacitor to withstand leakage of DC current. Variations in value of insulation resistance can occur due to:

A. Change in surface resistivity due to moisture or other absorbed contaminants.

B. Change in bulk dielectric resistivity due to intrinsic (material) or extrinsic (structure) causes.

In practice, it is found the value of the insulation resistance is also strongly related to the capacitor's design; number of dielectric layers, area of layers, thickness of dielectric, physical quality of structure, and test conditions. These are the extrinsic factors affecting IR and must be considered in the evaluation of capacitors.

Higher-capacitance value parts are made with more, or thinner, dielectric layers, or with larger electrode plates than units of lesser value, built with the same dielectric. The probability of the occurrence of the mechanisms that lower the resistance therefore increases with higher-value capacitors.

4.2.7 Dielectric Strength

Dielectric strength is defined as the voltage gradient that produces electrical breakdown of the capacitor. This voltage is a function of the material and thickness of the dielectric layer, usually expressed as volts per mil or volts per centimeter. In a properly applied capacitor, sufficient thickness of dielectric is provided for a margin of safety, that is, loading at "flash" voltage is well below the breakdown value for the dielectric. In addition to this consideration, the dielectric thickness selected must meet the voltage coefficient (VC) requirements.

4.2.8 Aging

Ferroelectric dielectrics exhibit a time dependence of dielectric constant and dielectric loss known as "aging." Theoretically, this behavior is caused by strain energy of the ferroelectric domains striving for orientation relative to one another after their formation. Domains appear in ferroelectrics of the $BaTiO_3$ (perovskite) structure after cooling past the Curie temperature, the point at which the structure goes from a cubic (paraelectric) to a tetragonal (ferroelectric) crystal habit.

4.3 CAPACITOR USAGE[29,30]

Capacitors are used as energy-storage components to accumulate energy through long periods of time and to discharge the energy over longer or shorter periods. Parallel RC circuits will maintain bias on the grid of a tube for long periods and, as in filter circuits, will smooth out pulsating direct current.

Blocking capacitors are used to prevent the flow of direct current without impeding the flow of alternating current and to attenuate low-frequency currents while permitting higher-frequency currents to pass. In combination with resistors, capacitors are used to reduce radio-frequency interference (RFI) caused by arcing contacts and to increase the operational life of the contacts.

Table 4.2 shows various types of capacitors and the corresponding applications and features that determine their particular selection.

Whenever a capacitor is designed into an electronic circuit it has been selected to do a specific job, and certain capacitors will perform certain jobs better than others. The following sections examine jobs performed by capacitors as well as analyzing the important factors leading to proper selection.

4.3.1 Blocking

Whenever direct current (DC) is applied to a capacitor, the capacitor will effectively "block" the current and will not allow it to pass through. Utilizing this property of the capacitor can isolate a circuit element from the DC sup-

Table 4.2 Capacitor Dielectric Comparison Chart[30]

Application	Mica	Ceramic	Electrolytic	Metallized Paper	Paper Oil	Mylar/Foil	Metallized Polycarbonate	Polystyrene/ Foil	Metallized Mylar
Coupling	High RF current	Yes High frequency	Yes Low impedance	Yes	No	Yes	Yes	Occasional	Yes
Bypass	Yes	Yes High frequency Low cost	Yes Up to 10 MHz	Yes	Occasional	Yes	Yes	No	Yes
Tuning (frequency net determination)	Yes	Yes	No	No	No	No	Yes	Yes	No
Timing	Millisecond range	Short time	Long delay	Not usual	Seldom	Seldom	Yes	Yes	Seldom
Ripple suppression	No	RF only	Yes Major use	Occasional	Occasional	Yes	Occasional	No	Occasional
Arc suppression	Seldom	Occasional	Seldom	Yes	No	Yes	Occasional	No	Occasional
Energy storage	Occasional high voltage	Fast rise time	Short life Portable light equipment	Occasional	Yes	Occasional	Seldom	Special cases	No
Power factor correction	No	No	No	No	Yes	No	No	No	No
Motor start	No	No	Yes	No	Yes	No	No	No	No
Motor run	No	Some special	No	Occasional	Yes	Occasional	No	No	No
Characteristics									
Capacitance range	1–100,000 pF	1 pF–10 μF	0.1 μF × 0.5 farad	0.001–10 μF	0.001–50 μF	0.001–12 μF	0.001–12 μF	10 pF–1 μF	0.001–12 μF
Voltage range	1 V–50 kV	3 V–40 kV	3 to 450	200–600	100–100 kV also to 75 kVAC	50–1000	50–100	50–1600	50–600
Temperature range	−60 to +125°C	−55 to +125°C	−55 to +125°C	−55 to +125°C	−55 to +125°C	−55 to +150°C	−55 to +125°C	−55 to +85°C	−55 to +125°C
Tolerance	±½ to ±20%	±1% to GMV	±5% to GMV	±5 to ±20%	±2 to ±20%	±1 to ±20%	±1 to ±20%	±0.1 to ±20%	±1 to ±20%
Relative cost	Modest	Lowest to high for high capacitance units	Low per microfarad	Modest	Low	Low	High	High	Modest
Military specifications	Mil-C-5 Mil-C-39001	Mil-C-11015 Mil-C-20 Mil-C-39014 Mil-C-55681	Mil-C-62 Mil-C-3965 Mil-C-39006 Mil-C-39903 Mil-C-39018	Mil-C-18312 Mil-C-39022	Mil-C-25 Mil-C-14157 Mil-C-19978	Mil-C-19978 Mil-C-55514	Mil-C-18312 Mil-C-39022 Mil-C-55514	Mil-C-19978 Mil-C-55514	Mil-C-18312 Mil-C-39022 Mil-C-55514

ply. A choice must be made among many dielectrics, each of which will permit a different amount of tiny "leakage current" to pass through itself. This leakage current is a normal result of the random passage of electrons through or around the dielectric (usually in the microampere range) and should not be confused with the normal DC current supply. Thus it is possible to match the dielectric with the critical value of leakage current that the circuit can tolerate without malfunction.

4.3.2 Coupling/Decoupling

Just as the capacitor will block out direct current, it will appear to pass alternating current (AC). The capacitor is really charging and discharging in opposite directions at each half-cycle as the impressed AC voltage alternates in polarity. For the rest of the circuit, the net effect is like having the capacitor allow the passage of the AC signal. By using this ability of the capacitor, it is possible to "couple" one portion of the circuit to another, or even to a different circuit. Here, the inherent dissipation factor of the capacitor style can be critical if the heating effect of the AC voltage is appreciable.

4.3.3 Bypassing

Bypassing is a combination of blocking and coupling functions. In this case, the capacitor is used to separate the DC and AC portions of a mixed-signal current. Here, the capacitor is used in parallel with a circuit element. The purpose is to ensure that the DC portion does appear on the circuit element, but not the AC portion. By properly selecting the capacitor rating, the AC signal sees an apparently low-impedance path "through" the capacitor compared to the circuit element. The DC signal "sees" a low-impedance path through the circuit element and travels that way. Both the leakage current and dissipation factor parameters are to be considered in any application where these factors are critical.

4.3.4 Frequency Discrimination

Somewhat related to the coupling function of the capacitor is the characteristic of a capacitor that allows circuits to discriminate between AC signals of different frequencies. In general, the greater the capacitance of a unit, the greater the current will be at any given frequency; and for a given capacitance, the higher the frequency, the more current will pass (up to the resonant frequency of the capacitor–the point of maximum current transfer). By utilizing this characteristic, it is possible to distinguish between AC currents of different frequencies. As in coupling application, the dissipation factor (heat produced under AC applications) can be critical.

4.3.5 Transient Voltage Suppression

This characteristic of a capacitor is widely used to provide voltage stabilization. Most unregulated power supplies are subject to transient peaks of voltage surges that can cause malfunction in the circuit. By utilizing capacitors, these transient peaks are absorbed or stored by the capacitor, and a steady voltage signal is supplied to the circuit. This principal is used to reduce AC ripple voltages in rectified power supplies. Parameter behavior of the various kinds of capacitors is usually of no critical importance for this application.

4.3.6 Energy Storage

By using the capacitor to store energy over a long period of time and then discharging this total energy within a very short period, high currents can be made available. For the case of fairly high pulsing service, the dissipation factor may be of concern from a heating consideration. The most critical factor here is the current-carrying capabilities of the unit, leads, and connections.

4.3.7 Arc Suppression

Make–break-type circuits that use mechanical relays or similar devices to interrupt a current flow periodically cause arcing at the contact points. This results in a "noise" signal that can interfere with nearby reception. The use of capacitor–resistor or inductor combinations, or both, will reduce or eliminate this interference and prolong the life of the contacts by absorbing the voltage pulse resulting from this current interruption.

4.3.8 Power Factor Correction

Part of the total power generated by, or supplied to, a circuit must be used to energize or activate certain components within the circuit. This power is effectively lost within the circuit and is therefore not available for other uses.

By using capacitors, this percent loss of power can be and is reduced, thereby "correcting" or improving the power factor of the circuit. The immediate benefit from this corrected power factor is the reduction in lost power from the original total power generated, or the ability to increase the total power generated without additional equipment.

4.3.9 Timing

Timing is really a control on the "speed of a response" of the charging and discharging time for a capacitor and its associated circuitry. The rate at which charge flows in and out of the capacitor is directly controlled by the capacitance (*C*) and series resistance (*R*) of the circuit. Therefore, the timing or speed of response of a circuit can be controlled by using various *RC* combinations. This combination of *R* and *C* is known as the constant (*TC*). Usually the most critical parameter to be considered in a timing application is that of capacitance change with temperature (the temperature coefficient). The variance of capacitance with time and the retrace capabilities of the capacitance after temperature excursions can be critical in certain applications.

4.4 APPLICATION DATA

The following should be used in conjunction with Table 4.3 in the selection and use of a capacitor type:

A. The capacitance tolerance that the designer uses in order to design a circuit that will operate satisfactorily for the desired time requires (1) acceptable tolerances according to the specification; (2) capacitance-temperature characteristics; (3) capacitance-voltage characteristics; (4) retrace characteristics; (5) capacitance-frequency characteristics; (6) dielectric absorption; (7) capacitance as a function of pressure, vibration, and shock; and (8) capacitor aging in the circuit and during shelf storage.

B. Capacitance between the capacitor terminals may be a consideration, as will stray capacitance and leakage currents. The terminal connected to the outside conductor is often identified by the manufacturer so that the circuit can minimize these effects.

C. The capacitance-temperature characteristic can be compensated for by using more than one type of capacitor to obtain the required capacitance. The characteristics of the other circuit components may also be used for compensation.

D. The peak voltage applied to the capacitor should not exceed the rating in the applicable specification. The safety factor between the peak applied voltage, in general, may decrease with (1) aging, (2) increased temperature, (3) increased area of dielectric, (4) higher frequencies of applied voltage, (5) decreased pressure, or (6) the entrance of moisture into the capacitor. In many applications, it is necessary to derate the capacitor from the specified voltage to provide the desired performance for the required length of time. It is to be emphasized that short-duration transient voltages cannot be neglected in capacitor applications.

E. The use of the self-healing properties of certain types of capacitors may not be desirable in circuits in which intermittent failures and noise would be troublesome. Some types are not self-healing at low voltages.

F. Operation of capacitors above the corona-starting voltage will reduce life and produce noise. Liquid-impregnated dielectrics have a higher corona-starting voltage than do dry solid dielectrics.

G. When a capacitor is operated at high voltages above ground, and when it is insulated from ground with supplementary insulation, one terminal should be connected to the case, since the division of voltage depends on capacitance between capacitor rolls and case and the capacitance between case and chassis.

H. The peak charge and discharge currents must be considered on the basis of the time constant of the circuit.

I. Internal heating and ambient temperature must be considered.

J. Capacitor surface temperature must be considered.

K. Environmental conditions such as humidity, pressure, corrosive atmospheres, fungus growth, shock, and vibration must be considered.

L. The insulation resistance must be considered, especially at high temperatures.

M. In series operation on DC, balancing resistors should be considered.

N. The effective inductance of a large capacitor can be reduced by shunting it with a small capacitor.

O. The inductance of various types of capacitors varies over wide limits.

P. Since capacitors have inductance, the operation of capacitors in parallel in circuits with fast rise times or transients may result in transient oscillations.

Q. Poor electrical contacts may open at low voltages and be noisy.

R. The stored energy in capacitors can be dangerous to personnel and equipment. Therefore, suitable precautions should be taken.

S. Extended-foil paper capacitors are generally considered superior to inserted-tab types, having less inductance and less series-contact resistance—important factors in low signal-to-noise circuits.

T. Oil- and acid-filled units should not be subjected to severe mechanical stresses. Leakage of the fluid can destroy the capacitor together with adjacent components.

U. Liquid-filled units should not be used inverted because internal corona may result.

V. Nonhermetically sealed capacitors may be pervious to moisture by the process of "breathing."

Table 4.3 Capacitor Applications and Features.[1]

Capacitor Type	Applications and Features
Mica	High frequency coupling, timing circuits, and filter networks. Low leakage and excellent stability. Good for high reliability and critical industrial use. Values from 1.0 pF to 0.1 μF: 100 to 2000 V. Tolerance from 0.5 to 5.0% TC from 20 to 200 ppm/°C.
Low-loss ceramics	High frequency coupling, timing, filters, and tuned circuit trimming. Good for circuits where temperature is a factor. Very small size for high packaging density. Values from 10 pF to 4.7 μF: 50 to 100 V. Tolerance from 1 to 20%. TC from 0 to 30 ppm/°C.
Standard disk ceramics	Low cost and small size for coupling and bypass use in IF and RF circuits. Values from 1.0 pF to 0.22 μF: 3 to 10,000 V. Tolerance from 20 to 100%.
High-voltage ceramics	Uses include filter, doubler, and bypass applications in high-voltage power supplies and radio transmitters. Values from 180 to 16,000 pF: 6 to 40 kV. Tolerance +50%, −20%.
RF and power ceramics	Used in power switchgear, microwave ovens, industrial air cleaners, and airborne military and commercial transmitters. Values from 10 to 10,000 pF: 6 to 40 kV. RF currents to 200 A.
Polycarbonates and Teflon	General-purpose and AC use to medium frequencies: approximately 1 MHz. Good for high reliability and critical industrial circuits and where high currents are present (e.g., motor starting, and other high current surge applications). Values from 0.001 to 22 μF: 30 to 400 V. Tolerance from 0.25 to 10%.
Mylar foil and paper	General-purpose and AC use (e.g., coupling, bypass.) Extremely low leakage, excellent for timing circuits, low-frequency filter networks, and voltage sensing. Values from 0.001 to 10 μF: 50 to 600 V. Tolerance: 5 to 10%.
Aluminum electrolytics	Low cost, used for filtering pulsating DC for power supplies, low-frequency bypass. Relatively high leakage. Values from 1 to 270,000 μF: 3 to 450 V. These capacitors are polarized.
Tantalum	Rugged, low leakage, small size high capacity. Good for high-reliability and critical industrial use. Generally polarized, but are also supplied as nonpolarized types. Can be operated to +175°C. These have the same uses as aluminum electrolytics but are superior in performance. Values from 1 to 1200 μF: 3 to 300 V. Tolerance from 5 to 20%.
Polysulfone	Same as the polycarbonates, but can opperate to +150°C with very low TC capability.
Variable air and variable ceramic	Used for tuning and trimming RF circuits in radiotransmitting and receiving equipment. Values range in various min–max levels. They are also used for tuning and trimming in microwave, VHF and UHF equipment. TC vary from 50 to 1500 ppm/°C.

W. Capacitors for AC and pulse operation require special ratings and tests not covered in most specifications.

4.5 CAPACITOR STYLES AND SHAPES[31]

Capacitors are available in many styles and shapes. Electrical capacitance and voltage ratings are usually specific values determined by circuit requirements. Capacitor style and the location and method of mounting, however, present a variety of choices. Selecting the most suitable possibility is simplified by using the answers to the following questions as a guide:

A. Must special environmental conditions be met? Some equipment specifications require certain types of dielectric materials, case materials, and so forth.

B. How much space is available? In most instances capacitor costs increase with decreasing case size.

C. Are lead lengths critical? A slight difference in lead length can cause unbalance in some circuits, while variations of a foot or more make no difference in other circuits.

D. Do maintenance problems exist? Depending on end use some equipment warrants plug-in capacitors that are easy to replace. Others can have the capacitor soldered into the circuit in relatively inaccessible locations.

E. Is the capacitor to be located near high-heat-emitting components? High ambient temperatures may require compensation for changes in electrical capacitance.

F. Must both sides of the capacitor be "floating," or can one side be grounded to the case? Two grounded types are available: one type consists of two leads, one of which is connected to the metal case while the other type has only one lead with the metal case providing the ground connection.

To satisfy various conditions, capacitors are available with either integral or separate mounting provisions. The principal features and arrangements of capacitors with integral mounting provisions (Figure 4.3) are the following:

A. ***Bathtub Capacitors*** with ear tabs for horizontal mounting (parallel to mounting surface). Lead connections can be located on side or on top. Side connections permit stacked mounting.

B. ***Tubular Capacitors*** with C clamp welded to metal case for horizontal mounting. Available with leads at one or both ends. Attached clamp saves assembly time.

C. ***Tubular Capacitors*** with angle bracket welded to metal case for vertical mounting (perpendicular to mounting surface). Bracket may be attached at any location along the length of the case to permit through-chassis mounting. This feature permits minimum lead lengths to be used for special circuits but requires two holes in the chassis.

D. ***Tubular Capacitors*** with threaded housing for vertical mounting. This capacitor is available with solder terminals or flexible leads and requires only one hole in the chassis.

E. ***Tubular Capacitors*** with two radially positioned threaded studs for horizontal mounting. This type of mounting is suitable for high vibration conditions and in cramped locations where side mounting cannot be used.

F. ***Specially Shaped Capacitors*** with molded plastic or ceramic cases. Capacitors can be fitted with flanges, tapped inserts, or through-holes for mounting. Case material and construction can provide a waterproof seal.

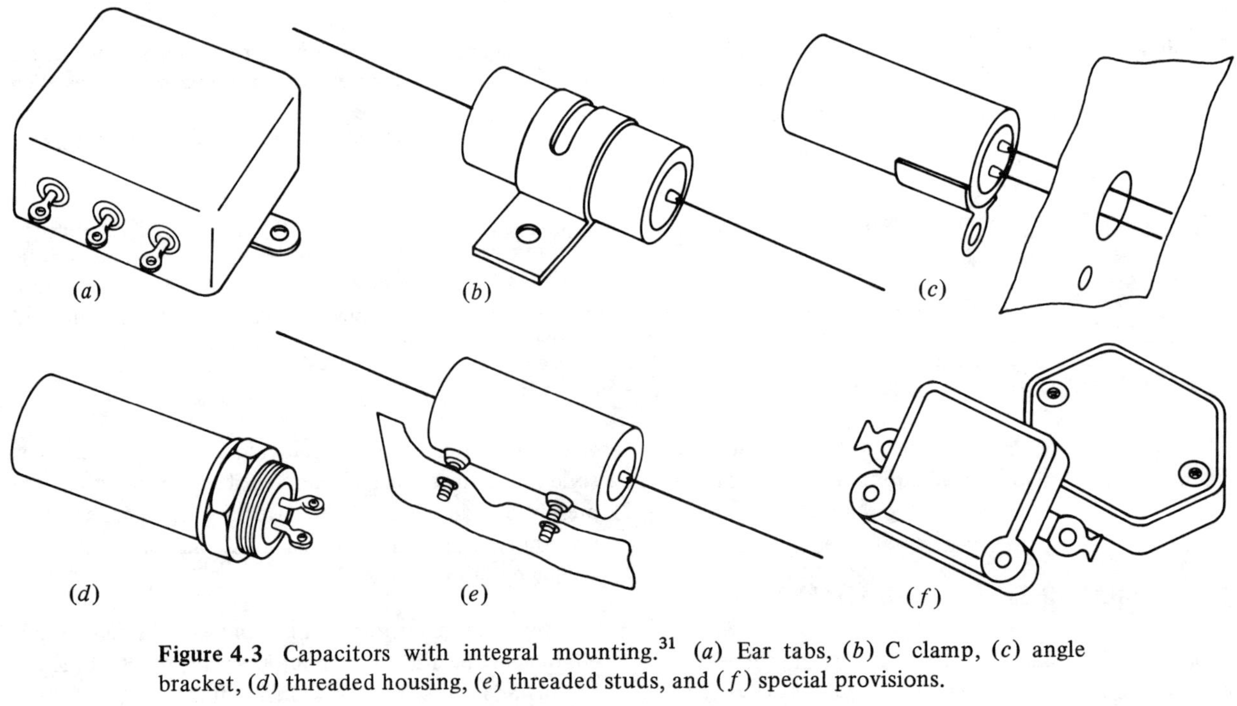

Figure 4.3 Capacitors with integral mounting.[31] (*a*) Ear tabs, (*b*) C clamp, (*c*) angle bracket, (*d*) threaded housing, (*e*) threaded studs, and (*f*) special provisions.

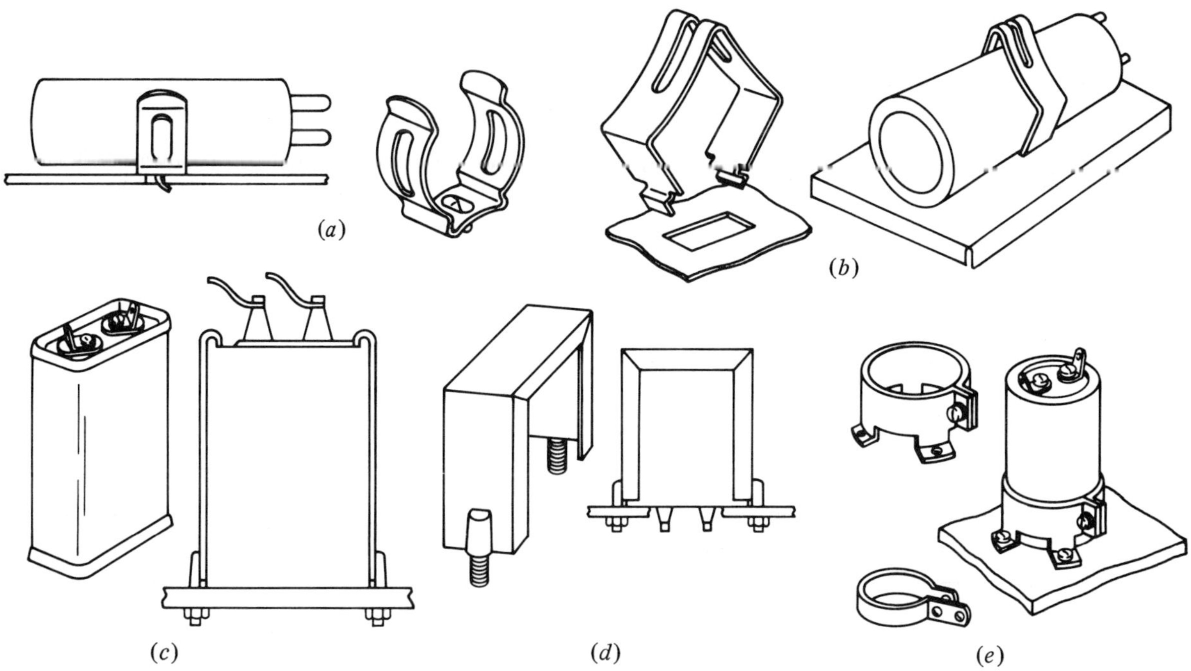

Figure 4.4 Capacitors with separate mounting.[31] (*a*) Spring-steel clip, (*b*) snapover clip, (*c*) hook strap, (*d*) inverted U strap, (*e*) strap and ring.

Separate mounting provisions for capacitors (Figure 4.4) usually consist of the following:

A. ***Quickly Installed*** spring-steel clips for horizontal capacitor mounting in various axial positions. The open cradle permits easy mounting and removal of capacitor. No bolts are required as the clip is rigidly held in a rectangular hole in the mounting surface.

B. ***Snapover Clip*** for horizontal mounting. No bolts are required. The clip fits into a rectangular hole in the mounting surface and provides a spring-loaded grip.

C. ***Hook Strap*** for vertical mounting of rectangular capacitor. A recess in the top of the capacitor is needed to provide effective gripping. Capacitors with terminals at the bottom require appropriate cutouts in the mounting surface.

D. ***Inverted U Strap*** for vertical mounting of rectangular capacitor. The strap has threaded studs attached. Holes are required in mounting surface for capacitor terminals and strap studs.

E. ***Strap and Ring*** for mounting tubular capacitors. Straps are used for horizontal mounting. For vertical mounting, ring permits capacitor positioning through hole in mounting surface.

4.6 CAPACITOR SELECTION

When selecting a capacitor type for a particular function to be performed, weigh numerous factors before coming to a final decision. Selection normally starts with the most important characteristics for the application, then selecting and compromising other characteristics. Here are the most important of these factors, with some of the reasons why they are important:

A. ***AC Ripple*** applies when a capacitor handles both DC and AC voltages, as in power-supply filters.

B. ***Dielectric Absorption*** causes inaccuracies and waveform distortion in timing, integrator, and nonsinusoidal oscillator circuits. When a charged capacitor is discharged, it fails to return all of the charge initially put into it. The unreturned charge (dielectric absorption) "soaks" into the dielectric.

C. ***Dissipation Factor*** varies with temperature, humidity, and frequency. Increasing humidity tends to increase the DF in unsealed capacitors. Changing the frequency and temperature causes highly nonlinear effects.

D. ***Humidity Coefficient*** of a capacitor is especially important with unsealed capacitors, in which water vapor can be absorbed by the dielectric and pockets and capillaries in the foil windings. The humidity coefficient becomes particularly significant in small capacitors in size up to about 250 pF, when the relative humidity exceeds 80%.

E. ***The Impedance*** of a capacitor varies with frequency.

F. ***Insulation Resistance*** determines the capacitor's leakage current for a continuously applied DC voltage. A fixed leakage-current value takes time to achieve. Electrolytic capacitors need about 15 minutes in which to stabilize. Insulation resistance can vary over several orders of magnitude with temperature changes.

Table 4.4 Capacitor Types by Dielectric[26]

DIELECTRIC	APPLICABLE SPECIFICATION	CAPACITANCE			DC rated voltage (Volts)
		Range	Tolerance	Stability after 2,000 hours life test	
GLASS					
Fixed	MIL-C-23269 (ER)	.5 to 10,000 pF	.25 pF to 5%	.5% or 0.5 pF whichever is greater	100, 300, & 500
Variable	MIL-C-14409	0.6 to 1.8 pF thru 1 to 120 pF	---	Cap. change vs rotation: ≤ 10%	250 to 1,250
MICA					
Button style	MIL-C-10950	5 to 2,400 pF	±1, ±2, ±5, or ±10%	≤1% or .5 pF whichever is greater	500
General purpose	MIL-C-5	47 to 27,000 pF	±1, ±2, or ±5%	≤5% or 1 pF whichever is greater	300 to 2,500
	MIL-C-39001 (ER)	1 to 91,000 pF	.5 pF to ±5%	≤1% or 1 pF whichever is greater	50 to 500
ELECTROLYTIC					
Aluminum	MIL-C-62	1 to 1,000 µF	-10, +50	±15%	400 & 450
Tantalum (nonsolid)	MIL-C-39006 (ER)	.5 to 1,200 µF	-15: +30, +50, +75, ±5% to ±20%	≤ 15%	6 to 450
Tantalum (solid)	MIL-C-39003 (ER)	.0023 to 330 µF	±5, ±10, or ±20%	≤ 2%	6 to 100
Aluminum oxide	MIL-C-39018	.68 to 220,000 µF	-10: +5, +30, +50, +75	≤ 15%	5 to 350
Tantalum (solid) chip	MIL-C-55365 (ER)	.001 to 4,700 µF	±5, ±10, or ±20%	---	2 to 50
PAPER					
Wax-impregnated	MIL-C-12889	.01 to .5 µF	+0, -30 or +50, -10	≤ 10%	100: 250 & 500 (AC-DC)
Metallized	MIL-C-39022 (ER)	.022 to 10 µF	±5 or ±10%	≤ 10%	50 to 400
PAPER-PLASTIC					
Polycarbonate	MIL-C-19978 (ER)	.001 to 1 µF	±5 or ±10%	≤ 6%	50 to 600
Paper & polyethylene terephthalate	MIL-C-19978 (ER)	.001 to 1 µF	±2, ±5, or ±10%	≤ 6%	200 to 1,000
Plastic or metallized plastic	MIL-C-55514 (ER)	.001 to 10 µF	±1, ±2, ±5, or ±10%	≤ 5%	50 to 600
Polyethylene terephthalate	MIL-C-19978 (ER)	.001 to 10 µF	±2, ±5, or ±10%	≤ 6%	30 to 1,000
CERAMIC					
Fixed, general purpose	MIL-C-11015	2.2 to 2,200,000 pF	.5 pF, ±10, ±20	≤ 10% or ≤ 20%	50 to 1,600
	MIL-C-39014 (ER)	10 to 1,000,000 pF	±10, ±20	≤ 20%	50 to 1,600
Temp compensating	MIL-C-20 (ER)	1.0 to 68,000 pF	.25 pF, ±1, ±2, ±5, ±10	±3% or .5 pF whichever is greater	50 to 200
Variable	MIL-C-81	1.5 to 7 thru 15 to 60 pF	---	---	200 to 500
Fixed, chip	MIL-C-55681 (ER)	10 to 180,000 pF	±5, ±10, or ±20	---	50 & 100
GAS OR VACUUM					
Fixed	MIL-C-23183	6 to 1,000 pF	±10%	---	30 kV
Variable	MIL-C-23183	5 to 3,000 pF	±10%	---	2 & 3 kV

1/ Where "C" = Capacitance and "V" = Voltage.

G. *Load Life* takes into account derating effects of excess working voltages and temperatures on life.

H. *Power Factor* for a good capacitor is represented by a low-power loss. For good capacitors, PF becomes nearly equal to DF (Z approaches X_c), and both factors are similarly affected by temperature.

I. *Surge Voltage* under worst conditions includes temperature, humidity, and frequency effects.

J. *Temperature Coefficient* of common capacitor types, identified by the dielectrics used, differ considerably and is crucial to selecting the proper capacitor.

TEMPERATURE		RELATIVE COST FOR EQUIV CV RATING1/	RELATIVE SIZE		DISSIPATION FACTOR (%)		
Operating temperature (in °C)	Temperature coefficient (in % or ppm/°C)		Varies as	For equiv CV rating	60 Hz	1,000 Hz	1 MHz
-55 to +125	140, 105, 0 ±25	Medium	CV^2	Large	---	<.001	---
-55 to +125 or +150	±50, ±100, ±150, and +50, -0	Medium high	C	Large	---	---	---
-55 to +85 or +150	±100, -20 to +100, and Not specified	Medium high	CV^2	Large	---	<.17	<1.2
-55 to +125 or +150	0 to +70, -20 to +100, ±100, ±200, and Not specified	Medium low	CV^2	Large	---	<.18	<.12
-55 to +125 or +150	0 to +70, -20 to +100, ±200, and Not specified	Medium low	CV^2	Large	---	<.1	<1,000
-40 to +85	Capacitance drops from 30 to 60% at -40°C	Medium	CV	Very small	---	15 to 18% at 120 Hz; varies with V	---
-55 to +85, derated to +125	Capacitance drops from 12 to 50% at -55°C	High	CV	Very small	---	10 to 32% at 120 Hz; varies with C and V	---
-55 to +85, derated to +125	Capacitance drops 10% max at -55°C	Medium	CV	Very small	---	3 to 8% at 120 Hz; varies with V	---
-40 to +85, derated to +125	---	Medium	CV	Very small	---	10 to 35% at 120 Hz; varies with C and V	---
-55 to +125	---	Medium	CV	Very small	---	.6 to 15% at 120 Hz; varies with V	---
-55 to +85	Capacitance drops <30% at -55°C	Medium	CV^2	Small	---	<1	---
-55 to +85 or +125	Capacitance drops <10% at -55°C	Medium	CV^2	Small	---	<1	---
-55 to +125	Capacitance change ±2% at -55°C	High	CV^2	Large	<.1	<.1	Higher
-65 to +125	±10%	High	CV^2	Medium large	<.1	<.1	---
-55 to +85 or +125	---	Medium	CV^2	Small	---	<2.0	---
-65 to +85	-7 to +5%	High	CV^2	Small	<.6	<.6	---
-55 to + 85 or +125	Capacitance change <+30, -80% at -55°C	Very low	CV^2+k	Small	---	<2.5	<2.5
-55 to +85 or +125	Capacitance change <+30, -80% at -55°C	Very low	CV^2+k	Small	---	<2.5	<2.5
-55 to +125	0 ±30, 0 ±60	Very low	CV^2+k	Small	---	0.15	<.10
-55 to +85	Capacitance change <-4.5, +2% at -55°C	Medium low	CV^2+k	Large	---	---	0.2
-55 to +125	0 ±30 or 0 ±15	Low	CV	Small	---	<2.5	<2.5
-55 to +85	---	High	---	Large	---	<0.001	---
-55 to +85	---	High	---	Large	---	<0.001	---

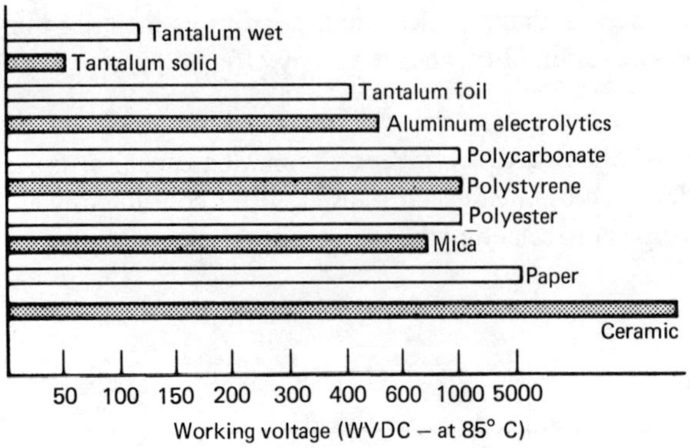

Figure 4.5 Capacitor working voltage.[25]

K. ***Temperature-Voltage Derating*** must be carefully evaluated if a temperature higher than specified, usually 65-85°C, is expected.

L. ***Quality Factor*** is a figure of merit that should be considered, especially in tuned-circuit applications.

M. An operating voltage higher than the specified *working voltage* will result in a shorter life, which may be permissible in some applications. Figure 4.5 compares the typical DC working voltage (WVDC) of different capacitor types.

N. ***Specifications*** pertaining to capacitors are listed in Tables 4.4 and 4.5.

4.7 CAPACITOR TERMS AND DEFINITIONS

Air Capacitor A capacitor in which air is the only dielectric material between its plates.

Aluminum Electrolytic Capacitor A capacitor with two aluminum electrodes (the anode has an oxide film) separated by layers of absorbent paper saturated with the operating electrolyte.

Anode Positive electrode of a capacitor.

Blocking Capacitor A capacitor that introduces a comparatively high series impedance for limiting the current flow of low-frequency AC or DC without materially affecting the flow of high-frequency current.

Bypass Capacitor A capacitor for providing an AC path of comparatively low impedance around some circuit element.

Capacitance A property of a capacitor that determines its ability to store electrical energy when a given voltage is applied, measured in farads, microfarads, or picofarads.

Capacitance Tolerance The maximum deviation (expressed in percent) from the specified nominal value at standard (or stated) environmental conditions.

Capacitance Reactance Opposition offered to the flow of an alternating or pulsating current by capacitance, measured in ohms.

Capacitor An electronic component part consisting essentially of two conducting surfaces separated by an insulating (dielectric) material. A capacitor stores electrical energy, blocks the flow of direct current, and permits the flow of alternating or pulsating current to a degree that depends on the capacitance and frequency.

Cathode Negative electrode of a capacitor.

DC Leakage (DCL) Stray direct current of relatively small value which flows through or across the surface of solid or liquid insulation when a voltage is impressed across the insulation.

Dielectric The insulating material (for example, air, paper, oil) between the plates of a capacitor.

Dielectric Absorption A property of an imperfect dielectric whereby all electric charges within the body of the material caused by an electric field are not returned to the field.

Dielectric Constant A property of a material that determines how much electrostatic energy can be stored per unit volume when unit voltage is applied. (It is the ratio of the capacitance of a capacitor filled with a given dielectric to that of the same capacitor having a vacuum dielectric.)

Dielectric Strength The maximum voltage that a dielectric material can withstand without rupturing. (The value obtained for the dielectric strength will depend on the thickness of the material and on the method and conditions of test.)

Disk Capacitor A small single-layer ceramic capacitor consisting of one dielectric insulator with the opposing surfaces conductively silvered.

Dissipation Factor The ratio of the effective series resistance (ESR) to the capacitive reactance (X_c).

Electrolyte A current-conducting solution (either liquid or solid) between two electrodes or plates of a capacitor at least one of which is covered by a dielectric film.

Equivalent Series Resistance (ESR) The sum of the resistance of the oxide film, electrolytic and paper, foils and the lead wires or terminals of a capacitor.

Feedthrough Capacitor A feedthrough insulator that provides a desired value of capacitance between the feed-

Table 4.5 EIA Standard RS-198[32]

The following is an abstract from EIA Standard RS-198 illustrating the codes currently used for capacitance, tolerance, and temperature characteristics.

1st & 2nd Significant Figure of Capacitance to be selected among the following series of preferred numbers			Multiplier	Numerical Symbol	Tolerance on Capacitance	Letter Symbol
5%	10%	20%				
10	10	10	1	0	±.5pF	D
11			10	1	±1%	F
12	12		100	2	±2%	G
13			1,000	3	±3%	H
15	15	15	10,000	4	±5%	J
16			100,000	5	±10%	K
18	18			—	±20%	M
20				—	+100%, −0%	P
22	22	22	.01	8	+80%, −20%	Z
24			.1	9		
27	27					
30						
33	33	33				
36						
39	39					
43						
47	47	47				
51						
56	56					
62						
68	68	68				
75						
82	82					
91						

Low Temp. Requirement	Letter Symbol	High Temp. Requirement	Numerical Symbol	Max. Capac. Change Over Temp. Range	Letter Symbol
+10°C	Z	+45°C	2	±1.0%	A
−30°C	Y	+65°C	4	±1.5%	B
−55°C	X	+85°C	5	±2.2%	C
		+105°C	6	±3.3%	D
		+125°C	7	±4.7%	E
				±7.5%	F
				±10.0%	P
				±15.0%	R
				±22.0%	S
				+22%, −33%	T
				+22%, −56%	U
				+22%, −82%	V

through conductor and the metal chassis or panel through which the conductor is passing.

Impedance (*Z*) The total opposition offered to the flow of an alternating or pulsating current measured in ohms. (The vector sum of the resistance and the capacitive reactance.)

Impregnant A substance, usually liquid, used to saturate paper dielectric materials in order to replace the air between its fibers.

Insulation Resistance (IR) The DC resistance between two conductors separated by an insulating material.

Liquid-Filled Capacitor A capacitor in which a liquid impregnant occupies substantially all of the case volume not required by the capacitor element and its connections.

Liquid-Impregnated Capacitor A capacitor in which a liquid impregnant is dominantly contained within the foil- and paper-winding, but that does not occupy a substantially portion of the case volume.

Load Life The expected life a capacitor under specified conditions of voltage, temperature, and ripple current.

Power Factor (PF) The ratio of resistance to impedance.

Quality Factor (*Q*) The ratio of capacitive reactance to resistance.

Radio Interference Undesired conducted or radiated electrical disturbances, including transients, which may interfere with the operation of electrical or electronic communications or other electronic equipment.

Ripple Current (or Voltage) The AC component of a undirectional current or voltage.

Solid Tantalum Capacitor A tantalum capacitor with a solid semiconductor electrolyte instead of a liquid.

Stability The ability of a part to resist changes in characteristic values or coefficients, or both.

Temperature Coefficient (TC) The change in capacitance of a capacitor per degree change in temperature.

Temperature-Compensating Capacitor A capacitor whose capacitance varies with temperature in a known and predictable manner.

Trimmer Capacitor An adjustable capacitor, associated with another capacitor, used for fine adjustment of the total capacitance of an element or part of a circuit.

Variable Capacitor A capacitor that can be changed in capacitance by varying the useful area of its plates, as in a rotary capacitor, or by altering the distance between the plates.

Wet Electrolytic Capacitor A capacitor in which the dielectric is primarily an anodized coating on one electrode, with the remaining space between the electrodes filled with a liquid electrolytic solution.

Working Voltage (WVDC) The maximum DC voltage that a capacitor can withstand continuously for a specified lifetime.

CHAPTER FIVE

Fixed-Value Electrolytic Capacitors

5.1 GENERAL

Electrolytic capacitors[33,34] come in an array of shapes, sizes, and characteristics. Sizes range from 0.1 to 3.0 in. in diameter, operating temperatures extend all the way from - 80 to + 175°C, and they work at frequencies into the megahertz region (see Figure 5.1). Whether an aluminum or tantalum electrolytic, the performance and operating life depend largely on three operating parameters: voltage, temperature range, and ripple current.

5.1.1 Ratings and Tolerances

The capacitance rating of electrolytic capacitors is usually stated as a specific value with a tolerance, for example, - 10% and + 50% for aluminum types. Low-voltage capacitors normally have a wider tolerance than do high-voltage units. For aluminum electrolytics ± 20% would be considered "tight," particularly at low voltages. Tantalum-foil electrolytics also have broad tolerances ranging from - 15% and + 30% (narrow) to - 10% and + 150% (wide). However, tantalum electrolytics made with sintered anodes, with either liquid or solid electrolytes, can be made with relatively narrow tolerances ranging from ± 5% to ± 20%. These "precision" capacitors can be used in critical RC-timing networks.

The voltage rating of a capacitor depends on the voltage at which the electrochemically deposited oxide film was formed, that is, the formation voltage. Normally, the rated voltage is considerably lower than the formation voltage. For a 6-V capacitor, a typical formation voltage would be 12 V; for a 100-V rating, it would be 150 V.

5.1.2 Operating Temperature Range

With electrolytic capacitors, high operating temperatures mean reduced life. Also, heating due to ripple current and DC leakage contribute significantly to operating temperature. The DC leakage, resulting largely from impurities in the foil, increases with operating temperature and so contributes to further heating. For voltage ratings of less than 100 V, however, heating from leakage currents is generally negligible.

5.2 TANTALUM ELECTROLYTIC CAPACITORS

Tantalum electrolytic capacitors are smaller than aluminums for equivalent ratings. Since tantalum oxide has a dielectric constant about twice that of aluminum oxide which serves as the dielectric, they generally have somewhat better characteristics and considerably longer oper-

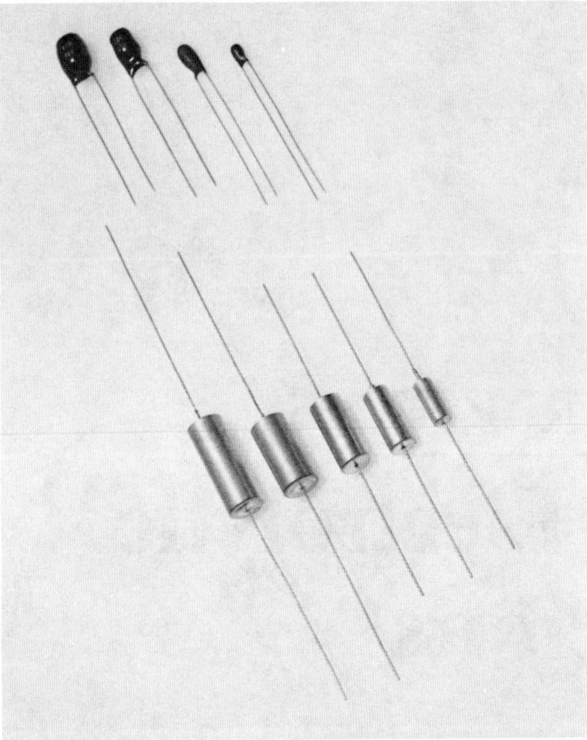

Figure 5.1 Electrolytic capacitors. (Courtesy of Mepco/ Electra Inc.)

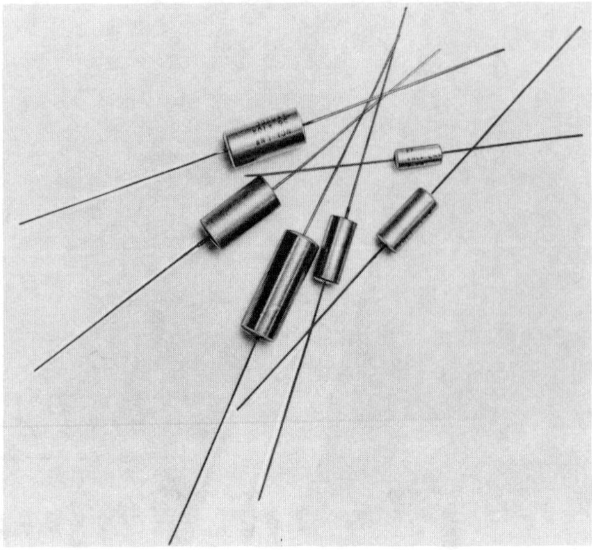

Figure 5.2 Tantalum capacitors. (Courtesy of Mepco/ Electra Inc.)

ating life, along with being more expensive. Because of their higher temperature range, higher reliability by life performance, and smaller-size tantalum electrolytic capacitors are common in military equipment.

These capacitors can be made with tantalum foil (plain or etched) or sintered anode structure with either a solid or liquid electrolyte. The liquid-electrolyte porous-anode types (wet slugs) have the best characteristics and highest volumetric efficiency of any electrolytic capacitor. Because they are generally cheaper, tantalum capacitors with anode slugs, containing a solid electrolyte, are most popular, especially for low-voltage applications, such as filtering, coupling, and bypassing. Foil types are available in plain and etched foil, polarized and nonpolarized (see Figure 5.2 and Table 5.1).

5.2.1 Tantalum-Foil Capacitors

Tantalum-foil capacitors are electrolytic capacitors utilizing tantalum foil for the electrodes and tantalum oxide as the dielectric. The foil electrodes may be plain or etched to provide greater surface area. They are electrochemically formed to the proper voltage and wound into rolls with a porous-paper spacer to provide separation and hold the electrolyte. Standard tubular units are assembled into metallic cases with roll-crimped Teflon elastomer end seals. Solderable nickel leads are externally butt-welded to the tantalum leads attached to the electrodes.

Tantalum-foil capacitors provide a wide range of microfarad and voltage ratings. Voltages up to 300 V DC at 85°C are available. New high-etched types have extended the capacitance range up to 1450 μF at 6 V in a standard tubular unit. Tantalum-foil capacitors are particularly well suited to nonpolar requirements or to any application in which there is some reverse-voltage requirement.

5.2.2 Solid-Tantalum Capacitors

Solid-tantalum capacitors are constructed of sintered tantalum-powder particles packed around a tantalum anode, which makes a rigid assembly or slug. They have a higher CV product per unit volume than do the aluminum electrolytics, are more temperature-stable, and usually have hermetic seals to eliminate humidity effects. Furthermore, both their shelf and operating lives are superior to aluminum electrolytics, but tantalum capacitors are also several times more expensive.

Tantalum capacitors can be obtained as either polar or nonpolar types. Although even a small reverse voltage applied to a polar tantalum can significantly increase its

Table 5.1 Tantalum Capacitors[34]

MIL Type		Voltage Range (D-c)	Microfarad Range	Temp. Range Deg. C	Polarity	Configuration	Terminals	Seal	Outline
MIL-C-3965	MIL-C-39006								
CL 20 thru 27 CL 30 thru 37	CLR 25 and 27 CLR 35 and 37	3-450	0.15 to 1450	−55 to +85/125	Polar and Non-polar	Tubular	Axial lead	Teflon Elastomer	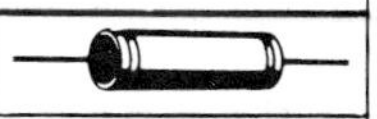
CL 51 thru CL 54		10-150	20 to 3500	−55 to +85/125	Polar and Non-polar	Rectangular	Solder lug	Glass-to-Tantalum Hermetic	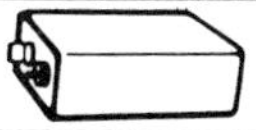
CL 70 thru CL 73	CLR 71 and CLR 73	10-150	0.75 to 920	−55 to +85/125	Polar and Non-polar	Tubular	Axial lead	Teflon Elastomer	
	CLR 25 and 27 CLR 35 and 37	10-200	0.15 to 580	−55 to +85/125	Polar and Non-polar	Tubular	Axial lead	Glass-to-Tantalum Hermetic	
		6-60	0.1 to 325	−55 to +85	Polar only	Subminiature tubular	Axial lead	Teflon	
CL 64 and CL 65	CLR 65	4-125	1.7 to 1200	−55 to +85/125	Polar only	Tubular	Axial lead	Teflon/ Double O-ring	
		6-60	3.3 to 450	−55 to +85	Polar only	Subminiature tubular	Axial lead	Teflon/ Viton	
CL 55		10-150	70 to 2400	−55 to +85/125	Polar only	Rectangular	Solder lug	Glass-to-Metal Hermetic	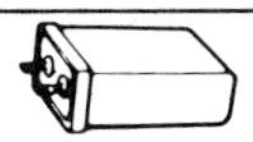
		4-75	22 to 1800	−55 to +85/125	Polar only	Tubular	Axial lead	Teflon/ Double O-ring	
CL 66 and CL 67		4-125	1.7 to 560	−55 to +85/125	Polar only	Tubular	Axial lead	Glass-to-Metal Hermetic	
	CLR 65	4-125	1.7 to 1200	−55 to +85/125	Polar only	Tubular	Axial lead	Glass-to-Tantalum Hermetic	

leakage current, nonpolars can sustain AC operation without appreciable degradation, if the rms voltage is within the ratings of the unit. Nonpolar tantalum capacitors are constructed with two polar capacitors, back to back, with their anode leads connected.

5.2.3 Wet-Slug Tantalum Capacitors

Wet-slug tantalum capacitors have the highest volumetric efficiency and lowest leakage per CV of any electrolytic capacitor. They use a porous anode or slug immersed in a liquid electrolyte. Since they are expensive, wet-slug tantalum capacitors are limited to high-reliability, small-space military aerospace and critical industrial applications.

5.2.4 Tantalum Chip Capacitors

Pellet-type, solid electrolyte, tantalum chip capacitors offer high capacitance ratings of 0.1-100 μF. Thin-film-type tantalum chips are available in the 100–1000 pF. pico-farad range. Both types exhibit low values of frequency and temperature coefficients and load-life drift. Some pellet chips can withstand 300°C for short periods and operate continuously at 175°C when derated to 25% of the 85°C voltage rating.

The very high dielectric constant (K) of tantalum gives capacitors a very high volumetric efficiency, which is much higher than with ceramic-based capacitors. Consequently, unlike ceramic capacitors, tantalum-chip units with large capacitances occupy relatively less amounts of real estate, and cost considerably less. However, tantalum chip capacitors with low capacitance values cost more than equivalent ceramic capacitors.

5.3 ALUMINUM ELECTROLYTIC CAPACITORS

Aluminum electrolytic capacitors are the most widely used electrolytic capacitors in electronic equipment. There are several grades to meet various requirements and costs. The failure mode of both aluminum and tantalum electrolytic capacitors is predominantly one of degradation through a decrease in capacitance, an increase in leakage current, and an increase in equivalent series resistance (ESR). Three types of aluminum electrolytic capacitors are available: polar, nonpolar, and semipolar (see Table 5.2).

Table 5.2 Aluminum Electrolytic Capacitors

Type	Relative Cost	Temperature Range (°C)	Capacitance Range (μF)	Maximum Rated Voltage at Rated Temperature	Maximum Ripple 120 Hz at 85°C	Size Range (diameter × length) (in.)
			Axial Leads			
Subminiature						
Computer grade	Low	−40 to 105	1 to 1500	150	20 to 1100 mA	0.26 × 0.56 0.38 × 2.25
Industrial grade	Low	−40 to 85	1 to 100	150	12 to 150 mA	0.26 × 0.56 0.38 × 1.31
Computer grade, low impedance	Low	−80 to 110	1 to 600	150	25 to 1000 mA	0.26 × 0.69 0.39 × 1.56
Military grade, high reliability	Moderate	−55 to 125	3.3 to 1000	200	90 to 2000 mA	0.41 × 0.91 0.41 × 2.78
Miniature						
Military grade, high reliability	Moderate	−55 to 105	16 to 13,000	200	0.2 to 6.9 A	0.66 × 1.19 1.04 × 3.69
Computer grade	Low to moderate	−40 to 85	1 to 18,000	450	0.2 to 3.6 A	0.51 × 1.16 1.01 × 3.66
Entertainment grade	Low	−40 to 85	1 to 45,000	700	10 mA to 7.8 A	0.26 × 0.56 1.44 × 4.31
Military and computer grade, 4 terminal, low impedance, high frequency	Moderate to high	−55 to 105	50 to 16,000	200	0.7 to 7 A	0.78 × 1.84 1.04 × 3.84
			Lugs			
High reliability, hermetic seal	High	−55 to 125	500 to 20,000	500	1 to 30 A	Not restricted
			Screw Terminals			
Large						
Computer and military grade	Moderate to high	−55 to 105	100 to 450,000	150	2.5 to 40 A	1.41 × 2.16 3.03 × 8.66
Computer grade	Low	−40 to 85	80 to 1,000,000	450	2.5 to 40 A	1.41 × 2.16 3.03 × 8.86
Computer grade, low impedance, high frequency	Moderate	−40 to 85	640 to 600,000	100	3 to 40 A	1.41 × 2.16 3.03 × 8.66

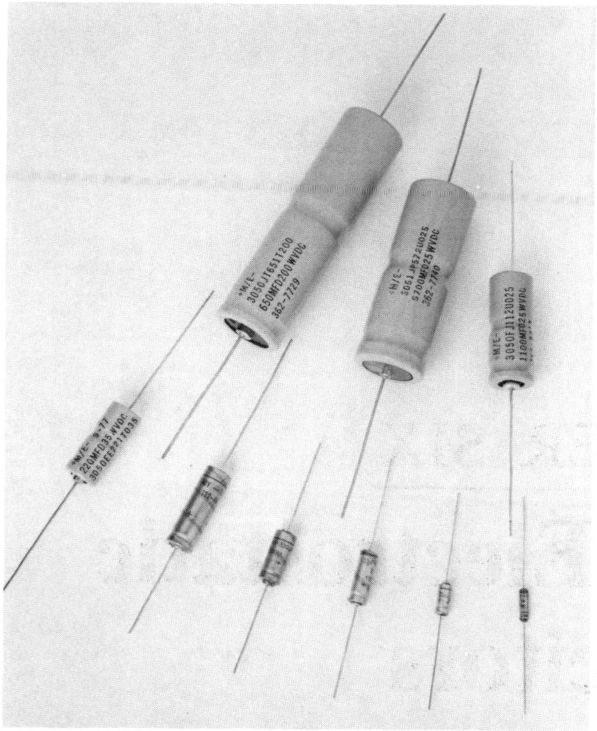

Figure 5.3 Polar aluminum electrolytic capacitors. (Courtesy of Mepco/Electra Inc.)

5.3.1 Polar Aluminum Electrolytic Capacitors

Polar aluminum electrolytic capacitors are used in applications where a constant DC voltage is applied in the forward direction and any AC voltage present is superimposed on this DC bias. The combination of the peak AC voltage and the DC voltage applied should not exceed the voltage rating of the device and the AC current caused by the AC voltage should not exceed the manufacturer's recommendations. A typical application for this type of capacitor, Figure 5.3, is power supply filtering.

5.3.2 Nonpolar/Semipolar Aluminum Electrolytic Capacitors

Nonpolar aluminum electrolytic capacitors are available for use where DC may be applied in either direction from time to time. These types are not suitable for use with unbiased AC. A special type of nonpolar device, commonly referred to as an AC-motor start capacitor, is available for use where the AC voltages are applied intermittently. Semipolar types are available for use where the level of reverse voltage is something less than the normally applied forward voltage.

These have a shelf life of several years under normal storage conditions. After this, the anodic film may deteriorate slightly, but after an application of rated voltage for a period of 30 minutes or more, the device should return to its initial leakage-current value.

CHAPTER SIX

Fixed-Value Electrostatic Capacitors

6.1 PLASTIC-FILM/METALLIZED-PLASTIC CAPACITORS

The present drive towards miniaturization, closer electrical tolerances, and higher operating temperatures is being met by the use of thin plastic-film dielectrics in the construction of capacitors.[35,36] Plastic-dielectric capacitors have insulation-resistance values far in excess of those for paper capacitors and, since they are nonabsorbent, their moisture characteristics are superior to those of mica. Also, it is better to use plastic-film dielectrics, instead of natural dielectrics, when high heat resistance is required.

There are several types of plastic films available for use as capacitor dielectrics. They may be used individually or in combination with other films and with paper in order to obtain the compromised advantages of the specific electrical characteristics of each individual film. The more-common films include polyethylene terephthalate and polycarbonate. When properly applied plastic-dielectric films lead to the solution of many special capacitor problems. (See Tables 6.1 and 6.2.)

Polymer films are available in thinner gauges, hence they have higher capacitance density than ceramic or glass capacitors. However, they cannot withstand the temperatures possible with the inorganic dielectrics. The capacitance of film devices is lower than that of electrolytics, but the capacitors are dry, nonpolar, and have low dissipation factors as well as excellent insulation resistance. As with paper units, both film-and-foil and metallized-film constructions are also used. In a film-and-foil capacitor, plastic film is interleaved with aluminum foil. (See Figures 6.1 and 6.2.)

The construction of metallized-plastic capacitors differs from conventional plastic capacitors in that instead of having separate layers of metal foil (capacitor plates) and plastic dielectric, the metal comprising the capacitor plates is imposed on one side of the plastic dielectric by means of a metallizing process. This technique results in an overall size reduction for metallized-plastic capacitors when compared with conventional plastic-foil capacitors of equal ratings. This space savings is the outstanding feature of the metallized-plastic capacitor. A 200-V metallized-plastic capacitor has a typical volume of 75% of the volume of a conventional plastic construction. At 600 V this increases to 80%; above 600 V there is no size advantage.

Another advantage resulting from the metallizing technique is that the capacitors are "self-healing." The metallic film imposed on the plastic is very thin and if breakdown occurs, a tiny area of the thin film surrounding the breakdown point burns away, leaving the capaci-

Table 6.1 Plastic and Paper Dielectric Capacitors[28]

Dielectric	Electrode	Volume Factor	Dissipation Factor (25C–1000 Hz) (%)	Insulation Resistance (MEG × MFD) 25°C	Insulation Resistance (MEG × MFD) High Temperature	Maximum Operating Temperature	Cost Factor	Outstanding Properties
Paper	Foil Metallized	0.30 0.10	0.60 0.65	20,000 2,000	10 5	125 C 125 C	1.0 1.1	Low cost Small size
Paper–polyester	Foil Metallized	0.46 0.17	0.45 0.50	20,000 2,000	50 10–40	125 C 125 C	3.0 1.3	Highest reliability
Polyester	Foil	0.25	0.30	50,000	10	150 C	2.1	High temperature Small size
Polyester	Metallized	0.14	0.45	25,000	20	125 C	1.7	Moisture resistant High insulation resistance
Polycarbonate	Foil	1.0	0.10	100,000	2000	125 C	2.7	Close tolerance High stability
Polycarbonate	Metallized	0.14	0.25	100,000	1000	125 C	1.8	Good high-frequency capability High insulation resistance
Polysulfone	Metallized	0.14	0.10	100,000	100	150 C	1.8	High insulation resistance
Polypropylene	Foil	1.2	0.05	100,000	5000	105 C	1.5	Low cost Good high-frequency capability
Polypropylene	Metallized	0.8	0.10	10,000	500	105 C	2.0	High insulation resistance
Cellulose acetate	Metallized	0.06	1.5	5,000	100	85 C	1.7	Smallest size
Polystyrene	Foil	1.4	0.025	250,000	40000	85 C	3.3	Close tolerance High stability
Polyester–polystyrene	Foil	1.5	0.20	100,000	100	125 C	2.8	High stability Small capacitance change with temperature

Table 6.2 Military Plastic Film and Paper Capacitors[26]

Specification	Type	Style	Capacitance Range (μF)	Capacitance Tolerance (±)	DC-Rated Voltage (V)	Operating Temperature Range (-55° to +__°C)
MIL-C-12889	Bypass, radiointerference reduction, paper dielectric, AC and DC	CA32	0.01	+0, -30	100	85
			0.1 to 0.5	+50, -10	100	85
		CA36	0.01	+0, -30	500 (AC-DC)	85
			0.1 & 0.25	+50, -10	500 (AC-DC)	85
		CA47	0.01	+0, -30	250 (AC-DC)	85
MIL-C-19978	Fixed, plastic (or paper-plastic) dielectric, established reliability	CQR07	0.001 to 1.0	5 or 10%	50 to 600	125
		CQR09	0.001 to 1.0	2, 5, or 10%	200 to 1,000	-65° to +125°C
		CQR12	0.001 to 1.0	2, 5, or 10%	200 to 1,000	-65° to +125°C
		CQR13	0.001 to 1.0	2, 5, or 10%	200 to 1,000	-65° to +125°C
		CQR29	0.001 to 10	2, 5, or 10%	30 to 1,000	-65° to +85°C
		CQR32	0.001 to 10	2, 5, or 10%	30 to 1,000	-65° to +85°C
		CQR33	0.001 to 10	2, 5, or 10%	30 to 1,000	-65° to +85°C
MIL-C-39022	Fixed, metallized dielectric, DC established reliability	CHR09	0.022 to 10	5 or 10%	50	85
		CHR19	0.10 to 8.2	5 or 10%	200 & 400	125
MIL-C-55514	Fixed, plastic (or metal-lized plastic) dielectric, DC established reliability	CFR02	0.001 to 1.0	1, 2, 5, or 10%	100 to 600	85 & 125
		CFR03	0.0047 to 2.2	2, 5, or 10%	100 to 400	85
		CFR04	0.001 to 1.0	1, 2, 5, or 10%	50 to 400	85 & 125
		CFR05	0.010 to 10	1, 2, 5, or 10%	200 & 400	125
		CFR06	0.010 to 10	2, 5, or 10%	200 & 400	85
		CFR08	0.001 to 0.39	5 or 10%	50, 100, or 300	85

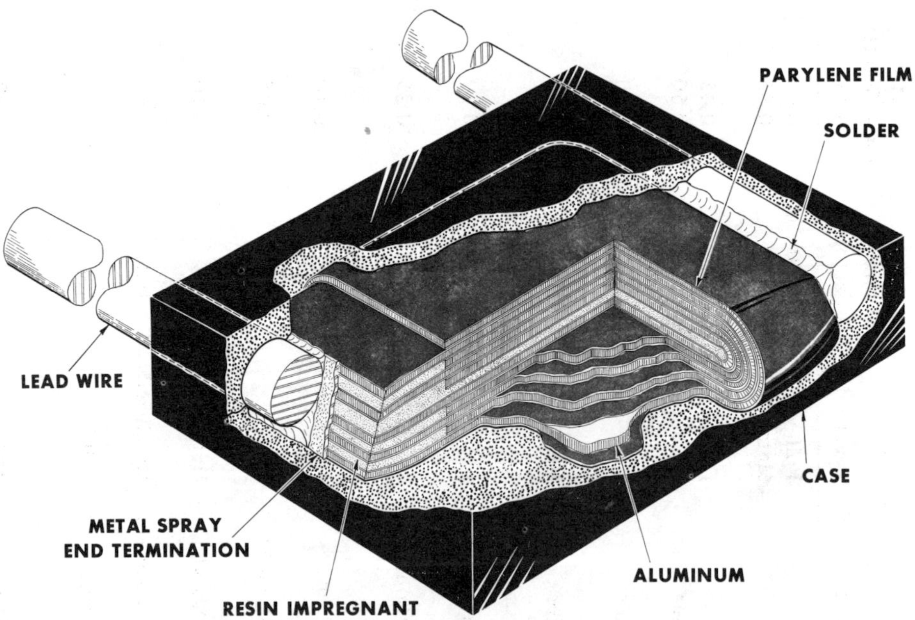

Figure 6.1 Typical radial-molded film capacitor. (Courtesy of Union Carbide Corp.)

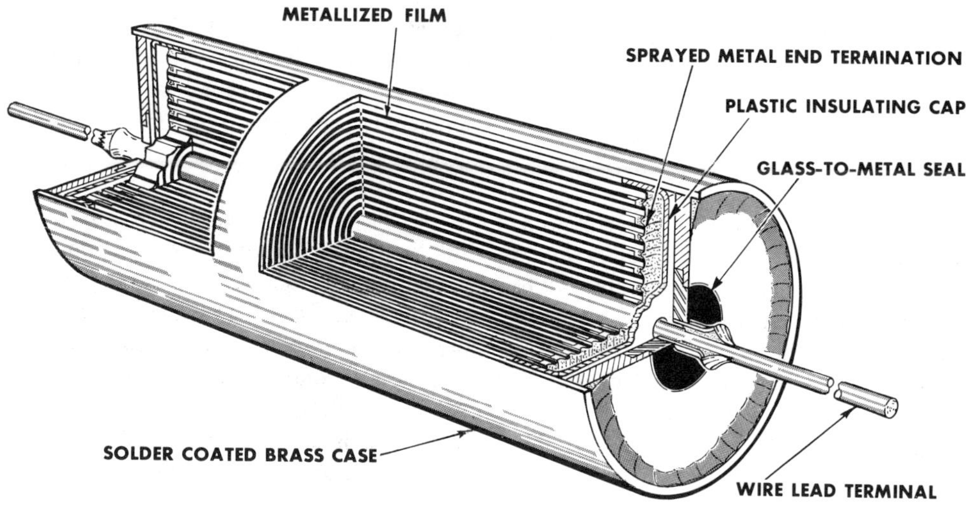

Figure 6.2 Typical hermetically sealed metallized capacitor. (Courtesy of Union Carbide Corp.)

tor operable, but with a slightly reduced capacitance. In the conventional plastic-foil type, where the foil is thicker, sustained conduction occurs on a breakdown causing a large area of plastic surrounding the breakdown to be carbonized, resulting in a permanent short circuit.

The breakdown of the metallized-plastic capacitor can be either of two types, that is, (1) complete breakdown lasting only for a moment, momentary breakdown; or (2) a sharp reduction in insulation resistance lasting for an extended period of time, but eventually returning to a normal period of low insulation. The general characteristics of the metallized-plastic type, aside from the breakdowns, are similar to the conventional plastic type except for a significantly lower insulation resistance.

6.1.1 Polypropylene[37]

The very low dielectric absorption, high insulation resistance, and low dissipation factor of polypropylene equal or exceed those characteristics of polystyrene. Although the temperature coefficient of polypropylene is not as low or as linear as that of polystyrene, its operating temperature rating of 105°C and its excellent stability and resistance to severe environmental conditions are among the reasons for its rapid increase in usage. Polypropylene has a high dielectric strength that permits small economical designs. In many instances it can be substituted for polystyrene or polycarbonate.

Capacitors utilizing metallized polypropylene are available in wrap-and-fill-type constructions, epoxy-case types, and in hermetically-sealed containers with tolerances of 1.0% and up for all capacitance values above 0.1 μF. Lower tolerances for capacitance value below 0.1 μF are available in hermetically sealed containers. For maximum stability, clamped or potted capacitors are recommended, which can be supplied in rectangular epoxy cases.

New plastic capacitor dielectrics, notably polypropylene, provide significantly greater efficiencies and reliability than electrolytics, especially at the frequencies used in switching power supplies (20–60 kHz). The high efficiency of polypropylene dielectric capacitors allows the use of lower rated-capacitance values than with electrolytic dielectrics. The effective-capacitance values of polypropylene units are higher than in electrolytics of the same rated value, when measured at high frequencies. Furthermore, polypropylenes have the very low equivalent-series-resistance (ESR) values need in filter circuits.

Though plastic-dielectric capacitors generally have a low dissipation factor (DF), polypropylene capacitors are among the lowest, being better than polysulfone, polycarbonate, polyester, and polystyrene.

Often overlooked is the current-carrying rating of capacitors. Most electrolytics are limited to under 1 A rms, although some large electrolytics over 100 μF can handle 2.5 amps. Polypropylene capacitors, however, are rated to carry up to 9 A of ripple current.

By weight polypropylene capacitors win over electrolytics, as polypropylene units typically weigh between 16 and 20 g/in.^{3}, while electrolytics typically weigh about 26 g/in.^{3}. Also, polypropylene capacitors are bone dry, whereas electrolytics contain pastes and liquids that can contaminate a circuit.

6.1.2 Polyimide

Polyimide film (Kapton®), often referred to as H film, capacitors have operating temperature capability up to 200°C. They find their major use in high-temperature applications. Polyimide-film capacitors also exhibit excellent electrical characteristics, such as high insulation resistance, low dissipation factor, and good stability.

6.1.3 Polystyrene

Polystyrene is a thermoplastic material with a melting point just over 90°C. Used only in film-foil configurations, polystyrene capacitors have a maximum temperature rating of 85°C, characterized by extraordinarily low dielectric absorption and a linear negative temperature coefficient of - 110/+ 30 ppm/°C. This film yields relatively high capacitance densities, because it is available in thin gauges. Polystyrene capacitors are ideally suited for computer and timing applications.

6.1.4 Polycarbonate

Polycarbonate film offers electrical advantages over polyester, while maintaining fine material properties. Polycarbonate-film capacitors are characterized by a low dissipation factor and dielectric absorption, high insulation resistance, and a very low temperature coefficient. The characteristics of polycarbonate approach those of polystyrene without the temperature limitation, as they can be used to 125°C. Polycarbonate is also available in the thinnest gauges of any polymer dielectric, that is, 0.00008 in.

Metallized-polycarbonate capacitors provide extremely small size, high-reliability capacitors. These capacitors have the same stability characteristics as the foil-type capacitors with reductions as much as 75% by volume in the lower-voltage ratings. They were primarily developed for compact construction where minimum weight and volume are required.

6.1.5 Polyester

Polyester is one of the most popular dielectrics, as it is inexpensive, has a good dielectric constant, and has good temperature stability for general use. Polyester films have largely replaced paper as the capacitor dielectric for electronic applications and have established reliability and dependability. They are available in a wide range of configurations in voltages up to 1000 VDC. The extreme versatility of polyester allows for many special designs and applications. Its major drawback is one of the highest temperature coefficients of capacitance (TCC) of all films in common use.

Metallized polyester has the same characteristics as mylar except that it has smaller size, tighter tolerances, and "self-healing" properties. The principal advantage of polyester-dielectric cpacitors is the high order of insulation resistance values available over the dielectric's temperature range of - 55 to + 125°C. However, for military applications the high-temperature limit is + 85°C. Polyester-dielectric capacitors have an insulation resistance that is normally about 100,000 MΩ/μF at room temperature and about 25,000 MΩ/μF at + 85°C. These insulation-resistance values decrease considerably when the capacitors are impregnated. However, a higher volt-per-mil rating is possible by impregnation and the possibility of corona and catastrophic failures due to pinholes in the dielectric is minimized.

6.1.6 Polysulfone

Polysulfone is one of the newest thermoplastic dielectrics used in wound capacitors. It exhibits electrical characteristics similar to those of polycarbonate, but having a temperature coefficient that is somewhat flatter, and temperature capabilities to 150°C.

6.1.7 Fluorocarbon

Fluorocarbon (Teflon) is perhaps the best polymer dielectric available, having both high-temperature capability to + 250°C and flat temperature coefficient of capacitance (TCC). This dielectric cannot be made in the very thinnest gauges, however, and has a low dielectric constant so that capacitance density tends to remain low.

6.2 PAPER DIELECTRIC CAPACITORS

Paper dielectric capacitors use a special thin paper impregnated with a dielectric wax or fluid, such as mineral oil or PCB, which bolsters the dielectric properties of the paper and seals out moisture. Metal foil is used in high-current and high-voltage paper capacitors. Good-quality units provide high capacitance and stability. However, the dissipation factor (DF) varies considerably as the temperature changes.

Metallized-paper capacitors, with the paper coated with a thin layer of zinc or aluminum, are considerably smaller than metal-foil units, but suffer from poor surge-handling capability. A tiny defect in the dielectric can result in an arc that will quickly and permanently short the capacitor.

Because the paper is an inexpensive material and large quantities are used, paper capacitors (foil or metallized) are relatively low cost (see Tables 6.1 and 6.2).

Paper/oil-filled capacitors consist of alternate layers of aluminum and one or more layers of paper in ribbon form that are rolled together. The paper may be saturated with an oil and the assembly mounted in an oil-filled, hermetically sealed, metal case. The addition of the insulating oil increases the paper capacitor's insulation resistance and temperature range and, primarily, the working voltage, since the oil fills the air gap, which would otherwise produce corona. The "AC" oil capacitor is designed primarily to provide large capacitance for industrial AC applications. As such, it must withstand large currents and voltages normally encountered in such applications at power-line frequencies.

6.3 MICA-DIELECTRIC CAPACITORS

Mica capacitors use the natural mineral mica as a dielectric. Metallized-film techniques are also applied to the mica dielectric. Originally, silver was the only metallization material used, but base metals are now employed without detracting from performance.

Mica capacitors are very reliable and stable and are

Table 6.3 Mica and Glass Capacitors[26]

Specification	Type	Style	Capacitance Range (pF)	Capacitance Tolerance (±)	DC Rated Voltage (V)	Operating Temperature Range (-55° to +__°C)
MIL-C-5	Fixed, mica dielectric	CM15	430 to 510	1 or 2%	500	150
		CM20	560 to 1,000	1 or 2%	500	150
		CM30	1,100 to 3,000	1 or 2%	500	150
		CM35	3,600 to 15,000	1 or 2%	300 or 500	125 or 150
		CM45	47 to 1,800	5%	2,500	150
		CM50	2,200 to 27,000	5%	600, 1,200, or 2,500	150
MIL-C-10950	Fixed mica dielectric, Button style	CB50	5 to 1,500	1, 2, 5, or 10%	500	85
		CB55	5 to 470	1, 2, 5, or 10%	500	150
		CB56	680 to 1,000	1, 2, or 10%	500	150
		CB57	1,200 to 2,400	1, 2, or 10%	500	150
		CB60	15 to 470	1, 2, 5, or 10%	500	150
		CB61	680 to 1,000	1, 2, or 10%	500	150
		CB62	1,200 to 2,400	1, 2, or 10%	500	150
		CB65	5 to 470	1, 2, 5, or 10%	500	150
		CB66	680 to 1,000	1, 2, or 10%	500	150
		CB67	1,200 to 2,400	1, 2, or 10%	500	150
MIL-C-23269	Fixed, glass dielectric, established reliability	CYR10	.5 to 270	0.25 pF to 5%	300 & 500	125
		CYR15	270 to 1,200	1 or 5%	300 & 500	125
		CYR20	1,500 to 4,700	1 or 5%	300 & 500	125
		CYR30	5,600 to 10,000	1 or 5%	300 & 500	125
		CYR13	0.5 to 270	0.25 pF to 5%	300 & 500	125
		CYR17	330 to 1,000	1 or 5%	300 & 500	125
		CYR22	1,200 to 1,800	1 or 5%	300	125
		CYR32	2,200 to 5,600	1 or 5%	300 & 500	125
		CYR41	0.5 to 1,000	0.25 pF to 5%	100	125
		CYR51	1 to 560	0.25 pF to 5%	300	125
		CYR52	620 to 1,000	2 or 5%	300	125
		CYR53	1,100 to 2,400	1, 2, or 5%	300	125
MIL-C-39001	Fixed, mica dielectric, established reliability	CMR03	1 to 400	0.5 pF, 2 or 5%	50 to 300	125
		CMR04	1 to 390	0.5 pF, 1, 2, or 5%	100, 300, or 500	150
		CMR05	1 to 390	0.5 pF, 1, 2, or 5%	500	150
		CMR06	430 to 4,700	1, 2, or 5%	500	150
		CMR07	5,100 to 20,000	1, 2, or 5%	500	150
		CMR08	22,000 to 91,000	1, 2, or 5%	100, 300, or 500	150

particularly good for high-frequency applications. Mica, a very stable material, provides high-Q, but capacitance values don't generally go above 0.1 μF, because the units become bulky and expensive (see Table 6.3).

Low-voltage micas provide great stability with moderate size at low cost. Mica also has very-low-loss features and is popular in high-voltage high-current applications in transmitters and other RF circuits.

6.4 GLASS-DIELECTRIC CAPACITORS

Glass-dielectric capacitors are intended for use in equipment where high orders of reliability are required and are primarily designed as a substitute for mica-dielectric capacitors. They are effective substitutes for mica-dielectric capacitors and can be employed in many applications where mica-dielectric capacitors are used, provided consideration is given to the differences in temperature coefficient and dielectric loss. They are capable of withstanding environmental conditions of shock, vibration, acceleration, extreme moisture, vacuum, extended life of 30,000 hours and more, and high operating temperature, such as those experienced in missile-borne and space electronic equipment (see Table 6.3).

6.4.1 Construction

Glass-dielectric capacitors are composed of alternating layers of glass ribbon and the electrode material. After assembly, the units are sealed together by high temperature and pressure to form a rugged monolithic block. Since the terminal leads are fused to the glass case, the seal cannot be broken without destroying the capacitor. Although these capacitors are of monolithic construction, they are not necessarily hermetically sealed, since the coefficient of thermal expansion of the terminals does not match that of the case. These construction features add up to the following advantages:

A. Fixed temperature coefficient.

B. High insulation resistance.

C. Low dielectric absorption.

D. Ability to operate in environments involving high humidity and high temperatures.

6.5 CERAMIC DIELECTRIC CAPACITORS[27, 38]

The wide range of dielectric constants gives ceramic-dielectric capacitors a notable advantage in microminiaturization. The ability to operate over wide temperature ranges and to possess either linear or nonlinear characteristics gives the ceramic capacitors significant capabilities. In addition, they can be formed into an unlimited variety of shapes and sizes, when cost is not a major factor, giving them both mechanical and electrical design possibilities not approached by most other materials.

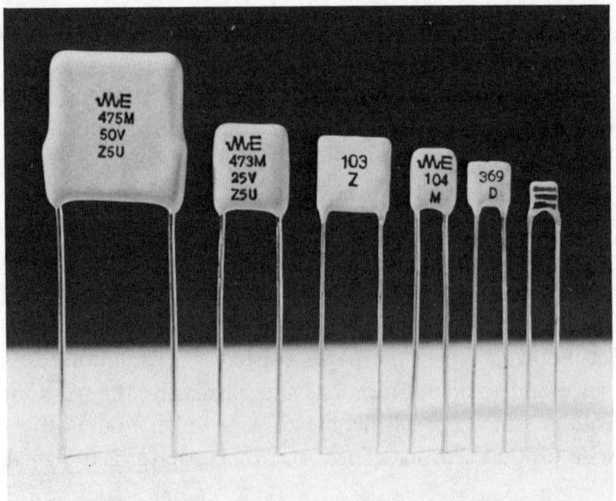

Figure 6.3 Ceramic capacitors. (Courtesy of Mepco/Electra Inc.)

General-application (GA) ceramic capacitors are used for bypass and coupling applications that require a capacitance tolerance of usually ±10 or ±20%. Available in the general range of the 100 pF to 0.15 μF, a GA-type unit will undergo a change in capacitance with temperature change within reasonable limits, making them suitable replacements for many paper- or mica-dielectric capacitors. (See Figure 6.3 and Table 6.4.)

6.5.1 Types

Disk ceramic capacitors are by far the most common type of ceramic-dielectric capacitor. They are produced on highly automated machines, resulting in minimum cost, and are usually furnished in 100-, 500-, and 1000-V ratings.

Disk and thin-plated subminiature ceramic capacitor types are extremely compact and have an inherent low series inductance owing to their construction. The placement of the leads facilitates making close-coupled, low-inductance connections suitable for printed-circuit applications. High insulation resistance allows these capacitors to be used in vacuum-tube grid circuits, while their extremely low leakage and small physical size make them suitable for use in transistor circuitry.

Buttonhead ceramic capacitors are designed for

Table 6.4 Ceramic Capacitors[26]

Specification	Type	Style	Capacitance Range (pF)	Capacitance Tolerance (±)	DC-Rated Voltage (V)	Operating Temperature Range (−55° to +__°C)
MIL-C-20	Fixed, ceramic dielectric (temperature compensating), established reliability	CCR05	1 to 10	0.25 & 0.5 pF	200	125
		CCR05	12 to 3,300	2, 5, or 10%	50, 100, or 200	125
		CCR06	390 to 10,000	1, 2, 5, or 10%	50, 100, or 200	125
		CCR07	2,200 to 47,000	1, 2, 5, or 10%	50, 100, or 200	125
		CCR08	3,900 to 68,000	2, 5, or 10%	50, 100, or 200	125
		CCR75	1 to 680	0.25 & 0.5 pF, 1, 2, 5, or 10%	50, 100, or 200	125
		CCR75	1 to 10	0.25 & 0.5 pF, 1, 2, 5, or 10%	200	125
		CCR76	82 to 1,000	0.25 & 0.5 pF, 1, 2, 5, or 10%	50, 100, or 200	125
		CCR77	150 to 5,600	0.25 & 0.5 pF, 1, 2, 5, or 10%	50, 100, or 200	125
		CCR78	820 to 27,000	0.25 & 0.5 pF, 1, 2, 5, or 10%	50, 100, or 200	125
MIL-C-11015	Fixed, ceramic-dielectric, general purpose	CK50	2.2 to 5,600	0.5 pF, 10 or 20%	50 to 200	125
		CK51	390 to 15,000	10 or 20%	50 to 200	125
		CK52	1,500 to 82,000	10 or 20%	50 to 200	125
		CK53	12,000 to 330,000	10 or 20%	50 to 200	125
		CK54	56,000 to 680,000	10 or 20%	50 to 200	125
		CK55	120,000 to 2,200,000	10 or 20%	50 to 200	125
		CK60	2.2 to 1,500	10 or 20%	500 & 1,000	85, 125
		CK62	22 to 6,800	10 or 20%	500 & 1,500	85, 125
		CK63	680 to 10,000	10 or 20%	500 & 1,000	85, 150
		CK65	4,700	20%	1,600	85
		CK66	5,600	20%	1,600	85
		CK67	6,800 & 7,500	20%	1,600	85
		CK68	10,000	20%	1,600	85
		CK69	15,000	20%	1,600	85
		CK70	10 to 1,500	10 or 20%	500 & 1,000	85
		CK80	10 to 1,500	10 or 20%	500	85
MIL-C-39014	Fixed, ceramic, dielectric, established reliability	CKR05	10 to 100,000	10, 20%	50, 100, or 200	125
		CKR06	1,200 to 1,000,000	10, 20%	50, 100, or 200	125
		CKR11	10 to 10,000	10, 20%	50 & 100	125
		CKR12	5,600 to 47,000	10, 20%	50 & 100	125
		CKR14	12,000 to 100,000	10, 20%	50 & 100	125
		CKR15	56,000 to 100,000	10, 20%	100	125
		CKR64	510 to 3,900	20%	1,600	85
		CKR72	100 to 1,000	20%	1,500	85
MIL-C-55681	Fixed chip, multiple layer, ceramic dielectric established reliability	CDR01	10 to 180,000	5, 10, or 20%	50 & 100	125
		CDR02	10 to 180,000	5, 10, or 20%	50 & 100	125
		CDR03	10 to 180,000	5, 10, or 20%	50 & 100	125
		CDR04	10 to 180,000	5, 10, or 20%	50 & 100	125

coupling, feedthrough, and bypass applications in very high and ultrahigh frequency applications. These capacitors use a ceramic disk or monolithic disk inside the head of a small metal button or screw. The design provides for a minimum lead inductance to ground and, therefore, the highest possible self-resonance frequency.

The "doorknob" capacitor so named because of its external appearance, is a disc capacitor with a dielectric of sufficient thickness to provide exceptional voltage capability, with ratings up to 40,000 VDC and 12,000 VAC.

Chip ceramic capacitors are leadless, unencapsulated, capacitor chips. They may be of either single-wafer or monolithic construction in many different ceramic compositions. (See Section 6.6 on chip capacitors for additional information.)

6.5.2 Construction

A ceramic capacitor consists of a ceramic dielectric on which a thin metallic film, usually silver, has been fired at a very high temperature. Terminal leads are attached to the electrodes by a pressure contact or by soldering. Ceramic capacitors are encapsulated to protect the dielectric from the environment and to electrically insulate the capacitor.

The disk types are covered by an insulating resin, plastic or ceramic; the thin-plated subminiature types may be in dipped, molded, or preformed cases. The feedthrough units are made of ceramic tubes modified for their required mounting.

Because the constituent materials have molecular polar moments, the dielectric constants of some mixes reach hundreds (even thousands) of times the value of paper, mica, and plastic films. This results in ceramic capacitors having the largest capacitance-to-size ratios of all the high-resistance dielectrics.

In monolithic ceramic capacitors, mechanical strength is improved by the multiple dielectric layers which provide a buildup to the required minimum thickness and maximum capacitance values. The material cost is somewhat higher since a more expensive electrode material is required. However, where minimum size is a prime requirement, very good high-frequency characteristics are desired, or capacitance values up to 4.7 μF are required, the monolithic construction is often the most desirable.

6.5.3 Material

For any capacitor, the capacitance C is determined by the formula $C \propto KA/d$, where A is the area of the metallic electrodes and d is the thickness of the material that separates the electrodes. For a given electrode size, the capacitance can be increased either by decreasing the dielectric thickness or by increasing the dielectric constant. The dielectric constants of some commonly used ceramic dielectrics are from 12,000 to 400,000.

A wide range of dielectric constants K can be obtained with various proportions of ceramic mixtures. This is part of the reason for so many different ceramic capacitor types. In addition, the stability of the dielectric constant with temperature is very important. In general, the lower the K value the more stable the capacitance value with temperature. High-K capacitors have a dielectric constant usually in excess of 3000 and are extremely small for any given capacitance value and voltage rating. They are used mainly for bypass or coupling and are usually in the range of 0.001 to 0.1 μF. The highest-K disk ceramic capacitors use a semiconductor type of ceramic to achieve K's of up to 400,00 over the operating temperature range. They provide capacitance values to 0.47 μF at low voltages in very small sizes.

Ceramic dielectric materials are nonhygroscopic, effectively imperable, and have practically no moisture absorption even after considerable exposure to humid conditions. Thus these units are intended to operate, through their full temperature range, at relative humidities up to 95%. Nevertheless, the termination materials under moisture conditions are subject to ionic migration, which can cause capacitor failure.

Where stability with temperature change is required, there are available semistable and temperature-stable types. These will exhibit a capacitance change of less than ± 15% or ± 5%, respectively, over the operating temperature range.

6.6 CERAMIC CHIP CAPACITORS[27, 39–41]

In the communications field chip capacitors (Figure 6.4) appear in radios, telecommunication products, and high-frequency microwave systems. In the medical field they are used in cardiac pacemakers and monitoring instruments. In the consumer field they show up in digital watches, appliances, and automotive equipment. In the commercial field they are employed in computers, industrial equipment, and avionic instruments. The military uses them in many applications. (See Table 6.5.)

6.6.1 Functions

Chip capacitors perform these more common functions:

A. Bypassing.

B. Blocking.

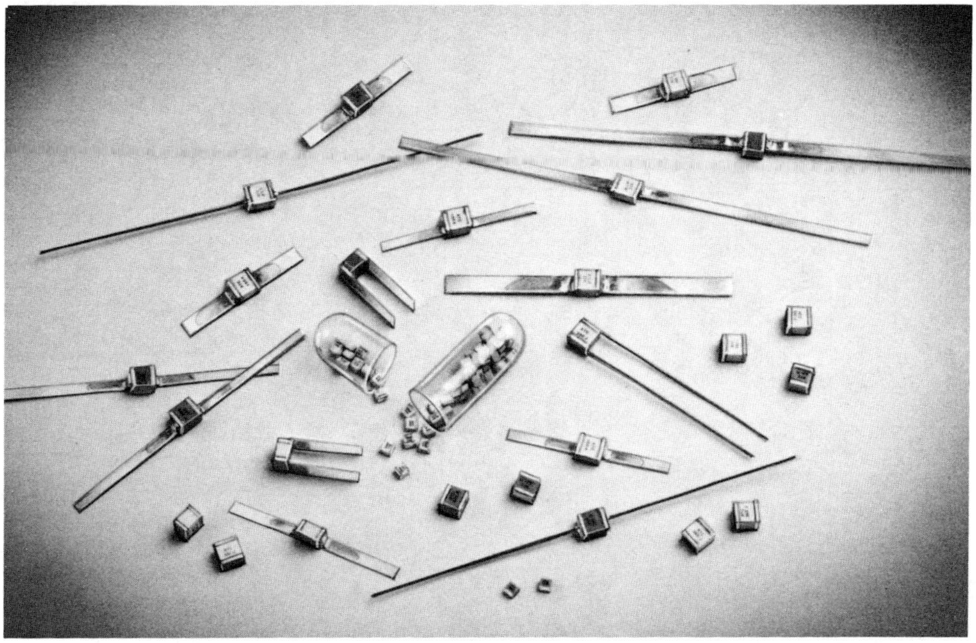

Figure 6.4 Typical chip capacitors. (Courtesy of American Technical Ceramics.)

C. Coupling.
D. Timing and delaying.
E. Frequency discrimination.
F. Energy discharge.
G. Arc suppression.
H. Voltage suppression.

6.6.2 Characteristics[42]

Almost an infinite number of ceramic materials can be formulated and used as capacitor dielectrics. The characteristics of the more commonly available materials are listed in Table 6.6.

6.6.3 Classes

Although an almost infinite number of ceramic materials can be formulated and used as capacitor dielectrics, current practice limits the mixtures to those that qualify to meet MIL-C-55681 and EIA-198 specifications (see Table 6.7).

A. *Class I Dielectrics* Class I specifications cover dielectrics with linear temperature coefficients (TCs) owing to the cubic crystal structure (paraelectric) of the ceramics. Class I materials also exhibit a low dielectric constant (K). K values range from 150 to 500 for Class I materials. They also provide a high Q factor, which measures the

Table 6.5 Applications for Ceramic Capacitors by Class[40]

Class	Dielectric	Applications
I	NPO/COG	Precision circuits that require stable dielectric characteristics with negligible dependence of capacitance and dissipation factor on time, voltage and frequency, low-loss or high-Q and linear TC; microstrip and hybrid microwave integrated circuits.
II	BX/X7R	Bypassing and coupling, filtering, frequency discrimination, DC blocking, and voltage transient suppressing.
II (High-K)	Z5U, Y5V	Best performance at, or near, room temperature with less than 25 VDC bias; high-capacitance density for units with space problems; replacements for tantalum capacitors; in bypass and decoupling elements.

Table 6.6 Ceramic Dielectric Characteristics[42]

Characteristics		Typical Data				
		NPO	SL	BX	AE	AF
Capacitance	Range	0.5 pF to 5600 pF	0.5 pF to 0.01 μF	47 pF to 0.15 μF	470 pF to 0.22 μF	2200 pF to 0.47 μF
	Tolerance (%)	±5, ±10, ±20		±10, ±20	+80 −20	+80 −20
Temperature characteristics @25°C reference	1.0 ± 0.2 Vrms	0 ± 30 PPM/°C −55°C to +125°C	+350 to −1000 PPM/°C −25°C to +85°C	±15% −55°C to +125°C	+20%, −55% −25°C to +85°C	+30%, −80% −25°C to +85°C
	Rated voltage applied (DC)	—	—	+15%, −25% −55°C to +125°C	+20%, −80% −25°C to +85°C	+30%, −95% −25°C to +85°C
Dissipation factor (or Q) @25°C		Q ≧ 1000 for capacitance 30 pF and over Q ≧ 400 + 20C* for capacitance under 30 pF		3.5% max. for capacitance 0.047 μF and over 2.5% max. for capacitance under 0.047 μF	4.0% max. for capacitance 0.1 μF and over 3.0% max. for capacitance under 0.1 μF	5.0% max.
Rated voltage (VDC)		50				
Insulation resistance		10,000 MΩ or 500 MΩ-μF minimum, whichever is less				
Voltage proof		Tested at DC V_R × 3, 5 sec. 1 = 50 mA max. . . . all values (V_R = rated voltage)				
Life	Test condition	1,000 hours @2 × DC V_R and 125°C	1,000 hours @2 × DC V_R and 85°C	1,000 hours @2 × DC V_R and 125°C	1,000 hours @2 × DC V_R and 85°C	
	Insulation resistance	1,000 MΩ or 50 MΩ-μF min., whichever is less				
	Dissipation factor (or Q)	Q ≧ 350 for capacitance 30 pF and over Q ≧ 275 + 5/2C* for capacitance 10 pF and over up to 30 pF Q ≧ 200 + 10C* for capacitance under 10 pF		5.0% max.		7.5% max.

Moisture resistance	Test condition	240 hours @DC $V_R \times 2$, 40 ± 2°C and 90 to 95% RH		
	Insulation resistance	1,000 MΩ or 50 MΩ-μF minimum, whichever is less		
	Dissipation factor (or Q)	Q ≧ 350 for capacitance 30 pF and over Q ≧ 275 + 5/2C* for capacitance 10 pF and over up to 30 pF Q ≧ 200 + 10C* for capacitance under 10 pF	5.0% max.	7.5% max.
Test frequency		1 MHz ± 1 kHz for 330 pF and under. 1 kHz ± 100 Hz for over 330 pF		

*C = Rated capacitance

Characteristics		Typical Data		
		RH	SH	UJ
Capacitance	Range	3 pF to 6800 pF	5 pF to 8200 pF	10 pF to 0.01 μF
	Tolerance (%)	±5, ±10, ±20	±5, ±10, ±20	±5, ±10, ±20
Temperature characteristics @25°C		−220 ± 60 PPM/°C −25°C to +85°C	−330 ± 60 PPM/°C −25°C to +85°C	−750 ± 120 PPM/°C −25°C to +85°C
Test methods, Specifications		Same as those of NPO except operating temperature range		

Table 6.7 EIA Dielectric Comparison[40]

Parameter	Dielectric Designation NPO	X7R	HIGH *K* Z5U
Dielectric factor (maximum)	0.1/0.2%	2.5%	2.5%
Temperature performance range	-55 to 125°C	-55 to 125°C	+10 to 85°C
ΔC tolerance	±30 ppm/°C	±15%	+22% -56%
ΔC (with rated bias)	±30 ppm/°C	N.R.	N.R.
Insulation resistance:			
at 25°C	1000 F	1000 F	1000 F
at upper temperature limit	100 F	N.R.	N.R.
Life test (maximum rated temperature)	1000 hr, 2X	250 hr, 2X	250 hr, 2X
Flash test	2.5X	2.5X	2.5X

N.R. = Not Required

performance of the capacitor when compared with pure reactance. The higher the Q, the more nearly the resistance and dissipation factor approach zero. Consequently, Class I dielectrics are low-loss materials.

Capacitors made with Class I materials change little with time (no aging), and their capacitance and dissipation factor depend very little on voltage and frequency. The most important Class I dielectric material is designated as a NPO or COG (see Table 6.8).

B. *Class II Dielectrics* Class II specifications for dielectric materials cover ceramics with *K* values greater than 500. They exhibit nonlinear characteristics and less-stable overall properties than Class I. Also, capacitors made from Class II materials have capacitance and dissipation factors that are time dependent (age), AC and DC voltage-dependent, and frequency dependent.

Class II materials are ferroelectric, that is, an alternating applied electric field polarizes the structure and causes charges to migrate toward the capacitor's electrodes. Ferroelectric ceramic structures change from cubic (paraelectric) to tetragonal (ferroelectric) crystals at the Curie temperature. The higher the *K* value, the lower the Curie termperature.

Class II materials are further divided into two groups. Mid-*K* ceramics (500–4000), the more stable of the two, meet military BX and EIA X7R specifications. HIgh-*K* ceramics (more than 4000) are less stable and must operate at or near room temperature in order to remain below the Curie Point to prevent severe degradation from aging effects (see Table 6.9).

6.6.4 Silicon

Silicon dioxide or silicon nitride are used to make MOS and NMOS chip capacitors, respectively. Because silicon

Table 6.8 Class I Dielectric (+55°C to +85°C)[41]

Ref.	Significant Figure	Ref.	Multiplier	Ref.	Tolerance (ppm/°C)
C	0.0	0	-1	G	±30
M	1.0	1	-10	H	±60
P	1.5	2	-100	J	±120
R	2.2	3	-1,000	K	±250
S	3.3	4	-10,000	L	±500
T	4.7	5	+1	M	±1,000
U	7.5	6	+10	N	±2,500
		7	+100		
		8	+1,000		
		9	+10,000		

Table 6.9 Class II Dielectrics[41]

Ref.	Low Temp. Limit	Ref.	Upper Temp. Limit	Ref.	ΔC Limit (%)
Z	+10	2	+45	A	±1.0
Y	-30	4	+65	B	±1.5
X	-55	5	+85	C	±2.2
		6	+105	D	±3.3
		7	+125	E	±4.7
				F	±7.5
				P	±10
				R	±15
				S	±22
				T	±22 to 33
				U	±22 to 56
				V	±22 to 82

capacitors are very high Q devices, they offer better insertion-loss characteristics for high-frequency applications than do ceramic-chip capacitors.

Silicon-capacitor applications include oscillator or tuned networks, padding, active filters, and feedback networks. One type of chip contains a number of geometrically arranged capacitance values. A user can substitute the unit for a trimmer capacitor in precision-oscillator circuits by bonding the leads between the appropriate sections to produce the required capacitance value.

6.6.5 Terminations

Chip capacitors are made from or coated with various types of metals and alloys. Capacitors without leads are called chips or pellets. Capacitor lead types include regular and narrow microstrip, regular and narrow ribbon, radial ribbon, radial wire, and axial wire.

The user chooses the type of termination, chip or lead, and the material on the surface of the termination to suit the application and the technique to be used to assemble the capacitor into the circuit.

6.6.6 Selection

The selection of chips and dielectric for any application is based on the following considerations:

A. ***Temperature Stability*** Dictates dielectric type.

B. ***Voltage-Temperature Coefficient*** Affects the dielectric and chip design requirements.

C. ***Capacitance/Size*** Capacitance value required dictates chip size. Chip size is also dependent on working voltage and dielectric-withstanding voltage requirements which determine dielectric-layer thickness.

D. ***Capacitance-Tolerance*** Determines chip and dielectric used.

E. ***Operating and Dielectric-Withstanding Voltage*** Can restrict dielectric type if voltage-sensitive high-K dielectrics are required to meet capacitance values.

F. ***Q-Factor*** Dictates types of dielectric material:

$Q > 1{,}000$ @1 MHz = NPO dielectric

$Q > 10{,}000$ @1 MHz = Hi-Q dielectric (K20)

G. ***Termination*** Standard termination materials and solder are available.

H. ***Safety Margin*** Minimizes potential failures and are usually included in chip-capacitor specifications.

I. ***Overdesign*** Can occur between the sequence of events that take place between original requirements and final chip, to the extent that costs and yields are affected. Delivery and costs are optimized when an understanding of true chip requirements and safety margins is reached.

CHAPTER SEVEN

Variable-Value Capacitors

7.1 DIRECT-ENGAGEMENT CAPACITORS

Direct-engagement, tubular, ceramic, variable-value capacitors[43] (Figure 7.1) consist of a ceramic tube, ground internally for a close fit with the metal spindle or rotor equipped with a tightly fitted metal tube on its free end and a mounting bushing on the opposite end. The spindle has a spiral groove which mates with a wire spring installed in the bushing; the spring guides in or out of the tube as it is rotated, and provides continuous electrical contact between spindle and bushing.

The variable capacitance is provided by the spindle and metal tube, separated by the ceramic dielectric. Rotation of the spindle, by means of a screwdriver inserted in a slot at either end, determines the degree of engagement of the spindle with the tube, and hence the capacitance, which reaches a maximum when the spindle is fully inserted.

In certain models, a plastic friction ring installed in the mounting bushing, or a locking nut threaded onto the top of the bushing, serves to lock the spindle in place at any setting, for preservation against the effects of shock and vibration after precise adjustment has been achieved.

Mounting bushings are designed for insertion into holes in a chassis and are provided with an annular flange that seats against the chassis. The solder type is secured by direct soldering of the bushing to the chassis: the tag-mounting type has an integral punched tag for rivet or machine-screw installation; nut-mounting types are equipped with threaded bushings and a retaining nut.

With the latter type a solder-lug terminal is furnished for installation on the bushing during mounting to provide for a soldered bonding connection to the chassis or signal ground. All types are equipped with a solder-lug terminal secured to the end of the metal tube for connection to the high circuit, though for very high-frequency applications, where strap conductors are used, the strap may be soldered directly to the silver-plated surface of the tube itself.

Tubular trimmer capacitors for printed circuit installation are similar to the others in construction, but are provided with terminal pins spaced to fit a 0.1-in. grid, for direct insertion and soldering. Vertical mountings, as well as horizontal-mounting types, are available.

7.2 FILM-DIELECTRIC TRIMMERS

Film-dielectric trimmer capacitors (Figure 7.2) are composed of a set of D-shaped plates rigidly mounted in a round of rectangular plastic, engaging a set of similar plates mounted on a central rotor shaft, interleaved with disks of plastic-film dielectric. Most types are single-stator designs, with a continuously rotatable rotor providing minimum-to-maximum capacitance adjustment over an effective angle of adjustment of 180°.

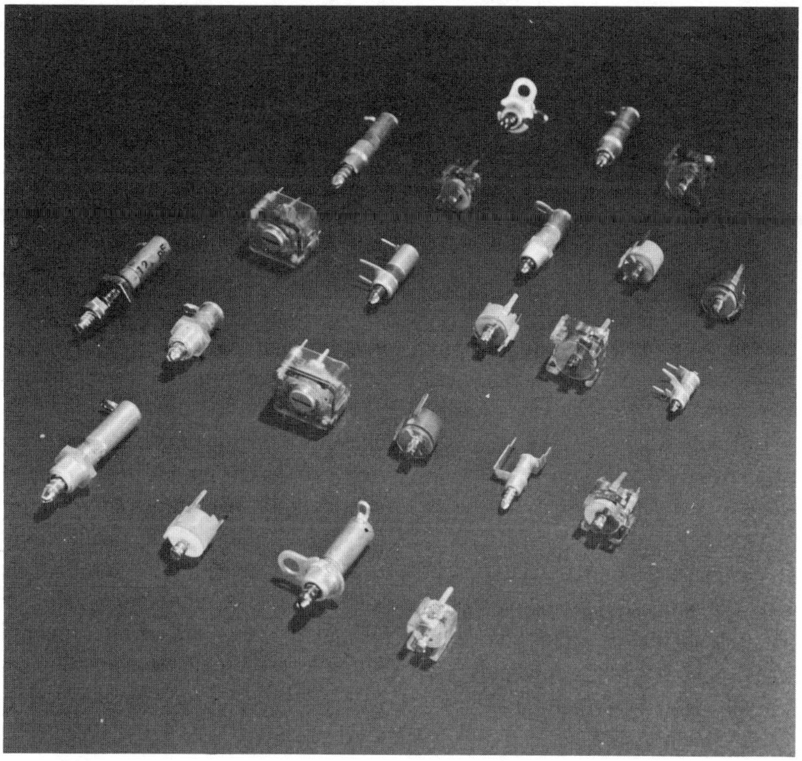

Figure 7.1 Tubular ceramic trimmer capacitor styles. (Courtesy of Mepco/Electra, Inc.)

Figure 7.2 Film-dielectric trimmer capacitor styles. (Courtesy of Mepco/Electra Inc.)

Table 7.1 Military Variable-Value Capacitors[26]

		Capacitance		Capacitance Change from 25°C						
				At −55°C		At +85°C			DC-Rated	Operating Temperature
Type	Style	Min (pF)	Max (pF)	Min (%)	Max (%)	Min (%)	Max (%)	*Q* or *DF*	Voltage (V)	Range (°C)
Variable, ceramic	CV11	1.5	7.0	−4.5	+2.0	−2.5	+2.0	⩽0.2%	500	−55 to +85
dielectric		3.0	12.0	−4.5	+2.0	−2.5	+2.0	⩽0.2%	500	−55 to +85
MIL-C-81		4.5	25.0	−4.5	+2.0	−2.5	+2.0	⩽0.2%	500	−55 to +85
		3.0	13.0	−1.0	+3.5	−2.5	−0.5	⩽0.2%	500	−55 to +85
		5.0	20.0	−1.0	+3.5	−2.5	−0.5	⩽0.2%	500	−55 to +85
		4.0	30.0	−1.0	+6.5	−4.0	−1.0	⩽0.2%	500	−55 to +85
		7.0	45.0	−1.0	+6.5	−4.0	−1.0	⩽0.2%	500	−55 to +85
		2.0	6.0	+1.5	+7.0	−5.0	−1.5	⩽0.2%	500	−55 to +85
		4.0	30.0	+1.5	+7.0	−5.0	−1.5	⩽0.2%	500	−55 to +85
		7.0	45.0	+1.5	+7.0	−5.0	−1.5	⩽0.2%	500	−55 to +85
	CV21	1.5	7.0	−4.5	+2.0	−2.5	+2.0	⩽0.2%	500	−55 to +85
		3.0	12.0	−4.5	+2.0	−2.5	+2.0	⩽0.2%	500	−55 to +85
		4.5	20.0	−4.5	+2.0	−2.5	+2.0	⩽0.2%	500	−55 to +85
		3.0	13.0	−1.0	+3.5	−2.5	−0.5	⩽0.2%	500	−55 to +85
		5.0	20.0	−1.0	+3.5	−2.5	−0.5	⩽0.2%	500	−55 to +85
		4.0	30.0	−1.0	+6.5	−4.0	−1.0	⩽0.2%	500	−55 to +85
		7.0	45.0	−1.0	+6.5	−4.0	−1.0	⩽0.2%	500	−55 to +85
		5.0	30.0	+1.5	+7.0	−5.0	−1.0	⩽0.2%	500	−55 to +85
		7.0	45.0	+1.5	+7.0	−5.0	−1.5	⩽0.2%	500	−55 to +85
	CV31	2.0	8.0	−4.5	+2.0	−2.5	+2.0		350	−55 to +85
		5.5	18.0	−4.5	+2.0	−2.5	+2.0		350	−55 to +85
		2.5	11.0	−1.0	+3.5	−2.5	−0.5		350	−55 to +85
		7.0	25.0	−1.0	+3.5	−2.5	−0.5		350	−55 to +85
		3.0	10.0	−1.0	+6.5	−4.0	−1.0	–	350	−55 to +85
		8.0	25.0	−1.0	+6.5	−4.0	−1.0		350	−55 to +85
		3.0	15.0	+1.5	+7.0	−5.0	−1.5		200	−55 to +85
		9.0	35.0	+1.5	+7.0	−5.0	−1.5		200	−55 to +85
		15.0	60.0	+3.0	+14.0	−10.0	−3.0		200	−55 to +85
Variable (Piston type,	PC25	0.8	10.0					>2000	250	−55 to +125
tubular trimmer)	PC26	0.8	10.0					>2000	250	−55 to +125
MIL-C-14409	PC38	0.6	1.8					>1500	750	−55 to +150
		0.6	5.5					>1500	1250	−55 to +150
		0.6	9.5					>1500	1250	−55 to +150
		0.8	16.0					>1500	1250	−55 to +150
	PC39	1.0	16.0					>750	1000	−55 to +125
		1.0	36.0					>550	1000	−55 to +125
		1.0	52.0					>350	1000	−55 to +125
		1.0	75.0					>250	1000	−55 to +125
		1.0	120.0					>250	1000	−55 to +125
	PC40	0.6	1.8					>1500	750	−55 to +150
		0.6	5.5					>1500	750	−55 to +150
		0.6	9.5					>1500	750	−55 to +150
		0.8	16.0					>1500	750	−55 to +150
	PC42	0.6	1.8					>1500	750	−55 to +150
		0.6	5.5	–	–	–	–	>1500	1250	−55 to +150
		0.6	9.5					>1500	1250	−55 to +150
		0.8	16.0					>1500	1250	−55 to +150
	PC43	1.0	16.0					>750	1000	−55 to +125
		1.0	36.0					>550	1000	−55 to +125
		1.0	52.0					>350	1000	−55 to +125
		1.0	75.0					>250	1000	−55 to +125
		1.0	120.0					>250	1000	−55 to +125
	PC48	0.8	5.5					>1000	750	−55 to +125
		0.8	11.0					>900	1250	−55 to +125
		0.8	16.0					>800	1250	−55 to +125
		0.8	23.0					>700	1250	−55 to +125
		1.0	38.0					>500	1250	−55 to +125
	PC52	0.8	5.5					>1000	750	−55 to +125
		0.8	11.0					>900	1250	−55 to +125
		0.8	16.0					>800	1250	−55 to +125
		0.8	23.0					>700	1250	−55 to +125
		0.8	38.0					>500	1250	−55 to +125

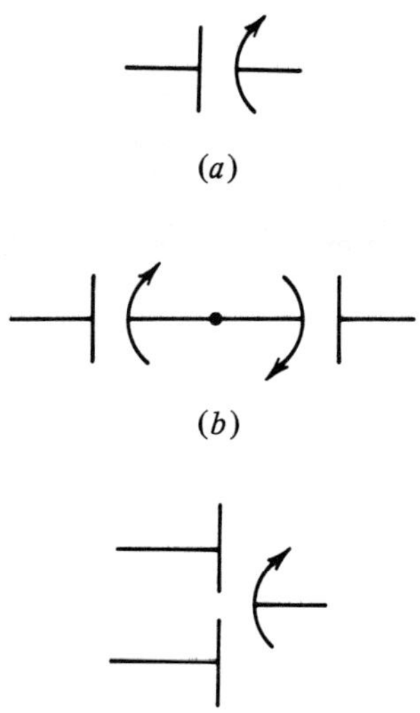

(*a*) (*b*) (*c*)

Figure 7.3 Electrical configurations.[43] (*a*) Single stator, (*b*) split stator, (*c*) differential.

These trimmers offer a choice of single-stator (Figure 7.3*a*), split-stator (Figure 7.3*b*), or differential (Figure 7.3*c*) configurations. In the split-stator design, rotation of the rotor provides progressive increase in capacitance coupling between the two halves of the stator by the rotor plates; capacitance measured between the two stator terminals increases from minimum to maximum over an angle of rotation of 90°.

The differential design provides reciprocal engagement of the rotor with two separate stators, so that the capacitance between one stator and the rotor increases as the capacitance between the other stator and the rotor decreases, over the effective angle of rotation of 180°. Connections to the rotor plates and to all stators are brought out as pins or tabs for direct insertion into printed-circuit boards.

7.3 MILITARY

Military specifications for variable-value capacitors are summarized in Table 7.1.

PART THREE

Electromagnetic Components

CHAPTER EIGHT

Transformers

8.1 GENERAL

A transformer[1,44,45] is a magnetic component that changes voltage, current, or impedance from one value to another by linking two electrical circuits with a magnetic field. (See Figure 8.1.)

The transformer consists essentially of two coils of wire, separated electrically, but wound on the same core. The coil connected to the source of the voltage is called the primary and the other is called the secondary. There can be several secondary windings to supply different levels of voltage from the common primary.

The alternations of the primary current set up an alternating magnetic field in the core and induce a voltage in the secondary coil. The relationship of the secondary voltage to primary voltage is the same as the number of turns between the two coils.

The transformer is not capable of producing energy. Therefore, the total power taken from the secondary cannot exceed that which is supplied to the primary. Since power is equal to voltage multiplied by current, voltage is affected by the turns ratio. (See Figure 8.2.)

8.1.1 Types

There are many variations on the type of construction, for example, shielded types, open-frame types, hermetically sealed types, potted or molded types, and plug-in types, to list only a few. (See Figure 8.3.) Some transformers are made to meet high-reliability military specifications and others are merely commercial or industrial grade.

A. ***Isolation Transformer*** An isolation-type power transformer consists of a minimum of two windings, where the secondary(ies) is isolated from the primary winding(s), ensuring limited voltage potential between the primary/input circuit and the secondary/output circuits. (See Figure 8.4.)

B. ***Auto-Type Transformers*** The autotransformer design is the least-expensive type available. It consists of one winding tapped, or two windings interconnected, to provide step-up or step-down voltages. The secondary winding(s) is common with the primary winding. Normally most isolation-type power transformers can be built as autotransformers, if isolation is not required. (See Figure 8.5.)

C. ***Audio Transformers*** Audio transformers change the level of voltage, current, or impedance (or a combination of all three). Usually operating over a band of frequencies from 20 to 20,000 Hz, with dependable output to input response; the audio transformer can be of the isolation or autotype. Normally a DC bias current flows through one or more of the windings.

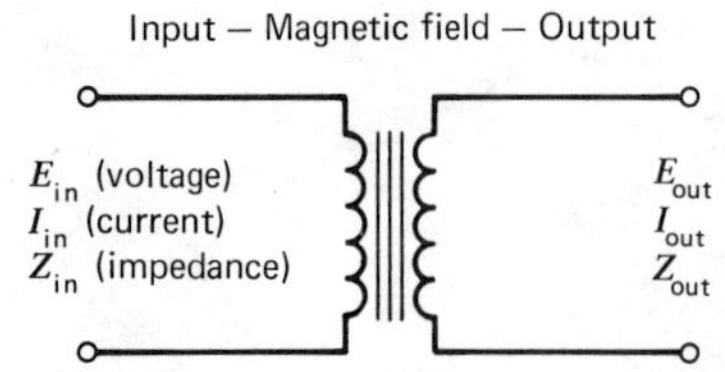

Figure 8.1 General transformer function.[45]

D. ***Constant Voltage-Transformers*** Constant-voltage transformers are designed to provide constant output voltage with input-line fluctuations. Formerly categorized as a power transformer, the unique characteristics of the constant-voltage transformer has resulted in its being recognized separately. (See Figure 8.6.)

E. ***DC Power-Supply-Transformers*** Cost savings can be achieved by including a rectifier and, if required, a capacitor as a part of the transformer. Thus DC power is provided from a single component instead of depending on the user to add these parts to the circuit. (See Figure 8.7.)

8.1.2 Applications

Generally, applications for transformers can be classified into the following uses:

A. Isolation of the power source (line voltage) from the circuit.

B. Shifting of voltage levels (step-up or step-down)

C. Coupling of AC signals when isolation or voltage matching is needed.

D. Special requirements (e.g., AC regulation, balanced-modulator circuits).

Also available are what are known as variable autotransformers, which are used to manually control, over a continuous range, the level of the line voltage being supplied.

8.1.3 Elements

The essential elements of a transformer are the coil, core, insulation system, termination, and hardware.

A. ***Coil*** Copper's excellent conductivity and availability are the reasons for its use in the manufacture of the transformer's coil. The selection of magnet wire size and insulation is dictated by the design requirements of the transformer application.

A large variety of winding equipment fulfill most coil-winding requirements, except for extremely large or ultrafine wire sizes. Bobbin, paper section, and hand-winding procedures are regularly utilized.

1 ***Bobbin-Wound*** This is a process of winding the magnetic wire randomly on a molded coil form. Because bobbin winding provides a randomly wound coil without layer insulation, the coil build is significantly reduced, thereby allowing more copper in the lamination window.

2 ***Paper Section*** (Multiple Wound) This method facilitates rapid production and provides a more-uniform reecision-wound coil. The coil is usually wound upon a kraft-paper tube. Interlayer support is provided by various sheet materials relevant to wire size. The tube is then cut, resulting in individual coils.

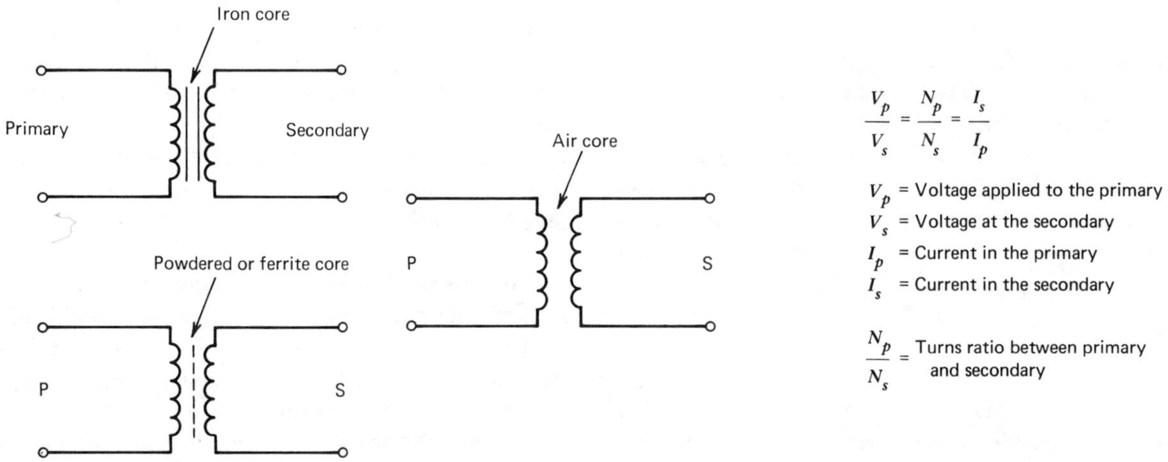

Figure 8.2 Turns ratio.[1]

Figure 8.3 Transformer types. (Courtesy of Staco Inc.)

3 ***Hand-Wound*** With this process winding and processing of coils which cannot be manufactured on multiple-winding machines, because of wire size or special constructional features, are wound individually. These can be similar in appearance to the bobbin, paper section, or the self-supporting type (which has no coil form or interlayer support.)

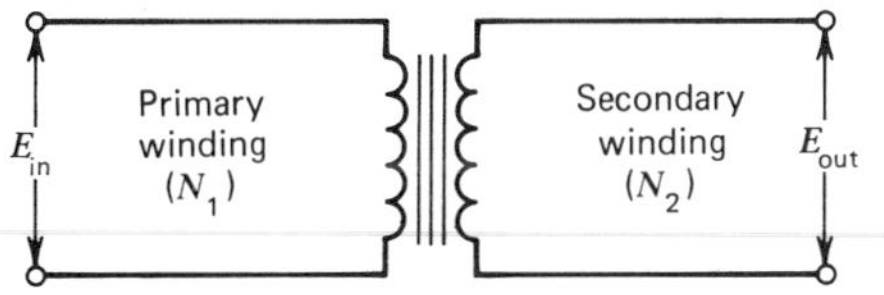

Figure 8.4 Isolation-type transformer.[45]

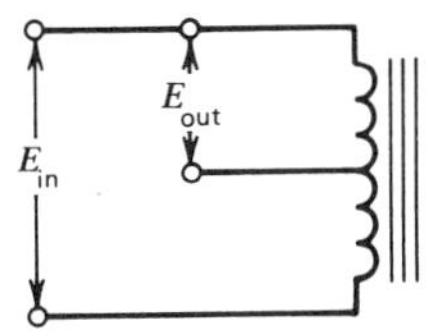

Figure 8.5 Auto-type transformer.[45]

B. ***Core*** There are three types of transformer cores used. Power and audiofrequency transformers use laminated iron cores. Higher frequencies use a powdered-iron or ferrite core, and very high frequencies in the RF region use air cores.

Many transformers are laminated with standard scrapless type E and I silicon-steel cores. Less common, but regularly used are the L, F, J, EE, and UI lamination types. (See Figure 8.8.)

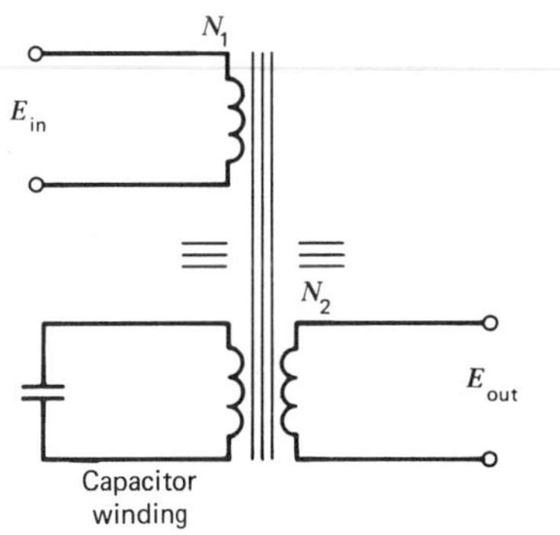

Figure 8.6 Constant-voltage transformer.[45]

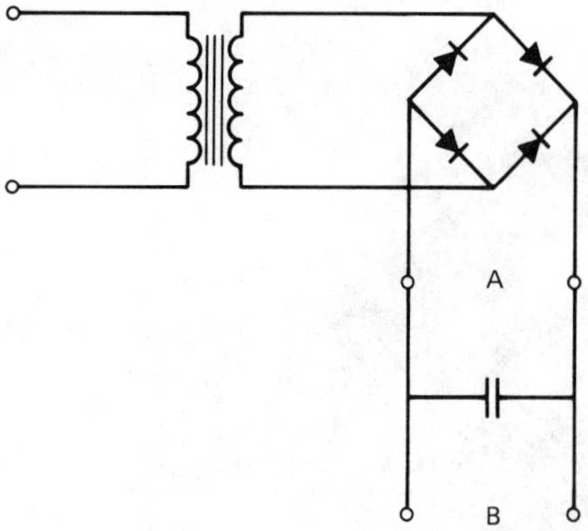

Figure 8.7 DC power-supply transformer.[45] A, Rectified output; B, filtered rectified output.

Core selection determines the overall transformer size and is a function of the permissible temperature rise, voltage-current requirements, efficiency, flux density, permissible core loss, frequency response (frequency of operation), and regulation. Typical silicon-steel laminations are available in 24, 26, and 29 gauge and in electrical grades ranging from M-6 through M-45.

C. *Insulation* The function of insulation includes thermal endurance, mechanical strength, moisture resistance, and corona-prevention. Table 8.1 designates the insulation classification system, which is based on maximum operating temperature and materials utilized.

Common materials used for impregnation include wax, varnish, epoxy, and heat-dissipating polyesters. These materials are applied by both dip- and vacuum-impregnation processes.

Equally important as the selection of the impregnation medium is the proper in-process controls for penetration. Viscosity levels must be maintained for adequate overall bonding and protection from various environmental conditions.

D. *Terminators*[46] The most common type of termination is lead wire cut and stripped to the required length. Spade quick connects, solder terminals, screw terminal boards, terminal blocks, and others are available as well. (See Figure 8.9.)

8.1.4 Electrical Parameters

As with all components, there are special types of transformers for unique requirements, but the basic primary-winding parameters that should be specified include input voltage and operating frequency(ies), DC current in primary(ies), and taps on primaries and loading on these taps. Secondary-winding parameters that should be specified are the number of secondaries, load on each secondary, and DC current and voltage per secondary. Also, if the secondaries feed rectified loads, the types of rectifiers and the filter type should be specified. Depending on the rectifier and filter configuration used, the rms secondary current is different from the DC output current.

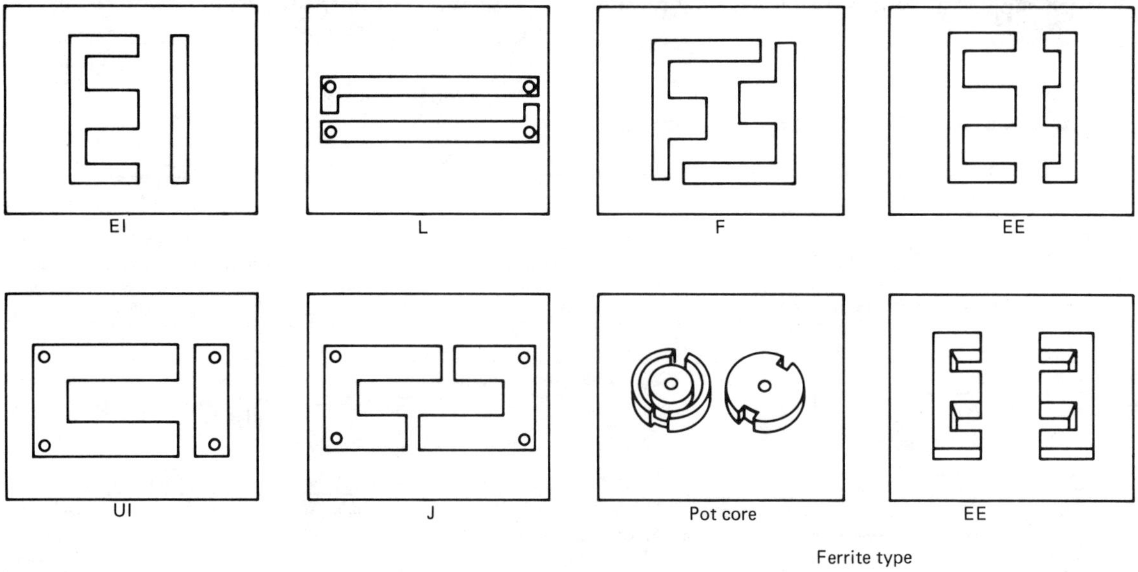

Figure 8.8 Types of lamination.[45]

Table 8.1 Thermal Classification of Insulation (Reproduced from Ref. 45)

CLASS	MAXIMUM OPERATING TEMPERATURE	MATERIAL
O	90°	Cotton, silk, paper and similar organic material neither impregnated nor immersed in a liquid dielectric.
A	105°	1. Cotton, silk, paper or similar organic materials either impregnated or immersed in a liquid dielectric. 2. Molded and laminated materials with cellulose filler, phenolic resins and other resins of similar properties 3. Films and sheets of cellulose acetate and other cellulose derivitives of similar properties. 4. Varnishes (enamel) as applied to conductors.
B	130°	Includes materials or combinations of materials such as mica, glass fibre, asbestos, etc., with suitable bonding substances.
F	155°	Includes materials or combinations of materials such as mica, glass fibre, asbestos, etc., with suitable bonding substances.
H	180°	Includes materials or combinations of materials such as a silicone elastomer, mica, glass fibre, asbestos, etc. with suitable bonding substances such as appropriate silicone resins.
C	OVER 220°	Consists entirely of mica, porcelain, glass, quartz and similar inorganic materials.

The specifying of some other electrical parameters should usually be left to the discretion of the transformer manufacturer, unless they directly affect the operation of the circuit. Examples include leakage inductance, winding inductance, distributed capacitance, and regulation. (Regulation is often specified when not necessary to the user's disadvantage.)

8.1.5 Mechanical Considerations

The transformer manufacturer must know the maximum temperature in which the transformer will operate. (The maximum temperature is the sum of the maximum ambient temperature and the transformer's permissible temperature rise.) This determines the class of insulation to be used in the transformer.

In general, overall dimensions, type of mounting (see Figure 8.10), and terminal type (see Figure 8.9) should be left to the discretion of the manufacturer. The manufacturer normally uses a military-size case. Be sure to inform the manufacturer of any limiting dimension arising from the chassis requirements. (What appears to be a simple limitation on height, for example, could prevent the manufacturer from using a standard method of construction.) Try to avoid limitations, as they often result in extensive preliminary design work on the part of the manufacturer as well as higher user costs.

The same is true for mounting requirements. Specify military studs or other standard mounting whenever possible.

If a standard case cannot be used and a nonstandard case results, it is a good idea to let the manufacturer

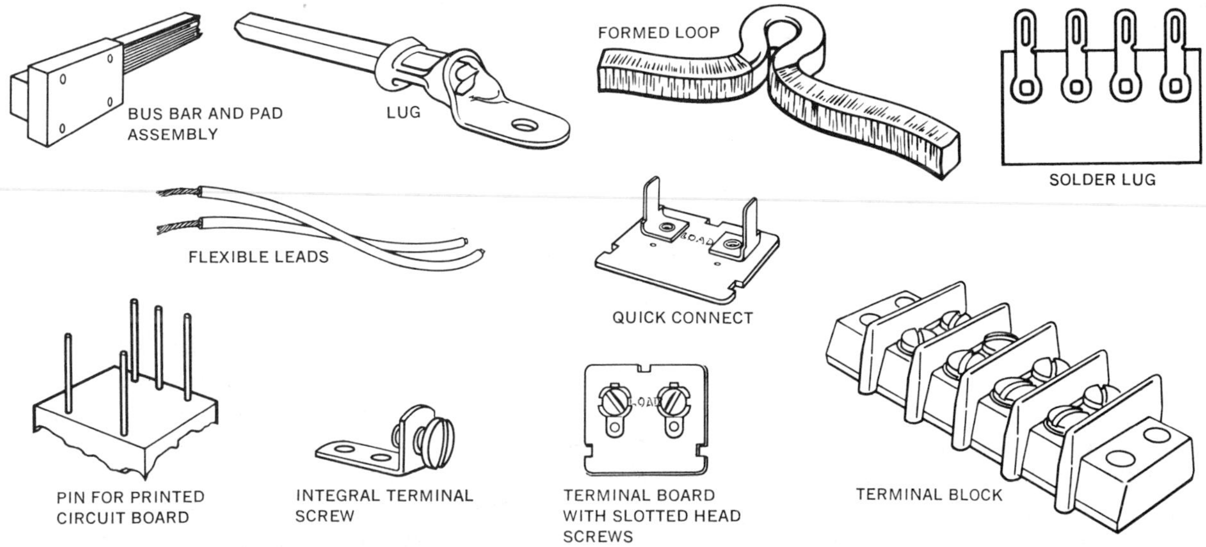

Figure 8.9 Transformer terminations. (Reproduced from Ref. 46)

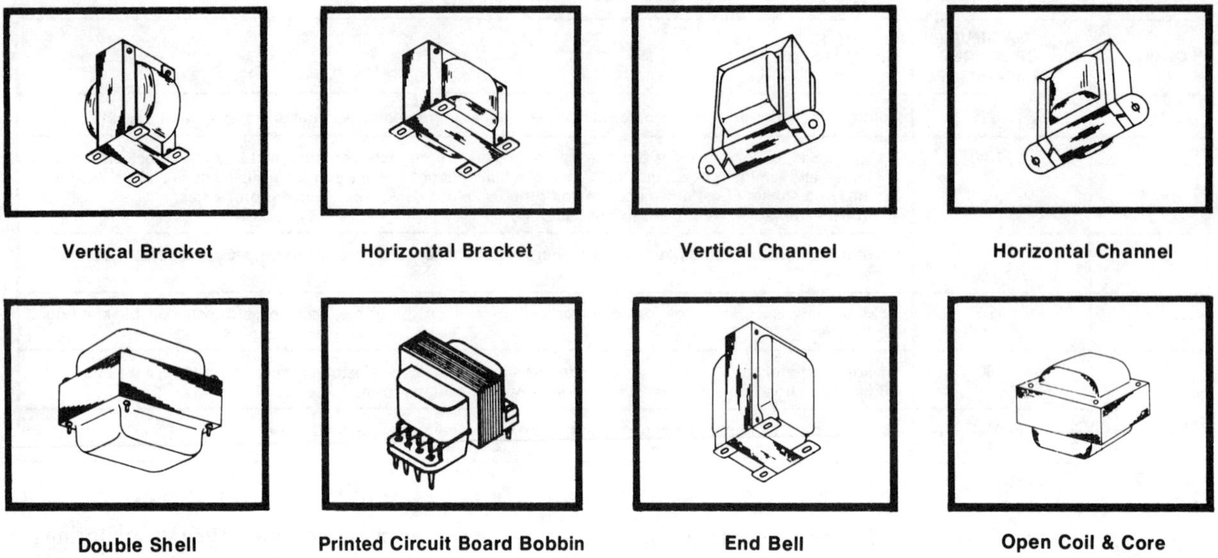

Figure 8.10 Transformer mounting styles. (Reproduced from Ref. 45)

choose the mounting dimensions and terminal locations on the basis of the internal construction of the transformer. Giving the manufacturer this freedom will almost always result in a more economical transformer.

8.1.6 Special Requirements

Special requirements include such items as magnetic shielding, electrostatic shielding, special tolerances for inductance, and self-resonant frequency, in addition to other items, such as special paint colors and special markings. Avoid these unless absolutely necessary, as they can be costly.

There are some items that should, and others that should not, be included if the specification is to be understood by all parties concerned. A schematic of the transformer showing terminal numbering, input and output voltages, currents, impedances, and load polarities (where applicable) make a transformer design/purchasing specification much easier to follow. References to methods of tests, type of plating, paint primers, paint, and marking should be avoided unless they are different from the methods and materials normally prescribed.

Electrostatic (Faraday) shielding between the primary and secondary windings may increase the size of the unit in some instances. Foil shielding is generally more effective than are wire-wound shields, but it also is more expensive because of the difficult manufacturing processes involved.

Electromagnetic shielding around the coil and core may increase the transformer's cost significantly. The circuit designer should attempt to locate the transformer so that the flux band is not necessary. Other types of stray-flux prevention are available by using enclosures of suitable materials.

8.2 AUDIOTRANSFORMERS[47]

The basic applications of an audiotransformer are impedance matching, output, driver, interstage, line, and voice coil. The audiotransformer represents components covering a frequency band from 20 to 20,000 Hz, at levels of megawatts to kilowatts and ranging in weight from $\frac{1}{20}$ oz. to 500 lb. Custom designs cover an even broader range of frequencies and power ratings.

8.2.1 Shielding

Audiotransformers usually require more shielding than any other type of transformer. Because of the low-power levels at which they operate, audiotransformers may be susceptible to radiated and line-coupled interference.

Electrostatic shielding is commonly used between the primary and secondary of a transformer to reduce line-coupled interference by reducing the interwinding capacity. This is accomplished using highly conductive materials, such as copper, silver, or aluminum, as a wraparound or between the coils of a transformer.

Magnetic shielding is used to reduce radiated-type interference from affecting the transformer. It is accomplished by encasing the transformer in a single, high-permeability, nickel-iron can or cans, depending on the intensity of the radiation.

8.2.2 Balance Winding

Many audiotransformer applications require two matched windings or winding halves. Depending on the parameters to be compared and the operating conditions, the type of balance required takes on a variety of names.

Low-frequency winding balance requirements are generally turned to accurate turn ratios, while high-frequency balance includes balancing of winding capacitances.

8.3 PULSE TRANSFORMERS[47,48]

A pulse transformer is a special type of wide-band transformer designed to transmit voltage or current pulses with specific requirements on wave-shape fidelity. Although it will provide the same electrical isolation as other types of transformers, a pulse transformer is more like a transmission line than a conventional power transformer. Load and source impedances, maximum peak output voltage, pulsewidth, duty factor, offset and bias voltage, and rise and fall times are associated with pulse transformers. In general, the turns ratio of a pulse transformer must be low to provide good wave-shape fidelity.

8.3.1 Selection

A pulse transformer is usually selected on the basis of a set of transformer equivalent-circuit parameters and saturation characteristics. As a minimum, manufacturers of pulse transformers need the following information:

A. Operating temperature range
B. Maximum and minimum pulsewidths
C. Maximum peak primary voltage
D. Turns ratio
E. Maximum pulse-repetition rate
F. Maximum physical dimensions
G. Desired terminations and locations
H. Maximum winding-to-winding peak withstanding voltage.

8.3.2 Applications

Application for pulse transformers range from thyristor triggers to laser triggers. The transformers can be used merely to provide isolation in computer circuits or to match the voltage and impedance of pulse-forming networks, to microwave tubes, such as klystrons, magnetrons, traveling-wave tubes, and cross-field amplifiers.

A thyristor-trigger transformer is a pulse transformer designed especially for gate-triggering applications. The triggering and impedance characteristics of the thyristors and the interwinding dielectric strength are the key factors in the design.

The relatively high operating voltage or thyristors set tough requirements for interwinding insulation. Most applications call for a step-down transformer, since the voltage required for effective triggering is relatively low. Several secondaries may be needed for simultaneous triggering of thyristors.

The characteristics required of a trigger transformer vary greatly with thyristor power rating, anode current waveform, operating voltage, and circuit function.

Trigger transformers differ from other pulse transformers in that the pulse rise time for selected values of usable volt-time products must be optimized with a typical thyristor circuit as a secondary load.

Since trigger transformers usually interface between low-power, sensitive, control circuits and highvoltage, high-power circuits, they need a high interwinding dielectric strength. An increase in dielectric strength causes an increase in transformer size, which results in an increase in pulse rise time.

Transformers used in most DC/DC converter applications are of the blocking-oscillator type. Selection of core material is very important here to ensure maximum efficiency. Blocking-oscillator transformers differ from other pulse transformers in that it is desirable to maximize the remnant flux density. This requirement contrasts with that for most other pulse transformers, which operate in a unipolar mode when pulses go only in one direction.

For a transformer to work in a unipolar application it must reset itself so that it does not look like a permanent magnet. The measure of a transformer's ability to reset itself is the flux density. The lower the flux density, the better for unipolar pulses. Therefore, the core of a regular pulse transformer can not be made from square-loop material. Converter, or blocking-oscillator transformers, however, can use square-loop cores because the pulses alternate in polarity.

8.3.3 Tradeoffs

With all pulse transformers there are limitations that make certain exact combinations of parameters unattainable. The most common tradeoffs are size versus inductance and operating temperature versus inductance stability. The present state of the art makes it possible, under favorable conditions, to achieve more than 4000 mH/in.[3] with a 1:1 turn ratio. Although high values of inductance can be achieved with only a few turns in a small package, it may not be possible to achieve simultaneously high inductance combined with the desirable inductance tolerance and turns ratio.

As core permeability increases, inductance stability tends to decrease. With cores that have the same permeability, the core with the higher Curie point, the temperature above which a ferromagnetic material becomes substantially nonmagnetic, will usually be more stable. Most core materials used for pulse transformers maintain inductance stability of ±10% from −10 to +80°C. This will degrade approximately 50% at −55°C and at +125°C from the initial value at +25°C.

Better stability can be achieved with a nonstandard material, but generally by sacrificing permeability or operating flux density and cost. Other tradeoffs that have to be considered include:

A. Rise-and-fall time versus load impedance.

B. Rise time versus maximum pulse width.

C. Size of the pulse transformer versus the product of the peak voltage and maximum pulse width.

D. Size versus peak voltage and power-handling capability.

CHAPTER NINE

Inductive Devices

9.1 RADIOFREQUENCY (RF) CHOKES

Although an RF choke is one of the least complex inductors, its effect can become quite complex. The usual data on RF chokes specify the value of inductance and Q measured at a nominal radiofrequency. In addition, the self-resonant frequency is given as an aid in determining the range over which the choke may be operated satisfactorily. Additional data usually include the DC resistance of the winding and the maximum DC current rating.

These data are generally quite satisfactory for use over a limited frequency range. The designer usually selects a choke with a parallel self-resonant frequency slightly higher than the highest frequency to be encountered in the circuit. Thus, the designer obtains maximum impedance with greatest voltage gain and avoids possible burnout at series-resonant frequency.

Some RF chokes are of the magnetically shielded series, which uses a ferrite bobbin and sleeve. In these inductors the winding is wound on the bobbin, the sleeve is placed over the bobbin, and the assembly is molded. This construction confines the magnetic field almost entirely within the magnetic material.

Two current ratings must be considered when designing one of these chokes into a circuit. First, the incremental-current rating is that current which causes an inductance to decrease 5% from nominal. In circuits where inductance is a critical parameter, the designer should specify a choke such that the incremental current rating is not exceeded.

Second, the maximum-current rating given is the least desirable operating condition because saturation can cause a wide variation in inductance value, and temperature rise may exceed specified values.

Variations of both inductance and Q should be watched very closely when using the magnetically shielded chokes in changing temperature environments. The specifications for this series of chokes gives maximum current ratings that can be greater than the incremental current. Whenever a current greater than the incremental current is used, the inductance will decrease more than 5% and the temperature-rise characteristics should be taken into consideration.

Chokes in the *MS* series are relatively low-power devices with maximum power ratings of between 0.25 and 0.55 W and are subject to derating above 90°C.

Although the inductance tolerance for chokes in the MS series has been established at ±20% for value of 0.1–0.82 μH and ±10% for 1.0 μH and greater in the standard molded-choke series, it is possible to wind these devices with tolerances as close as 1%.

netic path, thereby effectively decreasing the inductance and increasing the usable frequency range of any specific coil form. These slugs may also be used as a convenient means of lowering Q. The ratio of inductance change will be approximately 1.5:1.

9.2.6 Winding Configurations

RF inductors can be manufactured in a number of basic winding configurations. Each of these types has its advantages depending on its function in a circuit and the associated components.

A. *Solenoid (Single-Layer) Winding* Used in the simplest type of coil. (See Figure 9.2*a*.) This coil becomes a true air inductor when the winding is self-supporting. This type of winding provides the least amount of inductance in a given space. However, it produces a very low value of distributed capacitance, and allows excellent heat transfer, since air can move over the entire winding area.

Magnetic or nonmagnetic core material can be used to support the winding and to provide a means of mounting the start and finish leads. The core can also be made adjustable to vary the amount of inductance.

B. *Multilayer Wound Coils* (See Figure 9.2*b*.) Realize an appreciable increase in the amount of inductance over that of a single-layer wound coil. Turns are wound on top of each preceding layer to build up the coil diameter by adding layer upon layer. This permits the maximum amount of inductance.

Layer winding is satisfactory at low frequencies. However, even at the higher audio frequencies, the effect of the large distributed capacitance between turns can affect circuit operation significantly by limiting frequency response.

C. *Pi (Universal) Windings* Provide a larger value of inductance per volume than a solenoid, but not as much as a multilayer-wound coil. The main advantage of a universal coil over a layer-wound coil is a much lower value of distributed capacitance. If the coil is divided into a number of pi's, the value of distributed capacitance is lowered still further. This gives the same result as adding fixed values of capacitance in series. Generally, DC resistance of the inductor will increase, as more wire is required to produce the same inductance.

Pi or self-supporting universal windings must be done on a machine, while the single-layer or multilayer winding can be done by hand without the aid of a mechanical device. Basic design parameters of wire size, form diameter, winding width, or cam throw are entered into a number of basic formulas to give ratios that determine rotation of the winding form and wire movement on the coil form. These ratios establish an exact relationship to the wire at all points of the form rotation. Since the wire is carried from one side of the form to the other and back again as the form rotates, each turn should either progress or retrogress in relationship to the preceding turn in order to achieve a mechanically stable coil. (See Figure 9.2*c*–*e*.)

D. *Universal Progressive Windings* Developed to increase the inductance of the solenoid winding and to reduce further the distributed capacitance of the pi winding. This is accomplished by laying the pi winding along the form instead of allowing it to build up into a single pi. (See Figure 9.2*f*, *g*.) In addition to the calculations for the regular universal winding, gear ratios must be calculated for the progressive movement of the winding on the coil form used.

E. *Variable-Pitch Universal Progressive Winding* A special variation of the universal progressive winding. Whenever a piece of equipment is designed to use permeability tuning, such as most auto receivers, this type of winding is commonly used. The winding is layered on the form to obtain frequency distribution that is either linear or close to linear as the iron core is inserted. (See Figure 9.2*h*.) In normal progressive winding, a much greater increase in inductance is achieved during the initial movement of the core into the winding than is achieved after the core has entered the winding more fully.

F. *Pot-Core Winding* A highly specialized layer winding used with a cup-shaped powdered-iron or ferrite core to produce extremely efficient inductors. This type of coil exhibits most of the advantages of toroid coils, since the flux is confined almost entirely within the magnetic material.

G. *Custom-Wound Coils* Windings may be single-layer, close-wound, spaced, single-pi, or multi-pi according to requirements. In order to design and custom wind coils, the following information is required:

1 Desired instrument of measurement.
2 Inductance*: minimum, maximum, and at specified frequencies.

*Both electrical and mechanical characteristics may be specified. Unless otherwise indicated, electrical characteristics shall govern and mechanical specifications will be considered as reference.

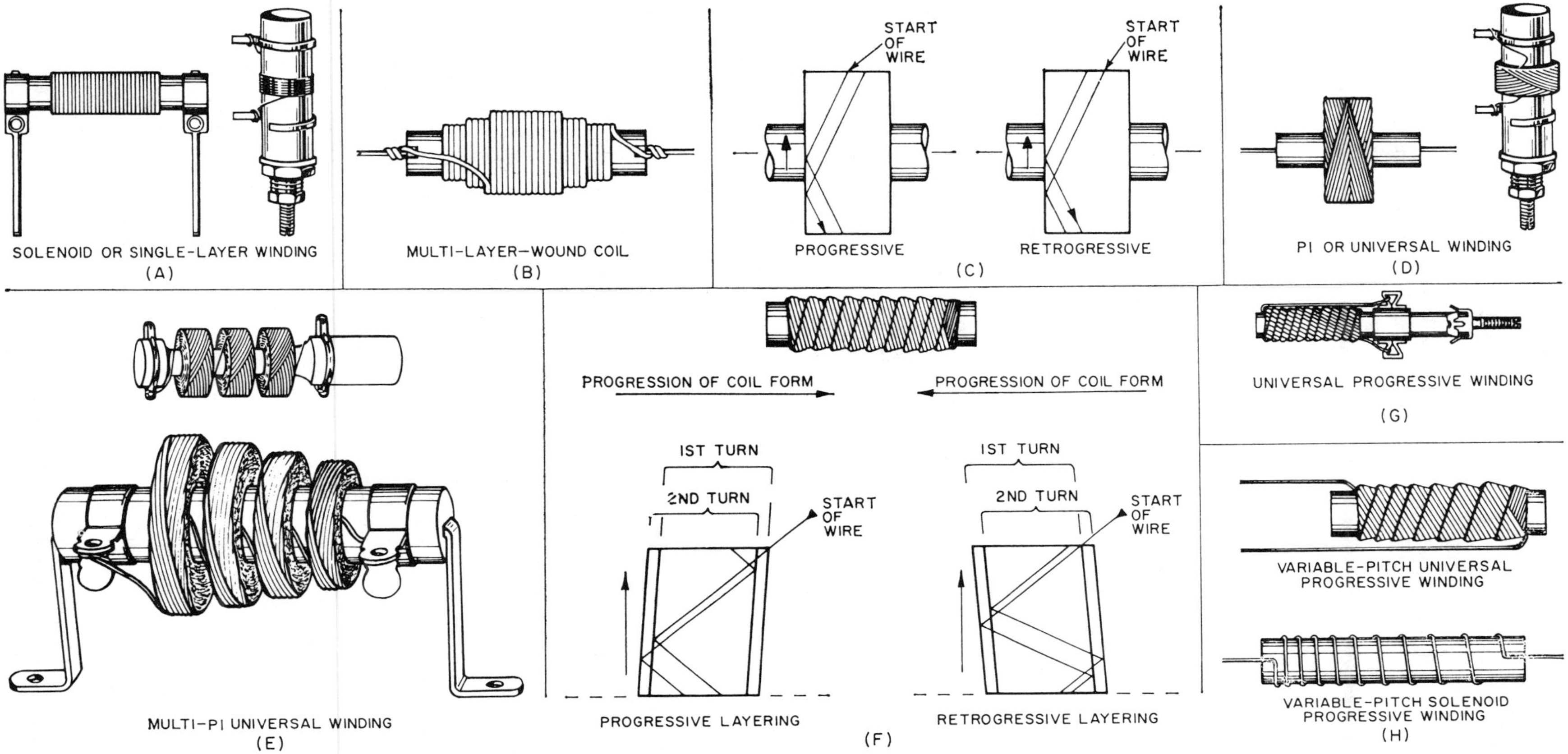

Figure 9.2 RF choke and coil-winding techniques. (Reproduced from Ref. 50.)

3 Q at a specified frequency(ies).
4 Self-resonant frequency, if applicable.
5 Preferred winding type, tap location, turns/inductance.
6 DC resistance, if applicable, or current capacity.
7 Wire size AWG.
8 Type of insulation.
9 Number of turns*.
10 Overall size.
11 Marking.

Addition information, if applicable, is also required, such as:

12 Turns ratio or impedance ratio
13 Loaded Q
14 Bandwidth (@ ___ F_0) (@ ___ dB down)
15 Test voltage
16 Shunt capacitance
17 Shielded or unshielded
18 Maximum or minimum temperature, or both
19 MIL-C-15305 grade and/or class
20 Desired test jigs

9.2.7 Magnet Wire[51]

The size of the magnet wire is limited by its current rating. The usual current rating for copper wire does not, however, apply to inductors. Current ratings for industrial and military inductors are made on the basis of maximum temperature rise. Since these ratings range typically from 5 to 20% of the wire's fusing current, rating the inductor wire at 10% of its fusing current is realistic, not to mention convenient, for calculation. This rating applies at maximum operating temperature. Current ratings for copper magnet wire are tabulated in Table 9.1, along with other selected magnet-wire design data.

9.3 CORES

9.3.1 *Q* Factor[52]

The Q of an inductor is defined as the ratio of reactance to resistance and is used to indicate the quality or figure of merit of the inductor. In filters, sharper cutoff, better-defined resonance, and higher attenuation ratios are realized as the Q factor is increased. Self-capacitance of the inductor winding will also have an effect on Q.

Each core size has an inherent Q versus frequency characteristic which is independent of the inductance to which it is wound. Any core of a particular size may be wound to any inductance without affecting Q, as long as the maximum size wire, physically permissible for winding, is used. This is only true at lower frequencies.

Actually the Q of a given inductor departs from the inherent characteristics curve above a certain frequency. The frequency at which the curves depart is called the "breakoff" frequency. Loss curves used for calculating R_{ac} are applicable only when the design frequency is near or below the breakoff frequency.

9.3.2 Core Materials[53]

Proper selection of materials for use in the core of a coil can improve several electrical parameters and reduce physical size. When a core is fixed in physical relationship to a coil winding, the concentration of the magnetic field permits a reduction in the number of turns necessary to provide a given inductance. Q is improved because less wire is required for a specific value of inductance. When the proper grade of magnetic material is selected, the losses introduced by the magnetic materials are less than the losses in the copper that would have been required for the additional turns on an air coil. However, distributed capacitance usually rises because the dielectric constant of the core material is usually greater than air. This has the effect of lowering the self-resonant frequency.

Table 9.2 lists a representative cross section of several popular core materials. To select the right core material, you should know the required frequency, operating temperature, desired core properties, cost, geometry, and winding limitations, among other things.

Each grade of powdered iron performs best over a different frequency range, and Q tends to decrease rapidly at higher frequencies. Optimum coil design depends on a compromise in physical size, inductance range, and stability of the device. If inductance range is the most important factor, ferrite or high-permeability powdered materials can be used. If stability is more important, lower-permeability material must be used.

A number of core materials are available in toroidal form for general application in the audio- and carrier-frequency range. In addition to molybdenum permalloy-powder cores, these include Sendust, iron powder, and ferrite cores. The range of initial permeabilities covered is from about 8 to 200.

*Both electrical and mechanical characteristics may be specified. Unless otherwise indicated, electrical characteristics shall govern and mechanical specifications will be considered as reference.

Table 9.1 Magnet Wire Design Data[51]

AWG Size	Bare Wire Max OD	Max I	Max Tensile Strength	Ω/ft	Single Film Max OD	ft/lb	Ω/lb	Tpi	Tpi²	Double Film Max OD	ft/lb	Ω/lb	Tpi	Tpi²
10	0.1024	33.6	82.4	0.00100	0.1047	31.6	0.0316	9.551	91.22	0.1061	31.5	0.0315	9.425	88.83
11	0.0912	28.2	65.3	0.00126	0.0935	39.8	0.0501	10.70	114.4	0.0948	39.7	0.0500	10.55	111.3
12	0.0812	23.7	51.8	0.00159	0.0834	50.3	0.0800	11.99	143.8	0.0847	50.0	0.0795	11.81	139.4
13	0.0724	19.9	41.2	0.00200	0.0746	63.3	0.1266	13.40	179.7	0.0757	62.9	0.1258	13.21	174.5
14	0.0644	16.7	32.6	0.00252	0.0666	79.9	0.2013	15.02	225.5	0.0682	79.3	0.1998	14.66	215.0
15	0.0574	14.1	25.9	0.00318	0.0594	101	0.3212	16.84	283.4	0.0609	100	0.3180	16.42	269.6
16	0.0511	11.8	20.5	0.00402	0.0531	127	0.5105	18.83	354.7	0.0545	126	0.5065	18.35	336.7
17	0.0455	9.94	16.3	0.00505	0.0475	159	0.8029	21.05	443.2	0.0488	158	0.7979	20.49	419.9
18	0.0405	8.35	12.9	0.00639	0.0424	201	1.284	23.58	556.2	0.0437	199	1.272	22.88	523.6
19	0.0361	7.02	10.2	0.00805	0.0379	253	2.037	26.39	696.2	0.0391	251	2.020	25.58	654.1
20	0.0322	5.92	8.14	0.0101	0.0339	318	3.212	29.50	870.2	0.0351	315	3.182	28.49	811.7
21	0.0286	4.95	6.42	0.0128	0.0303	402	5.146	33.00	1089	0.0314	397	5.082	31.85	1014
22	0.0254	4.15	5.07	0.0162	0.0270	508	8.230	37.04	1372	0.0281	503	8.149	35.59	1266
23	0.0227	3.50	4.05	0.0203	0.0243	633	12.85	41.15	1694	0.0253	625	12.69	39.53	1562
24	0.0202	2.94	3.20	0.0257	0.0217	806	20.71	46.08	2124	0.0227	794	20.41	44.05	1941
25	0.0180	2.47	2.54	0.0324	0.0194	1013	32.82	51.55	2657	0.0203	990	32.08	49.26	2427
26	0.0160	2.07	2.01	0.0410	0.0173	1282	52.56	57.80	3341	0.0182	1260	51.66	54.95	3019
27	0.0143	1.75	1.61	0.0514	0.0156	1608	82.65	64.10	4109	0.0164	1580	81.21	60.98	3718
28	0.0127	1.47	1.27	0.0653	0.0140	2033	132.8	71.43	5102	0.0147	1990	129.9	68.03	4628
29	0.0114	1.25	1.02	0.0812	0.0126	2525	205.0	79.37	6299	0.0133	2470	200.6	75.19	5653
30	0.0101	1.04	0.801	0.104	0.0112	3215	334.4	89.29	7972	0.0119	3140	326.6	84.03	7062
31	0.0090	0.874	0.636	0.131	0.0100	4065	532.5	100.0	10000	0.0108	3950	517.4	92.59	8573
32	0.0081	0.747	0.515	0.162	0.0091	5000	810.0	109.9	12076	0.0098	4880	790.6	102.0	10412
33	0.0072	0.626	0.407	0.206	0.0081	6369	1312	123.5	15242	0.0088	6170	1271	113.6	12913
34	0.0064	0.524	0.322	0.261	0.0072	8064	2105	138.9	19290	0.0078	7870	2054	128.2	16437
35	0.0057	0.441	0.255	0.331	0.0064	10210	3380	156.2	24414	0.0070	9940	3290	142.9	20408
36	0.0051	0.373	0.204	0.415	0.0058	12760	5295	172.4	29727	0.0063	12440	5163	158.7	25195
37	0.0046	0.319	0.166	0.512	0.0052	15800	8090	192.3	36982	0.0057	15300	7834	175.4	30779
38	0.0041	0.269	0.132	0.648	0.0047	19920	12908	212.8	45269	0.0051	19300	12506	196.1	38447
39	0.0036	0.221	0.101	0.847	0.0041	26040	22056	243.9	59488	0.0045	25100	21260	222.2	49383
40	0.0032	0.185	0.0804	1.08	0.0037	33110	35759	270.3	73046	0.0040	32200	34776	250.0	62500
41	0.0029	0.160	0.0661	1.32	0.0033	40100	52932	303.0	91827	0.0036	39500	52140	277.8	77160
42	0.0026	0.136	0.0531	1.66	0.0030	51000	84660	333.3	111111	0.0032	49800	82668	312.5	97656
43	0.023	0.113	0.0475	2.14	0.0026	65800	140.8k	384.6	147928	0.0029	63700	136.3k	344.8	118906
44	0.0021	0.0985	0.0346	2.59	0.0024	79400	205.6k	416.7	173611	0.0027	76300	197.6k	370.4	137174
45	0.00176	0.0756	0.0243	3.62	0.00205	104k	376.5k	487.8	237954	0.00230	99600	360.6k	434.8	189036
46	0.00157	0.0637	0.0194	4.54	0.00185	132k	599.3k	540.5	292184	0.00210	126k	572.0k	476.2	226757
47	0.00140	0.0536	0.0154	5.71	0.00170	162k	925.0k	588.2	346020	0.00190	153k	873.6k	526.3	277003
48	0.00124	0.0447	0.0121	7.29	0.00150	205k	1.494M	666.6	444444	0.00170	199k	1.451M	588.2	346020
49	0.00111	0.0379	0.0097	9.09	0.00130	258k	2.345M	769.2	591716	0.00150	252k	2.291M	666.6	444444
50	0.00099	0.0319	0.0077	11.4	0.00120	312k	3.557M	833.3	694444	0.00140	306k	3.488M	714.3	510204
51	0.00088	0.0267	0.0061	14.5	0.00110	416k	6.032M	909.1	826446					
52	0.00078	0.0223	0.0048	18.4	0.00100	555k	10.21M	1000	1.000M					
53	0.00070	0.0190	0.0038	22.9	0.00085	667k	15.27M	1176	1.384M					
54	0.00062	0.0158	0.0030	29.1	0.00075	859k	25.00M	1333	1.777M					
55	0.00055	0.0132	0.0024	37.0	0.00070	1.090M	40.33M	1429	2.041M					
56	0.00049	0.0111	0.0019	46.6	0.00065	1.380M	64.31M	1538	2.367M					
Units	Inches	Amperes	Pounds	Ω/ft	Inches	ft/lb	Ω/lb	T_{pi}	T_{pi}^2	Inches	ft/lb	Ω/lb	Tpi	Tpi²

Maximum ODs. Ω/ft and ft/lb are taken from Materials and Processes Handbook. Maximum rated current is 10% of the fusing current at 20 C. Maximum tension is based upon a tensile strength of 10.00 PSI. Ω/lb is derived from (Ω/ft) x (ft/lb). Tpi 1/Max OD. Tpi² (1/Max OD)².

Table 9.2 Common Core Materials[51]

Material Designation	Frequency Range (Hz)	Temperature Range (°C)	Initial Permeability (G/Oe) U_i	Saturation Flux Density (G) B_s	Residual Flux Density (G) B_r	Intrinsic Magnetic Field Strength (Oe) H_i
Electrical Steels[4]						
Silicon Iron	20 to 1 k	−55 to +300	500	17.5 k	12 k	5.5
Silectron	20 to 10 k	−55 to +375	1.5 k	16.5 k	14 k	0.83
Alloy 48	20 to 8 k	−55 to +250	1 k	12.5 k	10 k	1.25
HY-MV 80	20 to 25 k	−55 to +230	10 k	8 k	4.4 k	0.18
Supermalloy	20 to 25 k	−55 to +230	50 k	7.3 k	4 k	0.033
Supermendur	20 to 2 k	−55 to +460	1.5 k	21 k	19 k	0.67
Molypermalloy Powders						
MPP 14[2]	400 kto 1 M	−55 to +250	14	6 k	10	214
MPP 26[2]	400 to 650 k	−55 to +230	26	6 k	10	115
MPP 60	400 to 250 k	−55 to +200	60	6 k	10	49.8
MPP 125	40 to 100 k	−55 to +200	125	6 k	200	23.2
MPP 160	40 to 70 k	−55 to +200	160	6 k	200	18.1
MPP 200	40 to 30 k	−55 to +175	200	6 k	200	14.5
MPP 300	40 to 25 k	−55 to +150	300	6 k	300	9.5
MPP 550	40 to 20 k	−55 to +125	550	6 k	650	4.86
Powdered Irons						
Carbonyl SF[2]	2 to 50 M	−55 to +125	7.5	8 k	10	533
Carbonyl E[2]	200 k to 10 M	−55 to +125	10	8 k	10	400
Carbonyl C[2]	100 k to 2 M	−55 to +125	20	8 k	10	200
Carbonyl GQ4[2]	50 k to 1 M	−55 to +125	35	8 k	10	114
Carbonyl HA	1 to 100 k	−55 to +105	60	8 k	2200	48.3
75 Powder	400 to 50 k	−55 to +105	75	8 k	2200	38.7
90 Powder	400 to 10 k	−55 to +105	90	8 k	2200	32.2
Ferrites[3]						
F40	10 to 80 M	−55 to +250	40	2400	750	20.6
F125	200 k to 10 M	−55 to +250	125	2350	1200	4.6
F175	100 k to 5 M	−55 to +230	175	2550	1400	3.28
F250	50 k to 4 M	−55 to +200	250	2200	1100	2.2
F400	10 k to 2.5 M	−55 to +175	400	2700	1100	2
F750	1 k to 1.5 M	−55 to +125	750	4000	1800	1.47
F1000	1 k to 1 M	−55 to +125	1000	4200	1700	1.25
F1500	1 k to 650 k	−55 to +125	1500	4000	1100	0.97
F2000	400 to 500 k	−55 to +125	2000	4400	1500	0.72
F2300	400 to 300 k	−40 to +105	2300	4000	1200	0.61
F2700	400 to 250 k	−30 to +105	2700	4700	2000	0.50
F5000	400 to 100 k	−25 to +105	5000	4300	1200	0.31
F10000	400 to 80 k	−25 to +90	10000	4300	1200	0.16

Notes:
(1) For specific data and tolerances, refer to individual suppliers. (2) Excessive heating occurs before incremental effects are observed. (3) There are no standard designations for equivalent materials between ferrite suppliers. Designation used is for convenience. (4) Frequency range of electrical steel depends on material thickness. The thinner the material the higher the frequency range.

In the low-audio- and power-frequency range (below 500 Hz), permeability will be a prime consideration, and core losses generally will be secondary. In the audiofrequency range (500–15,000 Hz), both the hysteresis and eddy-current components of core loss become important, requiring only moderate permeability. Above the audiofrequency range, the eddy-current component of core loss predominates and lower permeability can be used.

At low-frequencies molybdenum permalloy powder cores and ferrite cores might be considered for applications requiring high Q, where stability to DC magnetization or modulation factor are important, however, ferrite cores will generally be unsuitable because of their relatively large change of permeability with level of excitation.

In the audiofrequency range, molybdenum permalloy powder, Sendust, iron, and ferrite cores would all be possible choices. In comparison with molybdenum permalloy powdered-iron cores, both Sendust and powdered-iron cores have substantially higher hysteresis loss coefficients, resulting in higher total core loss (and lower Q) in the audiofrequency range. They also have a somewhat higher modulation factor. The ferrite cores have an even greater hysteresis loss coefficient and modulation factor than do Sendust and iron powder.

In the high-audio and low-carrier-frequency range, the permalloy, Sendust and iron powder cores will tend to be more useful because of the lower eddy-current loss coefficients of the Sendust and iron powder.

In the high carrier-frequency and radiofrequency region, the Sendust, powdered iron, and ferrite cores would be favored because of their low eddy-current loss coefficients, with the Sendust and iron powder cores showing lower modulation factors than the ferrites. At very high frequencies, the powdered-iron and ferrite cores would be used almost exclusively, except where special consideration might dictate the use of one of the other materials.

9.3.3 Ferrites[54–56]

Ferrites are ceramic ferromagnetic materials whose electrical, magnetic, and mechanical properties are known and controllable to high degrees of reliability and repeatability. Many interesting and useful variations in electrical and magnetic characteristics can be obtained by the judicious mixture of two ferrites in formulating the raw-powder mix and by modifications in the firing process. The product, after firing, is a complex ferrite (actually a solid solution of the two ferrites) with a combination of two metallic ions in some fixed proportion.

The most significant advantage of using ferrite over laminated and powdered iron is its high resistivity, which affords a dense, homogeneous, magnetic medium with high permeability with stability, with respect to both temperature and time, but without the high eddy-current losses inherent in conventional core materials.

Linear-ferrite cores offer a unique combination of desirable electrical properties and great mechanical convenience in a broad range of sizes and shapes, at economical, mass-production prices. They are often the core material of choice, and at times the only means by which a set of design objectives may be met. Applications demanding the optimum combination of high Q, high stability of inductance, and minimum volume for frequencies up to 20 MHz in filters and tuned circuits show ferrite cores to be superior.

Most ferrite magnetic-core manufacturers supply core constants and an inductance factor (A_L), for particular core types and dimensions. (See Figure 9.3.) The inductance of a coil wound on a core can be figured to a respectable degree of accuracy with:

$$L = A_L N^2 \quad \text{(nH)}$$

However, the given A_L values are accurate only in regard to the specific applications and flux densities for which the core is tested and rated, that is, filters, pulse transformers, power inverters, and so forth. To calculate the operating, or effective flux, density (B) knowing the core's effective area (A_e), as well as effective volume (V_{ef}), simply use the following equation:

$$B = \frac{E(10^8)}{4.44\, fNA_e} \quad \text{(G)}$$

where E = rms (V)
F = frequency (Hz)
N = number of turns
A_e = effective core area (cm^2)

To get the effective strength (H_e) in the core or its air gap, if any, knowing the effective length of the magnetic path (d_e) in centimeters, use the following equation:

$$H_e = \frac{4\pi NI}{10 d_e} \quad \text{(oersteds)}$$

where I is the current in amperes through the coil.

In most cases, the size of the ferrite core can be chosen by a figure of merit, which is the winding area times the cross-sectional area of the core. The figure of merit required for a particular design can be determined from the basic equation for a transformer which

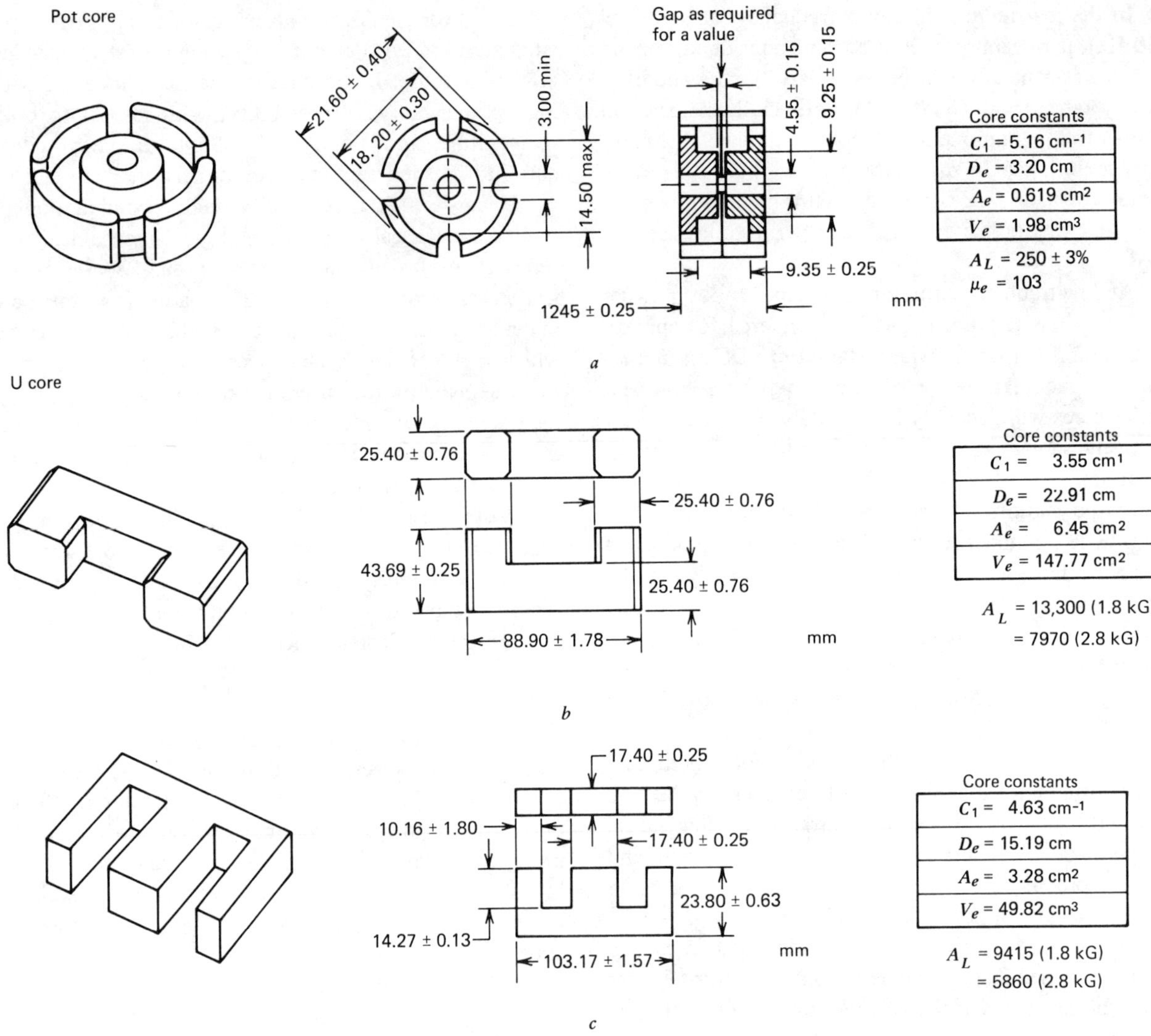

Figure 9.3 Ferrite coil and transformer cores.[55]

is then used to choose the proper core size. This number is usually listed along with the magnetic characteristics of the individual cores by the ferrite manufacturer. A general comparison of core geometries is given in Table 9.3.

A. ***Ferrite Pot Cores*** Offer combinations of controlled characteristics that achieve near-ideal performance, and to accomplish objectives that are impractical using any other technique. (See Figure 9.3*a* and 9.4.)

Inductors and transformers wound on ferrite pot

Table 9.3 Comparison of Core Geometries[51]

Selection Factor	E-U-I Cores	Pot Cores	Toroid Cores	Slug Cores	Other Shielded Cores
Core cost	Low	High	Low	Low	High
Winding cost	Low	Low	High	Low	Low
Winding flexibility	Excellent	Fair	Good	Good	Fair
Mounting flexibility	Good	Good	Fair	Fair	Good
Shielding	Fair	Excellent	Good	Poor	Good

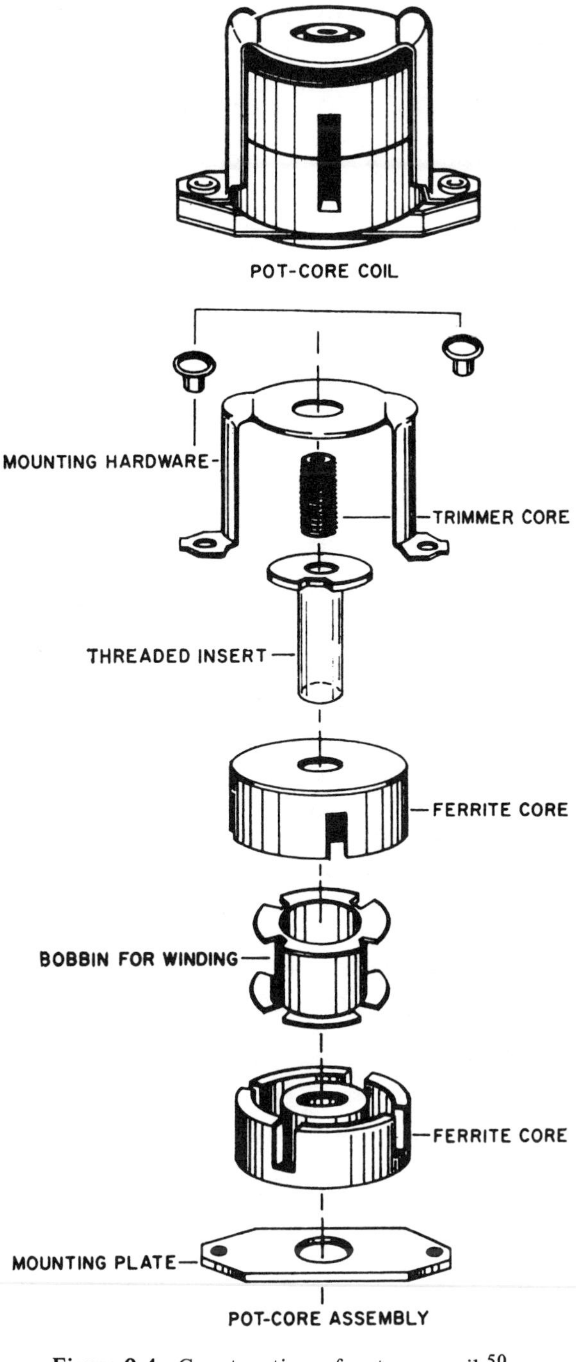

Figure 9.4 Construction of pot-core coil.[50]

cores, operating in the range from power frequencies up to tens of megahertz, are superior to conventional magnetic-core components and should be considered for small signal transformers, pulse and broad-band transformers, power transformers, from midaudio to the RF range, filters, and tuned circuits. A few of the advantages of the pot-core design are:

1 *Self-Shielding* Because the wound coil is enclosed within the ferrite core, self-shielding prevents stray magnetic fields from entering or leaving the structure.

2 *Compactness* Self-shielding permits more compact arrangement of circuit components, especially on printed circuit borads.

3 *Mechanical Convenience* Ferrite pot cores are easy to assemble, mount, and wire to the circuit.

4 *Low Cost* As compared with other core materials, ferrites are easier to make in unusual configurations (such as pot cores), resulting in a lower-cost component. Also, winding a pot core is usually quick and inexpensive because coils can be prewound on bobbins. When other costs of assembly, mounting, wiring, and adjustment are added, the total cost is often less than with other core materials or shapes.

5 *Adjustability* Final adjustment is simple by merely moving a threaded core in or out of the centerpost, and adjustment in the field is relatively easy as compared to any other type of construction. Fine to coarse tuning can be accomplished.

6 *Improved Temperature Stability and Q* Air gaps inserted between the mating surfaces of the centerposts provide good temperature stability and Q.

7 *Wide Core Selection* Many combinations of materials, physical sizes, and inductances offer a large number of choices in core selection.

8 *Low Losses* Since ferrites are ceramic materials with high resistivities, eddy-current losses are extremely low over the applicable frequency range and can be neglected. Hysteresis losses can be kept low with proper selection of material, core size, and excitation level. This makes it possible to reduce harmonic distortion in filter design.

B. *E Cores*[57] (See Figure 9.3*c*) The most popular of the ferrites for use in switch-mode power supplies. While specific requirements may dictate the use of another ferrite shape, the E core generally is the most economical one. Toroids, being less expensive, are sometimes used, but because of their solid construction must be wound by hand or on a toroid-winding machine.

E cores, on the other hand, can be wound on single- or multiple-bobbin winding machines more quickly and more inexpensively than an equivalent-size toroid.

Pot cores also are used for the transformers of switch-mode power supplies, but are considerably more expensive than E cores. The extra cost of the pot core is generally not warranted, except where shielding is needed to reduce the pickup or transmission of RF noise.

The round-centerpost E core has an advantage over the traditional square or rectangular-centerpost design because its conductor length is shorter and therefore creates less winding losses in the transformer. Other advantages include the absence of sharp bends in the conductors like those occurring in the winding of square or rectangular cores. The winding area in round-center-post is more fully utilized, while wasted space due to the inability of the wire to make a sharp bend is present in square posts. The problem worsens as the size of the transformer and the size of the wire increase. Flat conductors will alleviate this problem however.

C. *Square Cores* Provide a viable answer to the conflicting demands for higher performance, greater simplicity, and smaller volume. They feature all the advantages of the standard ferrite pot core, plus a few others.

Like the pot core, the square core is assembled from core halves, enclosing a bobbin on which the coils are wound. The coils are completely immersed in the magnetic medium, which provides magnetic shielding, as well as protection against mechanical interference or damage.

Also like the pot core, square cores are available in a choice of sizes and ferrite materials, and feature closely defined, reliable magnetic and mechanical properties because of the rigorous control exercised in the composition, fabrication, and dimensioning of the core. They are produced in ungapped, fixed-gap, and adjustable styles, providing a wide choice of inductance indices, and offering the circuit designer the same classic combination of high Q, high stability of inductance, and reliability and repeatability of characteristics.

D. *Balun Cores* Wide-band transformers are wound magnetic devices designed to transfer energy over a wide frequency range. Balun transformers are a special case of wide-band transformers, and are used to transfer energy from a balanced to an unbalanced line. Ferrite wide-band transformer cores for use at frequencies above 1 MHz are often simply called balun cores.

A main requirement of a wide-band transformer is that is should present a small insertion loss to the circuit over the full frequency range. This is best achieved by choosing the material that has the highest impedance value at the low-frequency end of the bandwidth of the transformer.

E. *Shield Beads*[58] (Figure 9.5) Simple, inexpensive, effective and reliable components for attenuating EMI and other unwanted signals in a circuit, without contributing to power losses at DC and low frequencies. The bead is integrated into the circuit simply by sliding it onto the lead of an involved component or a conductor carrying the unwanted signals. At DC or low frequencies such beads add little, if any, impedance to the lines. But at high frequencies, the impedance increases appreciably and will reduce or dissipate the undesired signals.

At high frequencies many portions of a circuit can generate unwanted signals. Consequently, ferrite beads

Figure 9.5 Typical shield beads. (Courtesy of Fair-Rite Products Corp.)

can be used in any part of a circuit which generates noise, or conducts it to the output. Such noise most often affects power-supply leads, connections to ground, and connections between circuit stages. Sometimes a single bead is insufficient to eliminate unwanted signals completely. Since the beads are relatively inexpensive and the assembly time is minimal, several can be used at each location to help achieve the desired impedance.

A main attribute of shield beads is that they provide an impedance that varies with the size and shape of the bead, as well as with the frequency of the unwanted signal.

When single-hold beads do not produce sufficient impedance, multihole beads can be used from a large selection of available sizes and shapes. Either two-hole or four-hole beads may be used, but generally they are bypassed in favor of six-hole beads. The use of six-hole beads with $1\frac{1}{2}$ or $2\frac{1}{2}$ turns of wire is now widespread throughout the industry.

Another method of increasing the absorption power of shield beads is to use a circuit comprising a bead and a capacitor in series, with the load in parallel with the capacitor. This combination yields a damping value equal to the bead impedance multiplied by the reactance of the capacitor. This technique offers many possible variations.

PART FOUR

Power Sources

CHAPTER TEN

Power Supplies

Every electrically driven load requires a stable power source. However, power supplies[59-61] are usually the last item considered when designing or purchasing equipment, despite the fact that it is impossible to operate any electrical or electronic load without them.

In addition, it is impossible to get optimum performance from a load coupled to a mismatched power supply. The relationship yields some undesirable phenomena, namely, overheating, arcing, spurious signal noise generation, overload shutdowns, and even blowing up of load or power supply. Worse, inadequate stability in an incorrectly specified power supply can completely mask the specified accuracy or sensitivity of the driven load.

The diversity of power supplies available makes it almost impossible to have a specific unit for any given load and space requirements. The plenitude of possibilities means that the user must decide which power supply most closely meets his needs. For such a decision, some foreknowledge is necessary.

Because only a few standard configurations exist, and because there is no standard for output power and regulation, each power supply can best be classified by its type of physical configuration (e.g., module, bench-type, stand-alone, rack-mount, open-frame, enclosed-chassis), by its type of power regulation (series, switching, ferroresonant, mag-amp, or simple-rectifier), and by its output-voltage range (low-voltage/under 1 kV or high-voltage/over 1 kV).

The regulator is the stabilizing heart of the power supply as the electronics that maintains the output voltage within its specified limits, no matter what is happening on the line or at the load. Regulator types generally used in low-voltage DC power sources are linear (series pass), switching, and ferroresonant.

10.1 SUPPLY REQUIREMENTS

When choosing a power source, it is most important to avoid overspecifying, as unrealistic tolerances or constraints are costly. It is better to specify realistic parameters that will match the load requirements to the ambient conditions. Also, specify output current at the highest operating temperature expected.

Realistic specifications for a power source can yield reliable clues to choosing an off-the-shelf unit or designing a custom supply. Having a realistic set of specifications also helps keep costs at a reasonable level.

The following is offered to help simplify terminology and help select an appropriate power source.

10.1.1 Type of Load

Identify the type of load to be operated and whether the load will be intermittent or constant.

10.1.2 Input

Characteristics of the line input have a large influence on how a given power source will perform. The power source must operate without derating or loss of efficiency over the entire input-voltage range tolerance and within its intended input-line frequency range.

Consequently, the power source should be able to compensate for normal line-voltage changes. In general, conventional linear power sources do not tolerate line-voltage brownouts without some loss of efficiency (derating), but switching power sources can sustain line-voltage drops of as much as 20% with no derating.

10.1.3 Line Regulation

Line power failure and input surge current always pose problems. If the load requires a specific shutdown sequence, the power source should have adequate storage time to permit this. On momentary loss of line power, switching power supplies have typical holdup times (carryover) of up to 30 msec for sequential shutdown. (For applications in which no power outage can be tolerated, a standby power source must be provided. These devices, in various capacities, comprise rechargeable batteries, oscillators, and other circuitry to provide power on an instant-takeover basis when line power fails.)

10.1.4 Output

Power-source output considerations are number of outputs, output voltage, output current, load regulation, stability, noise and ripple, temperature coefficient, holdup time, and recovery time. Output voltage and maximum output current must be determined first. Also do not overlook additional output requirements for remote sensing or for "marginal checking" of delicate circuits or loads.

10.1.5 Load Regulation

Load regulation is a measure of the maximum change in source output voltage due to a step change in load current, usually from no load to full load.

10.1.6 Noise and Ripple

Noise and ripple are undesirable (and unremovable) components of the output voltage. They are much higher in switching sources (typically 50 mV peak to peak) and can range from 1 to 50 mV peak to peak in linear sources (higher in proportion to the load-regulation percentage change increase).

Ripple may be a function of the filters and regulator-amplifier gain. In extremely low ripple units, it may be a function of harness layout within the supply. Certain wires are very sensitive to hum and noise, and lead placement is critical. Elimination of these spikes at an open-module supply's output terminals can be very important. Loss of stored data is a possible consequence. Ripple is specified more meaningfully as a peak-to-peak, rather than rms, value. Ripple frequency may also be a consideration.

10.1.7 Temperature Coefficient

Temperature effects can exceed the load regulation limits, depending on the magnitude of power-source temperature change. Consequently, particularly in lower-efficiency power sources, the output voltage can become solely a function of the source temperature, rather than of the load-voltage regulation, due to heat dissipation.

Allowance may have to be made for (1) a temperature coefficient that may not be exactly linear, although calculations often assume it is; and (2) a specification that refers to the long-term condition although, due to differences in thermal time constants of the various devices involved, the supply may go out of specification for the short term.

10.1.8 Stability

Stability is a measure of the drift (or lack of it) in the output voltage after the source has been operating steadily for a long time at constant temperature, line and load conditions.

10.1.9 Reliability

Reliability can be established by worst-case analysis and extensive high-temperature life testing. If this approach is too expensive, mean-time-between-failure (MTBF) or other statistical data may be used, or the use of Monte Carlo methods may be in order.

10.1.10 Recovery Time

Recovery time is a function of (1) the effective series inductance of the output leads, (2) the time constant of the output capacitor, and (3) the time delay of the internal error amplifier. All these items are affected by the rate of application and magnitude of the load change. For this, the rate of application of the load is 10 μsec and the magnitude of the load change is from no load to full load.

10.1.11 Protection[62]

The power source should be protected against input, output, and internal conditions caused by overheating, overvoltage, or overload (short circuit). Overheating and overload conditions (usually 120% of rated value) are related as the common protection against either is source shutdown, called current foldback, with automatic restoration or manual reset after the condition has been corrected.

Overvoltage protection usually provides source shutdown. This is effective against regulator failure in the supply or on unusually large rapid transients, especially where integrated circuits are involved.

An effective overvoltage protection system should contain the following:

A. ***Detector*** This element detects the presence of an overvoltage condition and activates the protection mechanism.

B. ***Protection Mechanism*** This device performs the clamping of the impending overvoltage to a safe value. To accomplish this, it must be designed to absorb all the stored electrical energy of known sources, thereby diverting the energy which may cause failure of the voltage-sensitive load. In general, a silicon-controlled rectifier, connected directly across the fragile elements of the load in the so-called "crowbar" configuration, is used as the protective device.

C. ***Turnoff Mechanism*** As a consequence of using a silicon-controlled rectifier as the protective device, the overvoltage protection system must include a means for disengaging the primary-energy source, once the protection has been accomplished. A fuse or a circuit breaker may be mounted externally. An electronic switch or the current-limiting feature of the power supply may be utilized to achieve the necessary turnoff.

However, a protective system which depends upon the current-limiting feature of the power supply to furnish the turnoff function may not operate if the power supply itself should fail. It is recommended, therefore, that a fuse or other current interruptor be always used in such a system.

D. ***Overload Protection*** This should be carefully specified before choosing a power supply. There are three types of overload protection, namely, current limit, constant current, and foldback.

10.1.12 Safety

If the power-source outputs are greater than 1 kV, safety becomes an important consideration. Most high-voltage power sources utilize a form of switching technology (DC-to-DC conversion) for efficiency and to avoid the potential hazards of huge power-storage capacitors found in so-called "brute-force" linear power sources. Many of the high-voltage output sources are encapsulated in an insulating medium.

10.1.13 Environmental Conditions

Ambient temperature, humidity, and altitude limits specified on each power source must be met to ensure a long service life. Ambient-temperature range is typically from 0 to +70°C for continuous duty and from 0 to +50°C for peak rating.

10.1.14 Mechanical Considerations

Limitations on size, weight, physical configuration, type of mounting, type of connector, cooling, and distance from load should all be determined.

10.1.15 Maintainability

The desirability of open construction, readily removable covers or readily removable cases should be evaluated. In contrast, inexpensive potted units may be more reliable and rugged for some applications.

10.1.16 Replaceability

System downtime must be kept to a minimum. Replaceability should be evaluated. A plug-in-type module may be more than one using terminal strips.

10.1.17 Peak Transient

The peak transient is caused primarily by the effective series inductance of the output leads and secondarily by the time delay of the internal error amplifier. Care should be taken to specify peak amplitude of the transient, especially where integrated circuits (ICs) are being used, since the transient itself may trigger the circuit just as transients due to ripple can induce undesired switching of ICs and cause crowbar firing, if not designed properly.

10.1.18 Output Impedance

Output impedance specifies what kind of variation in output may be expected from a varying load such as an audiopower amplifier or a series of ICs that are changing from one state to another and causing perturbations in the load current.

It is suggested to require a statement of the impedance of the power supply as a function of the frequency of variation of the load, so that as the load is varied in both amplitude and frequency, the output voltage excursion that will result from that variation will be known.

10.1.19 Warranties

On the best equipment, warranties are normally for five years. Two-year warranties are often given for some switching sources, and 60-day warranties are common for the least expensive open-frame power sources.

10.1.20 Options[63]

Some of the major power-source options available are remote source turn-on/off sequencing, AC line failure detection, reverse-voltage protection (insertion of DC for DC-to-DC conversion or reverse load connection), reverse output polarity switching, RFI and EMI filtering (for switching regulators used with sensitive loads or for CRT-power applications), auxiliary battery charging and crossover (for loads requiring fail-safe uninterrupted power at their terminals, such as in medical equipment and computer memories), remote output voltage or current control by remote digital-source programming (used extensively for automatic power testing and CRT-beam modulation or focusing), and "power-good" or "power-fail" indication.

Many options to improve performance can be added; some of the most important are:

A. *Crowbar* If the output of the power supply rises above a certain preset limit, the crowbar circuit shorts the output terminals and brings the output voltage to zero.

B. *Remote Sensing* If a critical load is located some distance from the power supply, extra leads are brought to the load to measure and control the actual voltage at the load rather than at the output of the supply.

C. *Constant Current* Some loads work best when supplied with a constant current rather than with a constant voltage. Some supplies can be operated either as a constant-voltage source or as a constant-current source.

D. *Current Limiting* When the current drawn by a load reaches a certain preset value, the power supply senses the condition and will not supply any more current.

E. *Foldback Operation* Similar to current limiting, except that when the load asks for too much current, the power supply reacts by reducing both its output current and output voltage.

F. *Programming* The output voltage or current of a power supply is set by commands from a computer or other input. Programming can increase the price of a supply by two times or more.

10.2 SERIES-REGULATED POWER SUPPLIES

Series-regulated power supplies probably represent the best compromise for a wide variety of applications. They excel in regulation capability, ripple and noise, and output voltage temperature coefficient. Medium-priced with wide application, the conventional linear series-pass regulator has many positive attributes. It provides excellent (0.1%) regulation, produces low ripple and output noises, and shows high reliability.

The series regulator usually contains paired collector-coupled Darlington transistors, which supply load current. Semiconductor technology has brought series regulators and overvoltage protectors to miniature size, but these must be mounted on large heat sinks, thereby limiting the ultimate output wattage of the device. A 50–60-Hz power transformer, the input to the series regulator, limits the minimum size of this type of supply.

10.2.1 Cost

The cost of series-regulated power supplies is sometimes the same as that for ferroresonants, although usually somewhat higher. Up to about 150 W, series-regulated supplies cost less per watt than switchers. At this power output level, switchers start to become more economical, and this economy increases significantly as power-output level increases.

10.2.2 Efficiency

Probably the most-important drawback of the series-regulated supply is its relatively poor efficiency. Efficien-

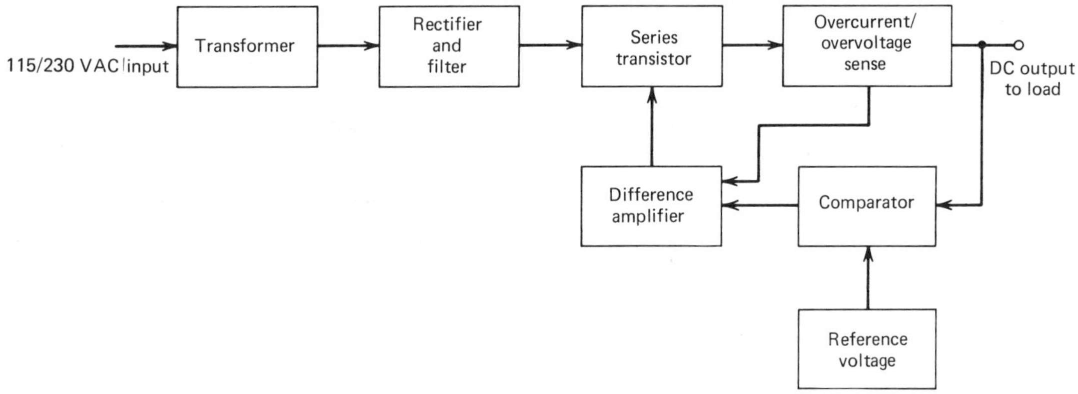

Figure 10.1 Series-regulated power supply.[61]

cies of 35-40% are typical. A few sophisticated and more expensive series-regulated supplies may rate a little higher, perhaps up to 50%. Lower efficiency usually means a larger power supply, hence more space as well as wasted power.

10.2.3 Operation

In a typical series-regulated power supply (Figure 10.1) the 120 VAC (or 230 VAC) power input is applied to the input transformer and stepped down to the desired voltage, typically 5, 9, or 12 VAC. This AC voltage is then rectified and filtered to provide a DC voltage, which is applied through the series transistor (in series with the load) and the overcurrent circuits to the power supply output.

Regulation is provided as follows: part of the DC output voltage is sampled and applied to a comparator circuit, along with a very stable reference voltage. As long as the sample voltage and reference voltage are the same, the comparator generates no output signal.

However, if the load demand changes, so too will the sample voltage. Since the reference voltage is constant, a different voltage (between the sample and reference voltage) will be produced by the comparator. This difference in voltage is amplified by the difference amplifier and applied as a bias signal to the series transistor.

The series transistor acts as a variable impedance (resistance) in series with the load. If the supply's output voltage starts to increase, the bias voltage from the difference amplifier causes the series transistor's impedance to rise, thereby causing the power supply's output voltage to drop until the reference voltage returns to "normal." If the supply's output tends to drop, the bias voltage from the difference amplifier causes the transistor's impedance to decrease, causing the supply's output voltage to increase to normal. In other words, the DC voltage at the output of the rectifier and filter is "shared" by the series transistor and load. Any change in the load voltage is absorbed by the series transistor.

10.3 SWITCHING POWER SUPPLY

Switching power supplies are characterized by their use of semiconductor elements to switch DC voltage on and off at a high rate, increasing or decreasing the resultant AC voltage through the use of a transformer and rectifying the output to produce the desired voltage.

10.3.1 Cost

Above the approximately 150-W level switching power supplies provide the lowest cost per watt. The switcher's cost per watt does increase dramatically in comparison to the other types in the high-power levels of 600 W and above.

10.3.2 Efficiency

When it comes to efficiency, the switcher excels. Efficiencies of 70-75% are typical. In general, when compared with other types of supplies, switchers offer the highest efficiency, the lowest size and weight, and superior operation over a wide input and frequency.

10.3.3 Reliability

The switching power supply's ability is not quite up to that of other technologies. Also, other parameters, such

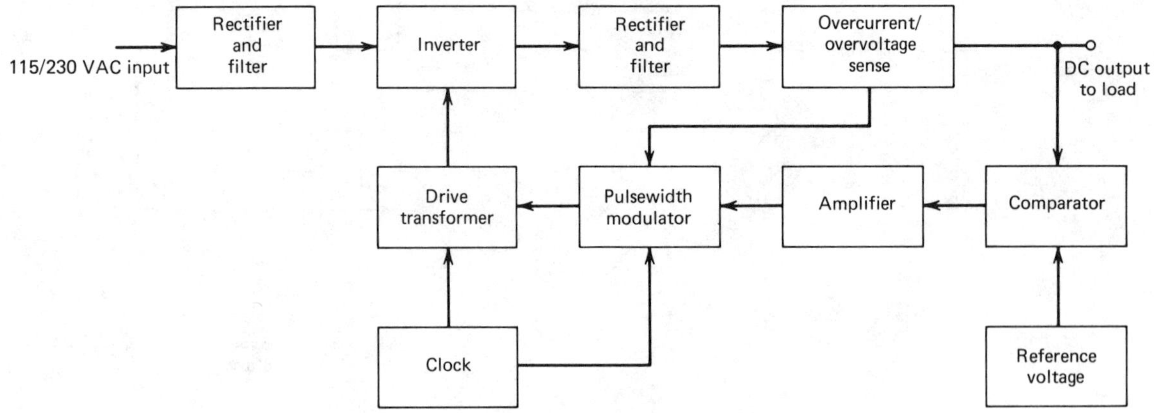

Figure 10.2 Switching-regulated power supply.[61]

as regulation, ripple, and noise are not quite as good as, or at least not any better than, the other types. The switcher's part count is also higher.

10.3.4 RFI/EMI

Switchers generate RFI/EMI because of their off-on switching operation. However, RFI/EMI can usually be handled by good design practices, such as locating the switcher away from sensitive circuitry, proper shielding of the switcher and its leads, short leads, good grounding, and so on. It is just as important to make sure that the switcher has enough input and output filtering built in to satisfy a particular application.

10.3.5 Operation

The switching power supply is much more complicated than the more conventional series-regulated type (Figure 10.2). AC voltage is immediately rectified to produce unregulated DC voltage, which is then applied to the inverter section of the circuit, thereby changing the DC voltage to a higher-frequency square-wave voltage. The output from the inverter is subsequently rectified and filtered. It passes to the regulator and the output terminals of the power supply.

In the inverter section, one side of the DC is connected to the emitter of two transistors and the other side to the center tap of the transformer. The bases of the two transistors are connected to a high-frequency switch, which at first turns one transistor on and the other one off. After a short time the switch reverses, the transistor that was conducting is turned off, and the one that was off begins to conduct.

The switch actually produces a constant flow of current to the transformer, first in one direction and then in the other, producing a square-wave output voltage.

10.4 FERRORESONANT POWER SUPPLY

As the least expensive of the three major power-regulator types, the ferroresonant regulator has few parts (meaning high reliability), good input/output isolation (4-5 kV), and built-in overload and short circuit protection.

10.4.1 Advantages

Ferroresonant power supplies, because of their great simplicity, relatively low cost, and excellent reliability, are probably a first choice where only moderate regulation is needed and system downtime must be avoided. Applications that rate reliability, maintainability, UL rating, and good efficiency are prime candidates for the use of ferroresonants. Ferroresonant regulators are best applied where load current demand remains relatively constant.

10.4.2 Disadvantages

Probably the most severe limitation of the ferroresonant power supply is its susceptibility to power-line frequency variations. Only very small frequency variations can be tolerated without upsetting the output voltage and current levels. Therefore, ferroresonants are designed for 50-, 60-, or 400-Hz operation.

Also on the negative side, the ferroresonant regulator has a bulky, heavy power transformer, limiting its minimum size; poor output-voltage regulation with line frequency change; poor load regulation with varying load-

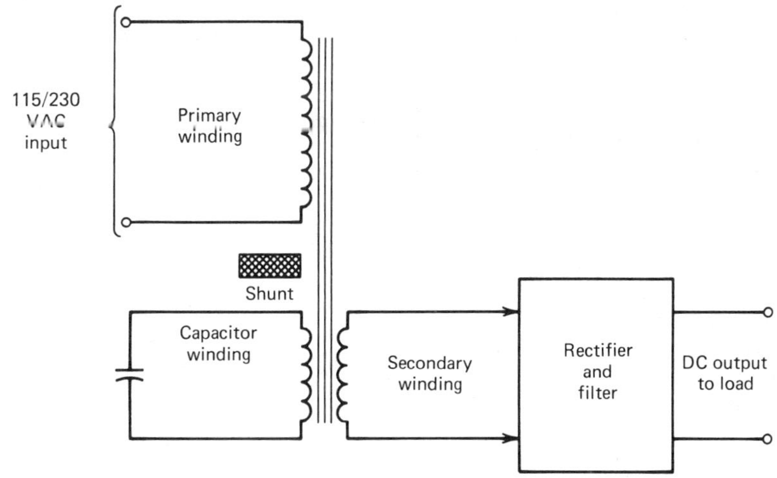

Figure 10.3 Ferroresonant power supply.[61]

current requirements; and a large leakage flux field that can create magnetic-field pickup problems.

10.4.3 Operation

The ferroresonant regulated power supply consists of a constant-voltage transformer, a rectifier, and a filter circuit. Regulation is provided by the constant-voltage transformer.

In its simplest form the constant-voltage transformer consists of a primary winding, secondary winding, and a third winding that has a capacitor connected across it (Figure 10.3). When the 115 VAC or 230 VAC power input is applied to the primary winding, the resulting magnetic flux is coupled through the shunt to induce a voltage across the secondary winding. The turns ratio of the primary to capacitor winding is such that the primary voltage is stepped down across the capacitor winding to a level that will produce the desired power supply output voltage after rectification. The AC voltage across the capacitor winding is coupled to the secondary winding, hence to the rectifier and filter circuit to provide the power-supply output voltage to the load.

Regulation is provided by the action of the capacitor and capacitor winding. The values of the capacitor and winding are such that they form a resonant circuit. Because of the high current flow in a resonant circuit, flux density in the core between the capacitor winding is at saturation level. Therefore, the supply's output voltage will remain fairly constant despite line voltage fluctuations at the input to the supply.

Ferroresonant regulators are sometimes used as inputs to other types of regulators. In these cases, they are often referred to as preregulators.

10.5 SELECTION[64]

Table 10.1 provides a general rating of regulated power supplies. Figure 10.4 presents a cost comparison of single-output DC power supplies, including series-pass and switching regulator devices. (CVT refers to a constant-voltage or ferroresonant supplies with rectifier and filter capacitor output. SCR Phase covers line frequency primary or secondary switching supplies with rectification and a passive LC filter.)

The series-pass or linear device is the logical choice for best electrical performance at low cost. Up to a power level of approximately 90 W, it costs less than a CVT-regulated supply, which is available generally for power levels beginning at around 100 W. The disadvantages of series-pass devices (size, weight, and efficiency) do not become objectionable until 250-W power levels are exceeded.

The CVT supply, which provides line regulation only, is a popular choice with a constant load current. It can deliver between 100 and 1000 W of line-regulated DC for less cost than other types.

Switching-regulated power supplies are more expensive because of their circuit complexity and the quality of other components used. These supplies look very attractive above 250 W because of their impact on system problems of size, weight, and cooling.

The series-pass power supply outperforms all other

Table 10.1[a] Power-Supply Ratings[61]

Characteristic	Series Regulated	Switching Regulated	Ferroresonant
Efficiency	3	1	1–2
Regulation	1	2	3
Ripple and noise	1	2–3	2–3
Output voltage T.C.	1	1–2	2–3
Relative cost	1–2	3	1
Relative size	2–3	1	2–3
Provable MTBF	1–2	2–3	1
Brownout capability	3	1	1–2
Line-frequency tolerance	1–2	1	3
Parts count	2	2–3	1

[a] 1, Best; 3, poorest.

types when comparing regulation, ripple, or transient response.

The high-frequency (20-kHz) switching-regulated power supply approaches linear regulation performance of better than 0.1% regulation, with even better specifications at higher-power levels (Figure 10.5).

The constant-voltage transformer (CVT) circuit provides reasonable line-voltage regulation, poor load regulation, and frequency regulation of 2.4% per hertz. This circuit is the lowest in cost but limited to a maximum ambient temperature of +55°C.

Figure 10.6 compares the power density of a power supply as a function of the type of regulation it uses with the output power range. (The data were derived for 60-Hz single-phase input to 1-kW and three-phase above 1-kW. Ambient temperature is + 40°C.)

The required output power has a great influence on the technique selected. The series-pass power supply has great appeal at power levels to 250 W because of simplicity, low cost, and high performance. Because of size, weight, and efficiency problems, however, this type loses appeal above 250 W. (Similar limitations are imposed on constant-voltage transformer supplies above 1 kW.)

Switching-regulated power supplies are available down to 50 W and become popular at levels above 250 W, with power levels presently limited to about 1500 W. The availability of new materials will, however, substantially extend power levels. This regulation technique offers the best power density because of its efficiency and high-frequency operations.

SCR phase-control has long been used to regulate output levels greater than 250 W. The availability of high-power SCRs is permitting the design of much greater power output supplies. It has good efficiency and yields reasonable power density, especially with three-phase input.

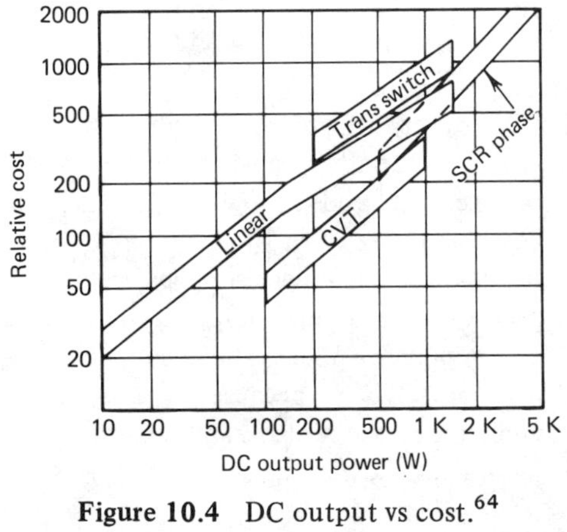

Figure 10.4 DC output vs cost.[64]

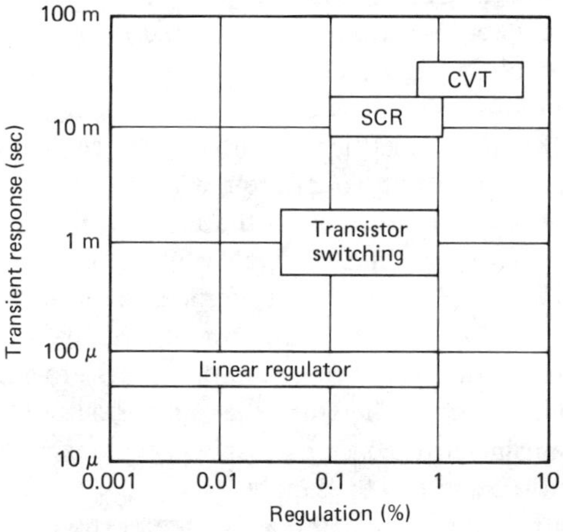

Figure 10.5 Regulation.[64]

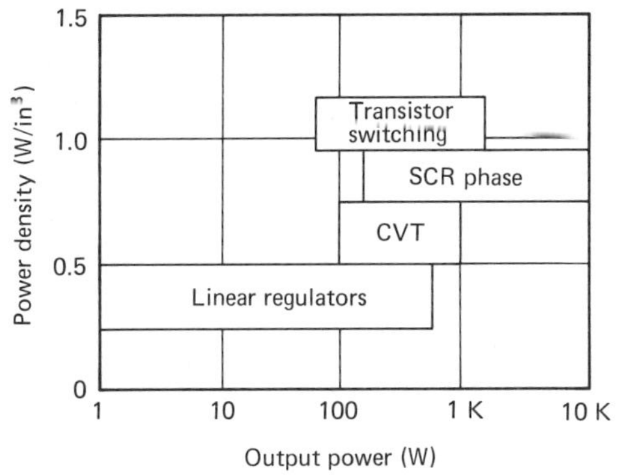

Figure 10.6 Power density.[64]

10.6 CUSTOM OR CATALOG—POWER SUPPLY?[65]

A number of factors on which to base a custom/catalog decision follow, along with criteria needed to evaluate these factors.

10.6.1 Physical Restraints

With respect to size and weight, either a catalog unit fits the allocated space or it doesn't. If it doesn't, the advantages of a custom design are obvious. A custom unit can optimize the power supply configuration for maximum system packaging efficiency.

If overall weight is not a factor, it is often possible to use a catalog power supply. Also, the use of two or three different power supplies in a small system or subsystem might result in less overall bulk and weight.

10.6.2 Environment

Evaluate a catalog power supply critically if a unique environment is involved. Whenever the unit must operate unprotected in an alien atmosphere, withstand temperature extremes, or undergo severe physical shock and vibration, it is usual to choose a custom power supply.

10.6.3 Electrical Specifications

There is an impressive selection of catalog units for commercial applications needing voltages of 500 V DC or less. As higher output voltages are required and the applications become more stringent, however, the choices narrow. Ideally, the power-supply requirements are dictated by the needs of the system. Many times, however, standardization causes system performance to be compromised to accomodate catalog power supplies. This should be carefully evaluated in the custom/catalog (standard) consideration.

Many catalog items were originally custom devices that found broader applications because of their price/performance characteristics. Ultimate performance is difficult, if not impossible, to obtain by trying to match specifications and can be obtained often by only working with the actual load during power supply design. In addition, specifications in a catalog are subject to change and might change without notification.

10.6.4 Options

Normally, the more that is paid for an item, the more features that are provided. Catalog or custom, add-on options cost disproportionately more than basic features. If protection circuits are needed, such as overvoltage, undervoltage, overcurrent, and overtemperature, it is best to avoid add-on circuits. If add-ons are needed, tooling costs, setup charges, and additional delivery time are often involved.

10.6.5 Approvals

Although a wide variety of catalog power supplies are capable of meeting UL, CSA, or equivalent commercial requirements, compliance with military specifications will probably dictate a custom-built, if not custom-designed, unit. If a catalog item is described as meeting military specifications, be sure that the costs of burn-in, screening of nonstandard parts, inspection, government certification, test data, and so forth, are included in the quoted price. Many catalog power supplies meet military-specification environments, but they do not comply with the parts/materials, inspections, and test criteria required by military applications.

10.6.6 Availability

For short-time availability, clearly an off-the-shelf catalog power supply is what is needed. This need eliminates not only all custom power supplies, but most catalog devices that are not economically maintained as inventory items as well.

The only definite way of being positive about delivery is to get a firm commitment from a reputable vendor. Don't be surprised if a particular custom design can be obtained as quickly as some catalog items, as the nature of custom business often results in competitive deliveries.

10.6.7 Engineering Service

If an application demands close liaison with the power-supply vendor for technical service, the advantage of direct communication with the designer is invaluable. The value of such coordination is underscored by the fact that as system specifications change during development, so can the power-supply requirements, hence the need for on-the-spot professional help to solve these change problems. Here, the custom-designed power supply holds an edge over the use of catalog items.

10.6.8 Cost

Although the common assumption is that custom-built units cost more than catalog packages, this is not always the case, especially if the volume requirements are large. High quantity alone is often sufficient to lower the price of a custom unit to below that of a catalog power supply, as you pay only for what you need with a custom supply.

10.7 HIGH-VOLTAGE POWER SUPPLIES

When specifying high-voltage power supplies, you should take into accout such major considerations as electrical performance, mechanical aspects, environmental stability, and testing.

10.7.1 Electrical Performance

Understanding the actual load should be the first concern, along with the definition of the power supply and the portion of the system in which it functions. Is the load resistive, capacitive, or inductive, and does it switch between these conditions during circuit operation?

If the load is not static, the dynamic loading for each output should be indicated in relationship to other outputs. Pulsed loads can shock-excite a power supply into oscillation or cause the output to grossly exceed the regulation limits. That is why it helps to know the load pulse width, repetition rate, and rise time. A timing diagram would also be helpful. (Define the load in terms of ohms, picofarads, microhenrys, and so forth, in addition to stating the application.)

A. ***Output Voltage*** Output-voltage performance characteristics are often overspecified. Aside from additional cost, there could be increased size and weight and a decrease in reliability due to greater complexity, all resulting from overspecified voltage tolerance. This appears to be more prevalent with high-voltage supplies because of the larger numbers involved.

Underspecification of voltage tolerance is not so common, but it is costly. Upgrading the tolerance on a finished supply could involve anything from minor design changes to an extensive redesign of the supply. Therefore, the best solution to voltage-range specifications is to use a respectable figure made on the basis of system-performance requirements and high-voltage power-supply characteristics.

B. ***Resolution*** Resolution of adjustable voltages is often neglected, as it is important to remember that all variable outputs have a finite resolution. However, output voltage range is not the only misspecified characteristic.

Overspecifying ripple adds size, weight, higher stored energy, and increases turn-on time and discharge time. The effect of underspecifying ripple is more complex, since it depends greatly on the application. It is often best to specify ripple in terms of peak-to-peak volts, rather than rms. Ripple, or noise, on high-voltage outputs will be more than likely to abrupt spikes of short duration with a fixed repetition rate.

C. ***Load Current*** Load current is often overspecified, leading to overdesign, higher weight, greater volume, and more power dissipation than necessary. In contrast, insufficient output current can lead to drastically reduced reliability, repeated overcurrent actuation, or component failure.

Fortunately, load-current requirements are easily calculated when various safety factors are known. In addition to knowing load current, output voltage, and allowing tolerance or adjustment range, each must be specified in relationship to the other.

10.7.2 Static Load Conditions

For static loading, output impedance can be determined directly from voltage and current. For dynamic loading, however, operating frequency is an important consideration, because impedance usually increases with frequency. Also, with dynamic loading, transient response becomes increasingly important, since the possibility of a random transient disruption is greater in a dynamic (especially a pulsed) load. Therefore, define acceptable peak excursion during load change and the allowable time for the output to return to a defined point, a point that may be outside the static regulation limits.

Failure to specify transient-load characteristics can result in an unusable power supply. Depending on the type of short-circuit protection utilized in the design, the supply could oscillate or shut down completely. A less important yet serious result could be gross deviations in the output voltage. Providing information on the dynamic load to the power-supply vendor will help avoid this predicament. Due to the voltage levels involved, it is often impractical to monitor high voltage directly.

When monitor points are needed, specify the division

ratio and tolerances required and the type of instrumentation to be used for monitoring. Otherwise, major measurement errors could result. A monitor circuit may not reflect actual ripple percentages unless it is designed to do so; therefore, to measure ripple at the monitor point, notify the power supply vendor.

A. *Input Voltage* Specifying an input voltage or frequency range greater than that actually encountered takes its toll in power dissipation and attendant temperature rise. Underspecifying obviously could avoid an expensive design.

If the input is AC, specify frequency, frequency range and tolerance, voltage range, and whether single or multiple phase. If DC, specify peak-to-peak ripple levels, characteristics of transients, and voltage range. For either type, indicate the source impedance, if known.

Be sure to mention any input power limitations that may exist, especially with portable, or hand-held, battery-operated equipment. And, unless indicated otherwise, most vendors will common the input and output returns for economy and simplicity.

B. *Regulation* When specifying regulation, indicate actual line and load-change ranges and the worst combination of the two. Also, costs can be reduced by specifying regulation (in percent) over the actual load-change range instead of the more common "zero-to-full-load" criterion. Where important, temperature coefficient should be specified in terms of percent of output regulation per degrees centigrade from nominal ambient temperature.

C. *Warmup* Indicate allowable warmup time, as all specifications tend to apply only after a finite warmup period. Normally, stability is specified over a fixed period of time under fixed line, load, and temperature conditions after the prescribed warmup period.

D. *Sequencing* If up-sequencing or down-sequencing is required, indicate the sequence-time delay and whether sequencing is to be internally or remotely controlled. If remotely controlled, specify the characteristics of the control signals.

Since a finite time is needed to turn off and bleed a high-voltage power supply, indicate the allowable time it will take for the high-voltage outputs to decay from their normal levels to a specified safe voltage upon turnoff.

10.7.3 Environmental Considerations

Specify the range of both operating and storage temperatures, as they can dictate materials, components, circuitry, and packaging. Also indicate the type of cooling, such as natural convection, forced air, and cold plate.

A. *High Altitude* Outgassing at high altitudes may be as detrimental to the system as arcing at sea level. For this reason, both operating and storage barometric pressures should be specified.

With respect to humidity, moisture absorption can alter the electrical properties of some insulating materials, and corrosion (especially galvanic) is sensitive to humidity. Therefore, both operating and storage ranges of humidity should be specified.

The major atmospheric contaminants present can also determine the materials required. Explosive environments warrant special consideration with high voltage present.

B. *Noise* High-voltage power supplies are often noisy because they contain oscillators. This noise is expensive to eliminate, but it is relatively easy to design outside of the audible range. It is usually sufficient to define tolerable audible noise in terms of the operating environment.

C. *EMI/RFI* Electromagnetic radiation is generated by the high-voltage power-supply oscillator. If this is of concern, indicate the acceptable levels of conducted emission on input and output leads, as well as radiated emission. This area is especially costly if overspecified. Also, if EMI and RFI are generated around the power supply, the radiated and conducted interference levels are characteristics that should be specified.

D. *Shock and Vibration* When specifying shock and vibration, be certain that it is indicated whether or not the unit must operate in, or merely survive, the specified levels. Note also the plane, frequency, and amplitude of the shock and vibration.

Special applications may expose the power supply to explosive decompression, thermal shock, random vibration, and so forth. Failure to specify these could prove catastrophic.

10.7.4 Mechanical Aspects

While there is an art to attaining high-power density and heat dissipation combined with minimum volume, it is generally true that specifying an inadequate volume for the available cooling will, in all likelihood, result in a hot-running, short-lived, piece of hardware. Therefore, when specifying the geometric configuration for the power supply, state the limiting dimensions and give some thought to whether or not the power supply must be in one piece or if a modular subdivision is possible, as the latter is often the best way to fit a power supply into a required space.

A. *Construction* Consideration should be given to whether the package can be open or closed. If closed, is hermetic sealing required, or can the package be of a hybrid construction? An ungrounded case can act as an antenna to pick up or radiate interference and can act as a capacitor that can change to and hold unsafe voltage levels. Therefore, it is normal to ground the power supply case unless otherwise specified.

B. *Mounting* Specify the method of mounting (studs, inserts, brackets, straps, clamps, and so on) as well as the mounting and operating position(s). This is especially necessary in the case of gas–oil-filled power supplies, as heat-dissipating components must be submerged in oil regardless of mounting orientation.

C. *Finish* When specifying the finish for the power-supply housing, keep in mind that dark colors are preferred where heat dissipation is a major consideration. Otherwise, it is sufficient to state the type of finish desired, whether it is insulating or noninsulating, and what marking and identification is required.

D. *External Connections* With the exception of the high-voltage lead or terminal, which should be as safe and corona-free as possible, external electrical connections present no special problem of their own, unless requirements peculiar to an application are involved.

10.7.5 Testing

It is important to specify if the power supply will be tested to military, UL, or CSA requirements, in addition to the methods, procedures, and instrumentation to be used, wherever possible. Also, indicate if a certificate of compliance is required, if performance must be documented for each unit, or if qualification testing is desired.

10.7.6 Other Considerations

It is often advisable to specify operating life in total hours, as the time period covered by a warranty often has no relation to operating time and has little bearing on the operating life of the power supply. Other items to consider include: terms of the warranty–whether the item will be repaired or replaced and in what turn-around time; first-article approval, if any; delivery schedule after approval; and so forth. And finally, it is necessary to indicate whether secret, confidential, or other security classification is applied to the power supply or its application.

10.8 TERMS AND DEFINITIONS[66]

Absolute Accuracy The correctness of the indicated value in terms of its deviation from the true or absolute value.

Ambient Reference The temperature range of the surrounding still air in which the power supply is specified to perform.

Ambient Temperature The temperature of the environment surrounding the power supply, generally assumed to be room temperature.

Automatic Crossover A term applied to bimodal power supplies (constant current/constant voltage) that prescribes the transferral from one operating mode to the other at a predetermined value of load resistance.

Bipolar Two output voltages of opposite polarity having different absolute values. A variable, single-output, supply in which the output amplitude and polarity can be linearly adjusted through zero to assumed positive and negative values.

Back Ripple Voltage (in DC-to-DC converters) The ripple voltage projected back through the input by disturbances caused from the converter's chopper, expressed in terms of peak-to-peak percent of the input voltage.

Calibration Accuracy Synonymous with absolute accuracy.

Case Reference The case temperature range over which the supply is specified to perform, normally measured in still air in a +25°C ambient.

Clamping A circuit action that keeps the power-supply output from exceeding a preset value.

Combined Line and Load Regulation (Combined Source and Load Effect) The maximum steady-state output voltage resulting in specified changes in input line voltage amplitude and output load condition, such change being applied concurrently and with directions chosen to make their effects additive.

Compliance Range The load-voltage capability of a power supply operating in the constant-current mode.

Constant-Current Power Supply A power supply that maintains a stabilized output current within prescribed limits with respect to changes of influence quantities (line, load, ambient, and so on).

Constant-Voltage Power Supply A power supply that maintains a stabilized output voltage within prescribed limits with respect to changes of influence quantities (line, load, ambient, and so forth).

Crowbar A load protection circuit designed to prevent excessive output voltage by rapidly placing a low-resistance shunt across the output terminals of the power supply, thereby initiating action to reduce the output voltage to a low value.

Current Foldback A self-resetting protective method that reduces the output current to less than full-rated output under overload and short-circuit conditions.

Current Limiting A general term applied to any of three kinds of short-circuit or overload protection and that describes the action of limiting the output current to some maximum predetermined value (fixed or adjustable) and automatically restoring the power supply to normal operation when the overload or short circuit is removed.

Current-Sensing Resistor A resistor of low value placed in series with the load to develop a voltage proportional to the output current.

Derating The reduction of output power of a power supply as a function of elevated temperature, referenced to either ambient or case temperature.

Drift The maximum change of an output parameter of a power supply during a specified period of time following a warmup period, with all external operating and environmental parameter influences maintained constant.

Efficiency The ratio of power supply output power to input power, expressed as a percentage, for a given set of input voltage and output load conditions.

Error Signal The output of the voltage comparison circuit, which is amplified to drive the series-pass elements.

Harmonic Distortion The distortion of a sinusoidal wave characterized by the presence of harmonics of the fundamental frequency.

Input/Output (I/O) Isolation The degree of electrical isolation between the input and output terminal of an electronic power source, which can be expressed in terms of voltage (breakdown), current (galvanic), as well as either resistance or capacitance (impedance), or both.

Input Reflected Ripple (Back Ripple) Transient currents generated back to the input by internal switching, expressed as a percentage of either the average input current or the DC input voltage (in DC/DC converters).

Input Voltage Range The maximum and minimum input voltages for which a power source will operate to specifications.

Inrush Current The maximum instantaneous value of the input current to the power supply when switching on the power supply under specified conditions.

Leakage Current Extraneous AC line current flowing between the primary and secondary winding in an isolated power source with rated input voltage applied to the primary.

Line Regulation A change in DC output voltage due to variation of input voltage with all other factors held constant, expressed as a percentage of the nominal DC output voltage.

Line Switchers A circuit that directly converts the AC line voltage to a lower DC value using switching techniques.

Load Regulation A change in DC output voltage due to variation of external load current with all other factors held constant, expressed as a percentage of the nominal DC output voltage.

Load Control The determination of the power-supply output by means of a control element integral with the power supply.

Load Effect See Load Regulation.

Noise A residual AC component of the output voltage caused by line or chopper harmonics and noise, specified as an rms value.

Operating Temperature Range The range of ambient temperature throughout which a power supply will meet all its ascribed output specifications.

Output Effect The change in the steady-state value of a stabilized output quantity (voltage, current, or power) due to a specified change in the steady-state value of one or more influence quantities with all other influence quantities maintained constant.

Output (Dynamic) Impedance The impedance presented by the power supply to the load, that is, the ratio of the change in output voltage (at the prescribed terminals) to the change in load current causing the change, specified from DC to a stated maximum AC.

Output Noise (Radiated) Electrical interference signals radiated by switching in DC/DC converters and by the transformer in line-operated power supplies.

Output Ripple The instantaneous periodic variations present in the output (see PARD).

Output Transient The non-steady-state part of a change in value of a stabilized output quantity that results from a change in one of the influence quantities or control quantity.

Output Voltage Accuracy The maximum deviation of the actual output voltage from the rated value when the input voltage, load, and temperature remain fixed at their typical values.

Overload Protection The protection of the power supply and associated equipment against excessive output current, including short-circuit current.

Overshoot A transient voltage or current rise above the regulation band caused by large steplike changes in the load or input line, or by disconnection of input power.

Overvoltage Protection The protection of connected equipment against excessive output voltage due to power-supply malfunctions.

Parallel Operation The interconnection of two or more power supplies in a manner as to obtain summation of their output currents.

PARD An acronym for Periodic and Random Deviation, that is, the deviation in the regulated output parameter (voltage or current) over a specified bandwidth with all external and evnrionmental parameters held constant, excluding deviations attributed to drift.

Pass Element An automatic variable resistance device in series with the source of DC power, which is driven by the amplified error signal to increase its resistance when the output needs to be lowered or to decrease its resistance when the output must be raised.

Perturbation Response The overvoltage spike a power supply generates during power turn on/off and step-load changes, including the transient response of the power source.

Phase Shift The difference between corresponding points on input/output signal wave shapes, not affected by magnitude, expressed as degrees lead or lag.

Power Factor The ratio of the source power, as measured by a wattmeter, to the apparent source power.

Recovery Time The time required for the output voltage to return to a value within the regulation specification after a step load or line change.

Regulation (1) *Line:* The maximum steady-state differential in the controlled output parameter resulting from a full-range change in source voltage. (2) *Load:* The maximum steady-state differential in the controlled output parameter resulting from a full-range load change. (3) *Line and Load Combined:* The maximum steady-state differential in the controlled output parameter resulting from concurrent full-range change in load and output voltage, both in the direction that represents the worst case.

Remote Control The determination of the power-supply output by means of a control element external to the power supply.

Remote Programming A power-supply feature whereby the controlled output parameter (voltage or current) may be controlled through the application of external resistances, voltage signals, or current signals to designated external terminals.

Remote Sensing A feature of a power supply that allows the user to shift the regulation point from the supply's output terminals to an external location.

Remote Voltage Adjustment Adjustment of the output voltage of a power supply remotely over a limited range by a variable resistor.

Repeatability The closeness in agreement among a number of consecutive measurements on the controlled output parameter, with each measurement following an identically programmed input command.

Reset A means by which a power supply is brought back into operation after a malfunction has been corrected.

Resolution The minimum discernible change in the controlled output parameter that can be made with the applicable controls.

Response Time The time required for the voltage excursion to be reduced to 37% of its peak value after a step load or line change.

Reverse Voltage Protection Protection of a power supply against a reverse voltage applied at its output terminals.

Ripple Those deviations in the controlled output parameter that are harmonically related in frequency to the applied input power and any internally generated switching frequencies.

Ripple Regulation A method in which the output ripple is sampled to maintain energy balance, hence voltage regulation, in an L-C circuit.

Series Operation The operation of two or more power supplies with the positive output terminal of one connected to the negative output terminal of another, so that the output voltages of the supplies are additive.

Series Regulator A device placed in series with a source of power capable of controlling the voltage or current output by automatically varying its series resistance.

Setting Effect The relatively low change of an output parameter that follows an initial variation of the same parameter and exists as an added output variation.

Short-Circuit Current The current delivered by the power supply when output terminals are short-circuited.

Short-Circuit Protection An output-power limiting system that enables a power source to sustain no damage under short-circuit load conditions.

Shunt Regulator A device placed across the output that controls the current through a series-dropping resistance to maintain a constant output.

Shutdown A circuit that prevents the power-supply output from exceeding a predetermined value by shutting down the supply.

Sinusoidal Frequency Response The maximum sinusoidal frequency to which a fast-programmable power supply will respond with negligible distortion of the output waveform.

Slewing Rate The response speed of a fast-programmable power supply.

Stability The variation in power-source output over a stated period of time with the input, ambient temperature, and load held constant.

Standby (Idle) Current The current drawn from the source by a switching regulator or DC/DC converter with no externally applied load.

Steady-State Value The value of a quantity that persists after all nonrecurring transients have decayed to an insignificant magnitude.

Step-Line-Voltage Change An instantaneous change in line voltage for measuring line regulation and transient response.

Step-Load Change An instantaneous change in load current for measuring the load regulation and transient response.

Storage Temperature The range of environmental temperature in which a power supply can be safely stored.

Switchers Power supplies that employ switching techniques in order to achieve higher efficiency regulation.

Switching (Oscillator) Frequency The rate at which the source voltage is switched or chopped in a switching regulator or DC/DC converter, respectively.

Temperature (Output Effect) Coefficient The percent change in the controlled output parameter due to a 1°C change in ambient operating temperature within the operating temperature range.

Tracking (Dual Output) A feature of a dual-output power supply that causes a change of one output to shift the other output, so that each maintains the same absolute voltage relationship, ignoring polarity, with respect to common.

Turn-on (Turn-off) Overshoot The overshoot resulting from the application (removal) of the source power or from the power supply input switch being turned on (turned off).

Voltage Reference A separate highly regulated voltage source used as a standard to which the output of the power supply is continuously referred.

Warmup Time That period, following the application of input power, before a power supply meets its rated performance characteriestics.

CHAPTER ELEVEN
Batteries

11.1 GENERAL

11.1.1 Cells

The terms "cell" and "battery"[67-70] are often used interchangeably but incorrectly. A cell may be used singly or two or more cells may be connected together to form a battery. Cells are generally classified in two major groups—primary cells, which are used until the voltage output is too low for useful work, and secondary cells, which are rechargeable.

In secondary cells, chemicals that provide the energy may be restored to their original condition by applying a direct current to the cell in a reverse direction to the flow of current during discharge.

Primary and secondary cells, although differently constructed, process the same basic components, that is, two electrodes and an electrolyte. One electrode chemically "reduces" when current flows and becomes the positive (cathode) terminal, while the other "oxidizes" and becomes the negative (anode) terminal. The process continues until either circuit is broken or until the cells are used up. (See Figure 11.1.)

Primary cells may be categorized into six significant subgroups as follows:

A. *EMF Standard Cells* These cells use chemicals of very high purity and are designed for use as voltage-reference standards.

B. *Solid-Electrolyte Cells* These cells use only solid materials. The principal advantage of such cells is their long shelf life. Their use, however, is restricted to low-current applications.

C. *Wet Cells* These cells handle relatively large currents, Two or more popular wet cells, the Lalande cell and the zinc "air" cell, are activated by adding a caustic-soda electrolyte just prior to use.

D. *Dry Cells* This type of cell is the type most widely used and the most familiar. In addition to the carbon-zinc (Lelanché) type, there is the zinc-mercuric oxide (Ruben or mercury) type, the zinc-manganese dioxide (alkaline) type, the zinc silver oxide (alkaline) type, and several newer types.

E. *Reserve Cells* These cells are designed for use in "one-shot" or "delayed-action" applications. A reserve cell remains essentially inactive in the standby state until required. It is then activated by the addition of a liquid or gas, or by the application of heat.

F. *Fuel Cells* The familiar process of burning fuel to obtain heat energy, which in turn is converted into mechanical energy and then electrical energy, can be replaced by a fuel cell. By adding carbonaceous or hydrogen fuels to these cells, a reaction takes place that converts chemical energy directly into electrical energy.

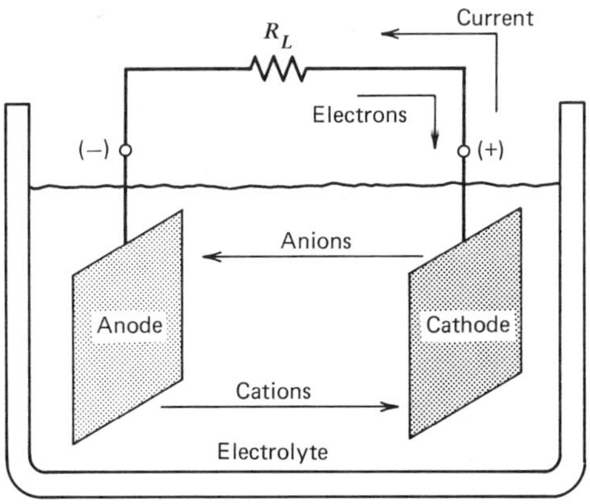

Figure 11.1 A basic cell.[68]

11.1.2 Batteries

A battery consists of a particular combination of anode, cathode, and electrolyte. Some common primary systems include carbon-zinc, alkaline-manganese, mercuric-oxide, and silver-oxide. (See Table 11.1.)

The anodic material used for the negative terminal is usually a metal such as lead, cadmium, manganese, or zinc. Each is characterized by the ease with which it gives up electrons and becomes positively charged ions (cations) in the electrolyte.

Cathodic materials, by contrast, are usually chemical compounds such as PbO_2, MnO_2, and Hg, and Ag_2O or AgO. Known as depolarizing materials, these compounds are also characterized by the ease with which they accept electrons. They reduce to a lower state of oxidation by electrolyte activity and, in so doing, form negatively charged ions (anions) in the electrolyte.

The electrolyte in a battery completes the electric circuit between the two electrodes with its ionic conduction path. Since it reacts chemically with both electrodes, the electrolyte must match the properties of the cathodic and ionic materials. The rate of chemical reactions occurring in the cell depends on the rate of the ionic diffusion, the temperature, the effective surface area of the electrodes, and the load connected across the cell. As the battery reaches the end of usefulness, the electrolyte grows progressively "weaker" and the electrodes are partially consumed.

11.1.3 Selection

Before a battery can be specified, the information that should be determined from the application includes voltage, current drain, operating schedule, required service life, service temperatures, environment, size/weight, and type of terminals.

If the equipment will not operate below a certain critical voltage, this end-point voltage should be specified. Both initial and operating current drains may need to be specified. These, along with the discharge schedule and the required service life, will determine the ampere-hour capacity for the battery.

Service temperatures must be known because they will affect battery capacity or life, or both. If the battery will be stored for any period of time before use, the length of storage time and temperature should also be checked carefully and specified.

Allowable size and weight will sometimes determine the battery selected in lieu of other requirements. A premium is usually paid for small size with high output capacity. Bulky and heavy batteries can be reliable, as well as economical, if their size can be accommodated in the application.

Shock or vibration may indicate the need for a rugged battery construction. Unusual rates of acceleration of high-altitude operation must also be considered, as well as time and temperature under any of these conditions.

Often the first decision involves whether to use a primary (nonrechargeable) or secondary (rechargeable) battery. Cost and battery capacity are the two most important factors influencing this decision. If low initial cost is the basic criterion for a product, then a primary battery is almost always the way to go, provided that a primary battery has the capacity required for the application.

When low operating cost is important, it is often advantageous to use a secondary battery, even if a primary battery has the capacity required for the application. In certain types of devices the higher initial cost of a secondary battery and its charger may be offset over the life of the equipment, eliminating continual replacement of primary batteries. Also, the power demands of an application often make a secondary battery the only choice.

11.2 PRIMARY BATTERIES/CELLS

Table 11.2 lists the major characteristics of the various types of available primary batteries. The carbon-zinc (Lelanché) dry cell is a major source of low-cost power, as this battery is available in a wide variety of styles, sizes, shapes, and terminal arrangements. However, carbon-zinc batteries have two major drawbacks–limited low-temperature performance and a substantial drop in efficiency at high current drains.

Table 11.1 Average Characteristics of Dry Batteries[70]

Usual Name	Carbon–Zinc	Carbon–Zinc (Zinc Chloride)	Alkaline–Manganese Dioxide	Mercuric Oxide	Silver Oxide	Nickel–Cadmium
Electrochemical system	Zinc–manganese dioxide (usually called Leclanche or carbon–zinc)	Zinc–manganese dioxide	Zinc–alkaline manganese dioxide	Zinc–mercuric oxide	Zinc–silver oxide	Nickel–cadmium
Type	Primary	Primary	Primary and rechargeable	Primary	Primary	Rechargeable
Rechargeability	Poor	Poor	Good	No	No	Yes
Number of cycles	10–20	10–20	50–60 rechargeable only			300–2000
Input if rechargeable			Approximately 100% of energy withdrawn rechargeable only			Sealed, minimum of 140% of energy withdrawn
Typical commercial service capacities	60 mA-hr–30 A-hr	Several hundred milliamp-hours to 9 A-hr	Several hundred milliamp-hours to 23 A-hr	16 milliamp-hours to 28 A-hr	35–210 mA-hr	Sealed: 20 mA-hr to 4 A-hr
Effect of temperature on service capacity	Poor low temperature	Good low temperature relative to carbon–zinc	Good low temperature	Good high temperature poor low temperature–depends on construction	Poor low temperature–depends on construction	Very good at low temperature–poor at high temperature
Leakage	Medium under abusive conditions	Low	Rare	Some salting	Some salting	No
Gassing	Medium	Higher than carbon–zinc	Low	Very low	Very low	Low
Reliability (lack of duds; 95% confidence level)	99% at 2 years	99% at 2 years	99% at 2 years	99% at 2 years	99% at 2 years	99% at 2 years
Shock resistance	Fair to good	Good	Fair to good	Good	Good	Good
Features	Low cost; variety of shapes and sizes	Service capacity at moderate to high current drains greater than carbon–zinc; good, leakage resistance; low temperature performance better than carbon–zinc	High efficiency under moderate and high continuous conditions; good low-temperature performance; low impedance	High service capacity/volume ratio; flat voltage discharge characteristic; good high temperature performance	Moderately flat voltage discharge characteristics	Excellent cycle life, flat voltage discharge characteristic; good high- and low-temperature performance; high resistance to shock and vibration; can be stored indefinitely in any charge state.
Limitations	Efficiency decreases at high current drains; poor low-temperature performance		Primary type expensive for low drains. Rechargeable-limited cycle life; voltage limited taper current charging	Poor low-temperature performance on some types		High initial cost; only fair charge retention

Table 11.2 Typical Characteristics of Primary Batteries[67]

Common Name	Carbon-Zinc (Leclanche)	Carbon-Zinc (Zinc Chloride)	Alkaline-Manganese Dioxide	Mercuric Oxide	Silver Oxide	Divalent Silver Oxide	Zinc Air	Lithium*
Electrochemical System	Zinc-Manganese Dioxide (usually called Leclanche or Carbon-Zinc)	Zinc-Manganese Dioxide	Zinc-Alkaline Manganese Dioxide	Zinc-Mercuric Oxide	Zinc-Silver Oxide	Zinc-divalent Silver Oxide	Zinc-Oxygen	*
Voltage per Cell	1.5	1.5	1.5	1.35	1.5	1.5	1.5	2.8 (also 1.9, 3.4)
Negative Electrode	Zinc	Zinc	Zinc	Zinc	Zinc	Zinc	Zinc	*
Positive Electrode	Manganese Dioxide	Manganese Dioxide	Manganese Dioxide	Mercuric Oxide	Monovalent Silver Oxide	Divalent Silver Oxide	Oxygen	*
Electrolyte	Aqueous solution of ammonium chloride and zinc chloride	Aqueous solution of zinc chloride	Aqueous solution of potassium hydroxide	Aqueous solution of potassium hydroxide or sodium hydroxide	Aqueous solution of potassium hydroxide or sodium hydroxide	Aqueous solution of potassium hydroxide or sodium hydroxide	Aqueous solution of potassium hydroxide	*
Typical Commercial Service Capacities	60 mAh to 30 Ah	Several hundred mAh to 30 Ah	Several hundred mAh to 23 Ah	16 mAh to 28 Ah	35 mAh to 210 mAh	45 mAh to 245 mAh	170 mAh to 300 mAh	30 mAh to 12 Ah
Energy Density:								
watt-hours/Lb.	20	40	45	50	50	65	125	100 to 150
watt-hours/in.^{3}	2	3	2 to 3	8	8	12	15	8 to 12
Practical Current Drain Rates:								
Pulse	Yes	Yes	Yes	Yes	Yes	Yes	Yes	Yes
High (more than 50 mA)	100 mA/square inch of zinc area ("D" cell)	150 mA/square inch of zinc area ("D" cell)	200 mA/square inch of zinc area ("D" cell)	No	No	No	Yes	Yes
Low (less than 50 mA	Yes	Yes	Yes	Yes	Yes	Yes	Yes	Yes
Discharge Curve (shape)	Sloping	Sloping	Sloping	Flat	Flat	Flat	Flat	Flat
Temperature Range:								
Storage	−40°F to 120°F	−40°F to 160°F	−40°F to 120°F	−40°F to 140°F	−40°F to 140°F	−40° to 140°F	−40°F to 140°F	−65°F to 160°F
Operating	20°F to 130°F	0°F to 160°F	−20°F to 130°F	32°F to 130°F	32°F to 130°F	32°F to 130°F	17°F to 130°F	−40°F to 160°F
Impedance	Low	Low	Very Low	Low	Low	Low	Low	Very Low
Shelf Life	Fair	Fair	Good	Fair to Good	Good	Good	Good to Excellent	Excellent
Cost:								
Initial	Low	Low to Medium	Medium	High	High	High	Medium	High
Operating	Low	Low to Medium	Medium to High at high power requirements	High	High	High	Low to Medium	Medium

* There are number of basic Lithium battery systems. Values given are typical for Lithium batteries.

The zinc-chloride type of carbon-zinc battery overcomes these limitations to a great extent. Its low-temperature performance is better than that of the Lelanché battery, and it provides better performance at moderate to high current drains.

Although more expensive than carbon-zinc batteries, alkaline batteries are very efficient under moderate and high continuous-drain applications. The alkaline battery gives much longer service than does the carbon-zinc battery in addition to providing very good low-temperature operation and excellent shelf life. However, they are relatively expensive, particularly in low current-drain applications.

The high concentration of energy in a mercuric oxide battery results in a package that is less than one-half the size of a comparable carbon-zinc battery. Mercuric oxide batteries are most often packaged in flat and button configurations for use in applications where space is at a premium. High cost and poor operation at low temperatures are two main drawbacks of these batteries.

Silver oxide batteries, which are packaged like mercuric oxide batteries, are used in many of the same applications. For some applications, silver oxide batteries have a distinct advantage in their higher cell voltage (1.5 V compared with 1.35 V for mercuric oxide). This better performance brings with it higher cost. The divalent silver oxide battery provides a much higher energy density and higher milliampere-hour rating than does the basic monovalent type, which costs about the same.

11.2.1 Cost

Lelanché batteries are low-cost and are often preferred for intermittent-duty applications. For heavy-duty, continuous, high-current applications, a better choice is the alkaline-manganese battery, as it has better shelf life, contains 50–100% more energy, and lasts three to seven times as long.

For high energy-to-volume ratios and a flat discharge curve, the traditional choice is the mercury battery. Its flat discharge characteristic makes it ideal for reference-voltage applications.

Like mercury cells, silver oxide cells provide a flat discharge curve. But, while they have a higher voltage than do mercury cells (1.55 V for silver versus 1.35 V for mercury), they have a lower ampere-hour capacity.

Table 11.3 gives cost ratios among the three most common primary cells and nickel-cadmium batteries. Because the prices of all these batteries vary widely, direct cost comparisons are not provided. Rather, the cost of the carbon–zinc (CZn) cell has been set at unity and the cost of other cells expressed as a ratio to the cost of the CZn cell. Numbers greater than one indicate that a battery type is more expensive than the CZn type.

Table 11.3 Battery Cost Comparisons[69,a]

	Size of Battery		
Battery Type	D Cell	AA Cell	C Cell
CZn	1.00	1.00	1.00
ZnCl	1.05	1.06	1.05
Alkaline	2.08	1.93	1.80
Nickel–cadmium	12.11	4.00	6.97

[a]These cost comparisons are ratios, not actual dollar figures. Although price of particular batteries may vary, ratios given are averages that express relationship between various types of battery that hold generally true regardless of price. Prices differ between types because of price differences in various materials used, and because of more complex cell construction in higher-priced batteries. All the batteries listed are primaries except for nickel–cadmium, which is secondary.

Table 11.4 Hours of Service for Batteries[69,a]

Battery Type	D Cell (Toy)	AA Cell (Photoflash)	AA Cell (Calculator)	C Cell (Cassette)
CZn	1.83	0.60	6.0	5.8
ZnCl	6.42	1.62	10.5	14.2
Alkaline	13.12	3.15	17.0	27.6
Ni–Cd[a]	1995	994	3384	3994

[a]Assumes up to 1000 or more charges.

From the ratios in Table 11.3, it is obvious that CZn cells are the least-expensive type on a first-cost basis and, on the same basis, that nickel–cadmium cells are the most expensive. However, over the life of the battery, nickel–cadmium cells are much less expensive to use than CZn cells. Table 11.4 gives the service life of the various cells to allow for making cost-per-hour comparisons.

11.2.2 Carbon–Zinc (CZn)

There are two major types of carbon–zinc (CZn) batteries available, the 1.5-V cylindrical cells, Figure 11.2, and

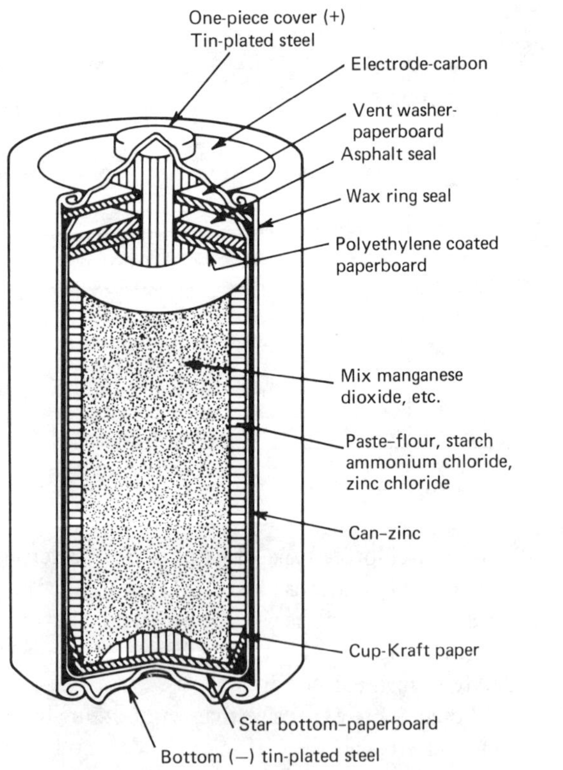

Figure 11.2 Carbon–zinc cell.[69]

multicell rectangular packages (such as the 9.0-V transistor radio battery).

Major uses for CZn primary batteries are in portable radios, clocks, flashlights, and toys. They are infrequently used in cameras, pagers, and calculators.

Although many improvements have been made in CZn cells, the tendency of the battery to leak electrolyte has not been fully resolved.

In the carbon-zinc (Lelanché) cell the zinc case serves as the negative electrode and the container for the cell contents. The material for the positive electrode of this cell is the cathode mix. Because the mix is a powder, it is not a mechanically suitable termination for the positive electrode. To overcome this problem a carbon rod with a large surface area is inserted in the cathode mix. The carbon rod is a good electrical conductor, is chemically inert, and its large surface area provides a low-resistance conducting path. The carbon rod is also porous enough to permit the escape of gas accumulating in the cell, but does not permit leakage of the electrolyte material.

11.2.3 Zinc Chloride (ZnCl)

The ZnCl cells are moderately priced, have good shelf life, and have good voltage characteristics at high current drain. They will give two to three times the service life of CZn cells in such devices as toys, camera flash units, and pocket calculators.

Because of their ability to operate at high electrode efficiencies, the current output of zinc chloride cells is usually higher, and they will operate at high current drains for longer periods of time than will Lelanché cells of the same size. The construction of the zinc chloride cell (Figure 11.3) is similar to that of the conventional carbon-zinc Lelanché cell, but the electrolyte system is different. In a zinc chloride cell the electrolyte contains only the soluble salt, but a Lelanché cell also contains a saturated solution of ammonium chloride, along with the zinc chloride.

Without ammonium chloride in the electrolyte, ZnCl cells can give longer service at higher working voltages than can CZn cells. ZnCl cells are more tightly sealed than CZn primary cells and have a good shelf life despite the absence of ammonium chloride in the electrolyte. The tighter sealing also helps prevent bulk electrolyte leakage.

When all chemicals but zinc chloride are eliminated, electrode blocking by reaction products is avoided, and electrode polarization at high current densities is minimized. Because of the electrochemical reactions in the zinc chloride cell, water is consumed along with the electrochemically active materials, so that the cell is nearly dry at the end of its useful life. Also, the zinc chloride battery operates much better at low temperatures.

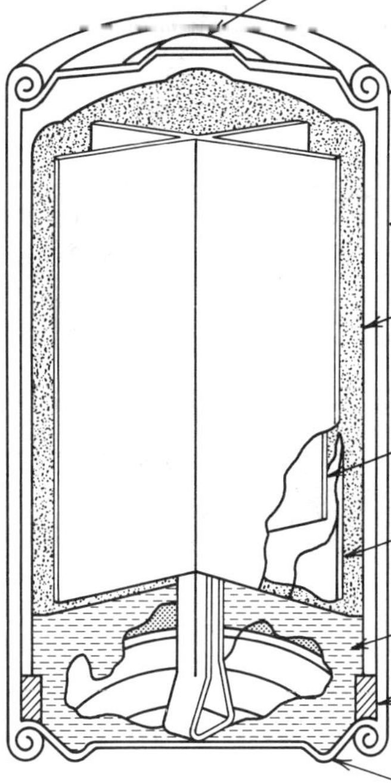

Figure 11.3 Zinc chloride cell.[69]

11.2.4 Zinc-Manganese Dioxide (Alkaline)

Alkaline cells find their greatest use in flashlights, cameras, tape recorders, and calculators. Occasionally they are used in portable radios, clocks, toys, pagers, and small appliances. When used in toys, alkaline cells last about seven times longer than CZn cells; in pocket calculators the ratio is about 3:1; and in cassette recorders about 5:1.

Alkaline cells differ from conventional carbon-zinc cells in that their electrode structure and in their electrolyte material, which is an aqueous solution of potassium hydroxide saturated with zinc oxide (Figure 11.4).

The negative electrode is a tightly packed mixture of manganese dioxide and graphite. The positive electrode is powdered zinc, heavily amalgamated, and packed as a loose powder at the center of the cell.

Because of a complicated construction and the use of more expensive materials, the alkaline battery is significantly more expensive than its CZn counterpart. However, for its premium price the alkaline battery delivers premium performance. It packs more power into a given

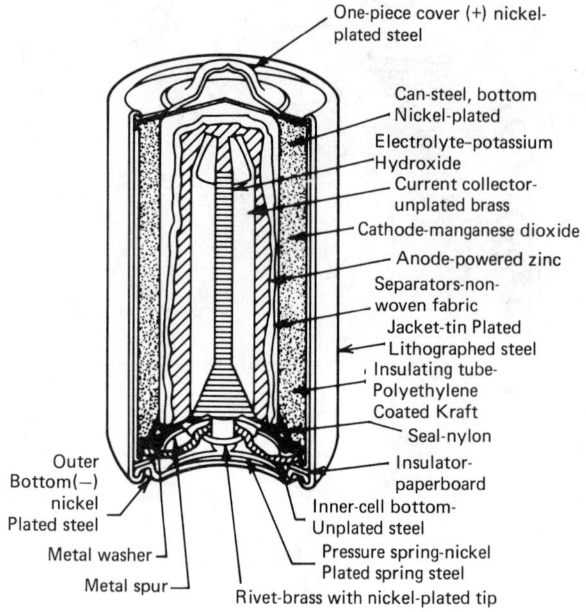

Figure 11.4 Alkaline cell.[69]

physical package than either the CZn or ZnCl types, can discharge at high rates with less significant voltage drop, has good shelf life and good low-temperature performance, and its ability to supply power is more independent of the duty cycle of the device it powers.

The basic alkaline cell has a 1.5-V open-circuit voltage. As the cell is discharged, the closed-circuit voltage of the cell gradually drops. This decrease in working voltage occurs at a slower rate than that of a comparable carbon–zinc cell. For a given size, alkaline batteries will have longer service life than will carbon–zinc batteries when used in similar applications.

Alkaline cells have relatively low internal impedance. The low internal impedance and rapid depolarization characteristics enable them to operate efficiently at high current drains, even in continuous-discharge applications, such as photoflash usage.

The alkaline cell has a very good shelf-life characteristic that approaches that of mercury cells. The shelf life of alkaline cells can be increased by storing them at low temperatures. Alkaline cells can also operate well at low temperatures.

11.2.5 Zinc Mercuric Oxide (Mercury)

Small equipment, such as electronic watches, exposure meters, cameras, hearing aids, and pacemakers, are powered by button-size mercuric oxide cells, are typically 0.125 in. high and 0.220-in. in diameter. These cells come in cylindrical, flat-pellet, and button configurations.

Mercury cells are divided into two basic types—one with a nominal voltage of 1.35 V, and the other with a 1.4-V rating. In general, the 1.35-V cells and batteries are recommended as voltage-reference sources for scientific and medical applications. The 1.4-V cells are used mainly in consumer applications where a flat voltage characteristic is not needed. Mercury cells offer a flat voltage discharge characteristic, good performance at high temperatures, good resistance to shock and vibration, and a shelf life of as long as 2.5 years.

In all these cells an ion-permeable barrier separates an anode, formed from high-purity amalgamated zinc, from the cathode of mercuric oxide and graphite. The ions of the electrolyte, which is a solution of alkaline hydroxide, act as carriers for the chemical action in the cell. To minimize corrosion the cell containers are made of nickel-plated steel.

In the mercury cell shown in Figure 11.5 the powdered-zinc electrode makes contact with a plated-steel cap that is insulated from the outer steel case. This arrangement makes the cap, or the terminal top, negative with respect to the steel casing. Consequently, the polarity of the mercury cell is the reverse of that of the carbon–zinc cell.

The basic mercury cells differ in their internal construction and external case design. However, electrochemically they are identical.

Because the mercury system is chemically balanced and highly efficient, the capacity of a mercury cell or battery is relatively constant for both intermittent and continuous discharge cycles within the recommended current drain. The average hours of service that can be obtained from a particular mercury cell or battery will

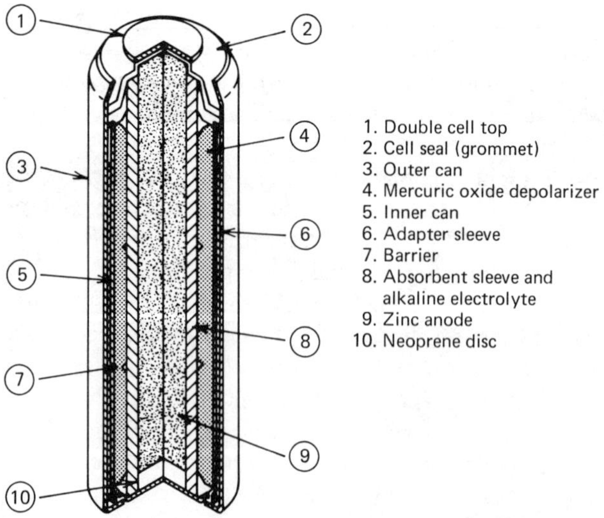

Figure 11.5 Cylindrical zinc mercuric oxide dry cell.

depend on the application and environmental conditions to which the battery has been subjected prior to and during use.

11.2.6 Silver Oxide

Although a battery system using monovalent silver oxide has about the same energy density as a mercury cell, its 1.5-V output is higher. This makes it desirable for light-emitting-diode readouts in digital watches, as well as cameras and hearing aids. However, chiefly because of material costs, the silver oxide cell is more expensive than the mercuric oxide cell. Unlike the packaging of the mercury cell, silver oxide cells are supplied chiefly in the button configuration.

A. ***Cadmium Silver Oxide Cells*** These cells employ silver-oxide as the positive electrode for high watt-hour capacity and cadmium negative electrodes for long operating life. Although cadmium silver oxide cells do not have as high a watt-hour capacity as do zinc silver oxide cells, they have longer operating life and more usage in low-current applications.

B. ***Zinc Silver Oxide Cells*** These cells, Figure 11.6, have greater current-handling capacity and higher watt-hour capacity than do most secondary cells. Because of their relatively high cost and shorter operating life, zinc silver oxide cells are not used as extensively as are the lead-acid and nickel-iron cells.

The zinc silver oxide cell is chemically similar to the alkaline cell. Zinc is used for the anode and potassium or sodium hydroxide for the electrolyte, depending on the intended application. Silver oxide acts as the depolarizer and cathode.

The silver oxide cell has an open-circuit voltage of 1.6 V. When the cell is discharged within its suggested current range the closed-circuit voltage of the cell will be approximately 1.5 V, as compared with 1.3 volts for mercury cells. The higher output voltage under load is a result of design that permits a high energy density. The cell also exhibits a relatively flat voltage discharge characteristic which contributes to long operating life.

The impedance of silver oxide cells is low and fairly constant and does not change appreciably until an end-of-life condition has been reached.

The average hours of service that can be obtained from a silver oxide cell will vary depending on the application, discharge cycle, and environmental conditions. But the silver oxide cell will deliver a high percentage of its useful capacity even if it is used to a relatively high end-point voltage of 1.1 V. Because of the large surface area of the anode, these cells perform well at low temperature. The shelf life of silver oxide cells is approximately 1 year.

11.2.7 Zinc–Air

An exceptionally high power density, at least two to five times that of most other primary sources, is the main advantage claimed for zinc–air batteries. Zinc–air batteries have a long shelf life. However, they are relatively untried and may be adversely affected by high humidity and high carbon dioxide concentrations.

Zinc–air batteries find most of their uses in nonconsumer areas. Although relatively expensive, typical applications are railway and marine signaling, telephone and telegraph circuits, and mining and highway equipment power.

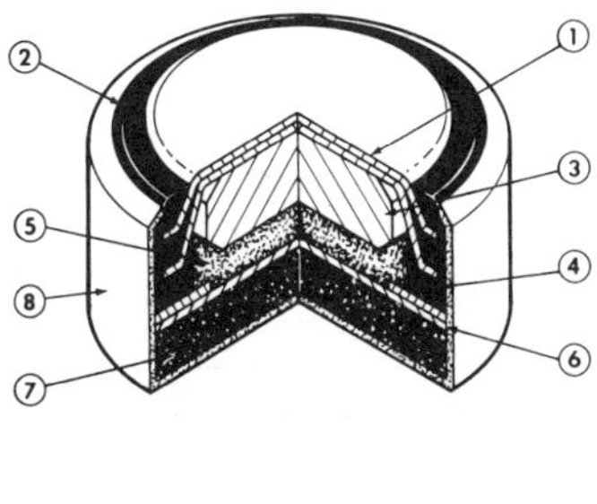

1. Double Cell Top
2. Cell Seal (Grommet)
3. Zinc Anode
4. Insulating Ring
5. Absorbent and Alkaline Electrolyte
6. Barrier
7. Silver Oxide Depolarizer
8. Cell Container

Figure 11.6 Zinc silver oxide dry cell.

Table 11.5 Lithium-Cell Systems[68]

Reactants	Open Circuit (V)
Soluble Reactants	
Thionyl chloride ($SOCl_2$)[a]	3.6
Vanadium pentoxide (V_2O_5)	3.4
Sulfur dioxide (SO_2)	2.9
Molybdenum trioxide (MoO_3)	2.9
Solid Reactants	
Copper fluoride (CuF_2)	3.4
Silver chromate (Ag_2CrO_4)	3.0
Copper sulfide (CuS)	2.2

[a]Inorganic solvent; all others organic.

Table 11.7 Primary-Cell Energy Densities[68]

Primary-Cell Systems	Nominal Voltage	D-Cell Weight (oz)	Energy Density W-hr/lb	Energy Density W-h/in.^{3}
Lithium/inorganic electrolyte	3.4	3.5	190	13
Lithium/organic electrolyte	2.8	3.3	140	8
Silver oxide	1.5	–	60	8
Magnesium	1.8	3.5	55	4
Mercury	1.35	5.9	45	6
Manganese–alkaline	1.5	4.5	35	3
Carbon–zinc	1.5	3.3	25	1.5

11.2.8 Lithium[71]

Different cathode systems develop different open-circuit voltages, but even the lowest lithium-cell voltage is higher than that of other primary-cell systems. (See Table 11.5.)

Lithium cells have greater energy density than do other primary units. And in addition to lithium's high electrode potential, its light weight contributes to the high-energy density of lithium batteries, as shown in Tables 11.6 and 11.7.

Lithium batteries operate over an extremely wide temperature range of from −53 to +71°C. Because the electrolyte is nonaqueous, conductivity at cold temperatures is far superior to that of other primary cells. Even at −55°C some lithium systems will provide about 50% capacity. (See Table 11.8.)

To some applications the long shelf life of lithium batteries makes a significant difference, as in many uses

Table 11.6 Lithium Battery Comparison[71]

System	Lithium–Iodine P2VP[a]	Lithium–Manganese Dioxide	Lithium–Sulfur Dioxide	Lithium–Carbon Fluoride	Lithium Thionyl Chloride	Lithium Lead Iodide	Nickel–Cadmium[d]	Lead–Acid[d]
Energy density (W-hr/cm^3)	0.60	0.44	0.45	0.33	0.50	0.47	0.09	0.08
Seal	Hermetic	Crimp	Crimp or hermetic	Crimp	Crimp or hermetic	Hermetic	Crimp	Crimp
Voltage (V)	2.8	2.9	2.9	2.8	3.6	1.9	1.2	2.0
Electrolyte	Solid salt	Liquid organic	Liquid organic	Liquid organic	Liquid inorganic	Solid salt	Liquid inorganic	Liquid inorganic
Self-discharge rate	Low	Moderate	Moderate	Moderate	Moderate	Low	High	High
Internal resistance	High	Moderate	Moderate	Moderate	Moderate	High	Low	Low
Separator	Not required[b]	Plastic	Plastic	Plastic	Plastic	LiI[a]	Paper/plastic	Paper/plastic
Initial unit cost	Moderately high	Moderate	Moderately high	Moderate	Moderately high	Moderately high	Moderate	Low
Relative life-cycle costs (5+ years)	Low	Low to moderate	Low to moderate	Low to moderate	Low to moderate	Low to moderate	High	High

[a]Poly-2-vinylpyridine.
[b]Chemical reaction of lithium anode and iodine P2VP cathode forms its own self-healing lithium iodide separator.
[c]Added during fabriaction process; not self-healing.
[d]Secondary batteries require periodic recharging. Energy density shown is based on small sealed cells.

Table 11.8 Low-Temperature Performance[68]

Temp. (°C)	Lithium V_2O_5	Lithium SO_2	Mer-cury	Magne-sium	Alka-line	Carbon-Zinc
	Capacity at 20°C (%)					
- 7	88	96	0	58	15	5
-30	78	85	0	23	3	0
-40	73	60	0	0	0	0

a device is "on the shelf" longer than it is in use. Alkaline systems are better, but lithium systems far exceed any other primary, with 10 years of life projected at +20°C. Also, at elevated temperatures lithium batteries have a substantial advantage, as shown in Table 11.9.

Besides their use as primary power sources in portable equipment applications, lithium batteries can also be used as standby power sources for volatile semiconductor memories. Their small size, very low discharge rate, and high power density make lithium units suitable for use in microcomputers and other small computer-type systems.

The lithium battery, with its spirally wound core packed in a steel case, Figure 11.7, looks like any commerical D cell, but the performance differs greatly. One standard lithium D cell can power a flashlight continuously for 18 hr, whereas two comparable zinc-carbon cells would last much less time.

Lithium cells operate efficiently at high discharge rates, which is needed for applications having high current drain. A lithium D cell's energy with a 1-A current drain at +21°C is equivalent to four mercury-zinc D cells or 30 carbon-zinc D cells. A comparison of a lithium cell's flat voltage/time profile with other types of cells shows that the lithium unit is clearly superior.

11.3 SECONDARY BATTERIES/CELLS

In secondary cells the chemical reactions that produce electrical energy are reversible. The materials used in most commercial secondary cells are lead, cadmium, iron, and zinc for the negative electrode and lead dioxide, nickel oxide, silver oxide, and manganese dioxide for the positive electrode. This major group of cells may be classified into the following subgroups: lead-acid, nickel-iron, nickel-cadmium, zinc silver oxide, cadmium silver oxide, and zinc manganese-dioxide cells.

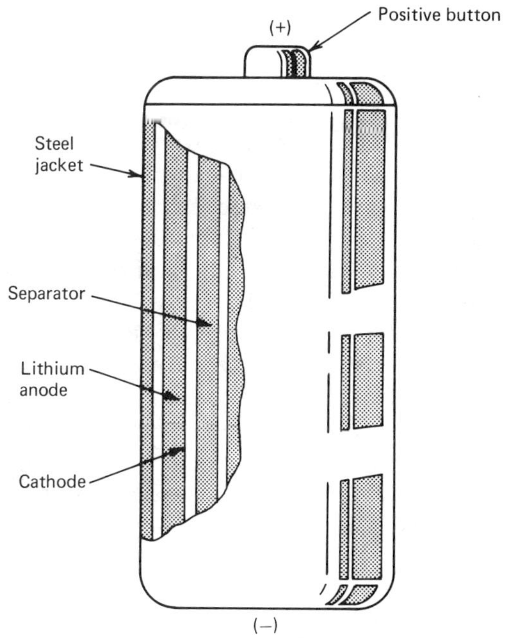

Figure 11.7 Lithium cell.

The proliferation of battery types makes it possible to closely match a battery to a battery-operated device. On the other hand, the great number and variety of batteries available can complicate the selection process. Matching a battery to a battery-operated device involves many possible choices. The first is between primary (throwaway) and secondary (rechargeable) batteries.

The nature of the reaction and physical changes within the battery determines whether or not it can be recharged. The useful reaction in a primary battery is one-way. When the chemical reaction has gone to completion and the battery is fully discharged it must be discarded.

However, the chemical reaction in a secondary battery

Table 11.9 Storage Life[68]

Storage Temp. (°C)	Lithium	Mercury	Magnesium	Alkaline	Carbon–Zinc
20	10 years +	3–4 years	5–7 years	2–3 years	1–2 years
65	12 months +	4 months	7 months	2 months	1.5 months

is reversible. By reversing current back into a discharged secondary battery, it is possible to restore the chemicals to their original charged state.

The number of charge/discharge cycles a secondary battery can undergo depends on many factors. For example, automotive lead-acid batteries can be recharged several thousand times because each cycle is shallow and the batteries are rarely fully discharged.

Sealed nickel-cadmium and sealed lead-acid secondary batteries are also used for consumer applications. Both are used in small emergency lights, alarm systems, cassette recorders, and so forth, and offer good sealed life. Sealed nickel-cadmium secondary batteries can be recharged up to 1000 times and sealed lead-acid secondary batteries up to 500 times.

Table 11.10 lists the characteristics of the major types of secondary batteries. By far, nickel-cadmium batteries are used in most applications. However, lead-acid cells are challenging them in many applications.

Silver-cadmium and silver-zinc secondary batteries are more suitable than both the nickel-cadmium and lead-acid units in those applications that require a small, lightweight source of high power. The exceptional-high-power densities of both silver cadmium and silver-zinc batteries makes them well suited for powering airborne electronics, portable television cameras, and electronics for many military applications. However, they also cost considerably more than nickel-cadmium and lead-acid batteries.

The silver-cadmium battery is the most expensive secondary battery, followed by silver-zinc, which is slightly higher than a nickel-cadmium system. The least expensive secondary battery for powering electronic equipment is the gelled-electrolyte lead-acid cell, largely

Table 11.10 Typical Characteristics of Secondary Batteries[67]

Common Name	Nickel-Cadmium (Sealed)	Silver-Cadmium	Lead-Acid (Gelled Electrolyte)	Lead-Acid (Gates cell)	Silver-Zinc
Electrochemical System	Nickel-Cadmium	Silver-Cadmium	Lead Acid	Lead-Acid	Silver-Zinc
Voltage Per Cell	1.2	1.1	2.0	2.0	1.5
Negative Electrode	Cadmium	Silver Oxide	Lead Calcium/Antimony	Sponge Lead	Silver Oxide
Positive Electrode	Nickel Hydroxide	Cadmium Oxide	Lead Dioxide	Lead Dioxide	Zinc
Electrolyte	Aqueous solution of Potassium Hydroxide	Potassium Hydroxide	Gelled solution of Sulphuric Acid	Sulphuric Acid	Potassium Hydroxide
Cycle Life	300 to 2000	150 to 300	200 to 400	300 to 500	25 to 100
Typical Commercial Service Capacities	20 mAh to 10 Ah	100 mAh to > 50 Ah	30 mAh to 25 Ah	2.5 Ah to 25 Ah	50 mAh to > 50 Ah
Energy Density:					
Watt-hours/Lb	12 to 16	22 to 34	8 to 10	10 to 12	40 to 50
Watt-hours/in.^{3}	1.3 to 1.7	1.5 to 2.7	1.1	1.4	2.5 to 3.2
Discharge Curve	Flat	Flat	Sloping	Sloping	Flat
Temperature Range:					
Storage	−40°F to 140°F	−85°F to 165°F	−40°F to 100°F	−40°F to 100°F	−85°F to 165°F
Discharge	−40°F to 140°F	−10°F to 165°F	−76°F to 140°F	−76°F to 150°F	−10°F to 165°F
Charge	32°F to 113°F	32°F to 115°F	32°F to 113°F	−40°F to 150°F	32°F to 115°F
Impedance	Low	Very Low	Low	Low	Very Low
Shelf Life	Fair	Fair to Good	Good	Good to Excellent	Fair to Good
Cost:					
Initial	Medium to High	High	Low	Low	High
Operating	Low	Medium to High	Low to Medium	Low to Medium	Medium to High

because of the relative abundance of its materials. In general, sealed nickel-cadmium cells are more economical when purchased in large quantities than other rechargeable types in the small size (AAA, AA, and C).

However, a silver-zinc cell with the equivalent energy density of 45 W-hr/lb provides four times the energy of a nickel-cadmium cell and five times the gelled-electrolyte lead-acid cell's energy density. At 22-34 W-h/lb, a silver-cadmium cell has slightly less energy density than does a silver-zinc unit, but its maximum life is three times as long as the typical 100 cycles for the silver-zinc cell.

At ratings of 1 A-hr and higher the lead-acid cell costs only one-third to one-fourth as much as a comparable nickel-cadmium cell. But a silver-zinc cell is somewhat more expensive than its nickel-cadmium counterpart with 4 A-hr of nickel-cadmium power costing more than the same amount of power from a lead-acid battery.

Neither the silver-zinc nor the lead-acid cell has the "memory effect" that can be a drawback for nickel-cadmium batteries that are operated at a certain low-level discharge level for short repetitive periods. In this situation, the battery becomes conditioned to that level of operation and "forgets" its design capacity, so that it delivers only the customary output when the demand for power is increased. However, nickel-cadmium cells supply a high rate of current, have a high-energy density, are available in a wide range of sizes, operate over a wide temperature range, and have a life expectancy of nearly 1000 cycles (as compared with 300 cycles for silver-cadmium cells).

11.3.1 Nickel-Cadmium[72]

Although available in either sealed or vented types, it is the sealed nickel-cadmium battery type, Figure 11.8, that finds greater application. This is because it is maintenance-free and can be used in any position without danger of electrolyte leakage.

Sealed nickel-cadmium batteries are designed with greater capacity at the negative (cadmium) electrode than at the positive (nickel hydroxide) electrode. During charging and overcharge, the positive electrode reaches full charge before the negative electrode, and oxygen formation occurs at the positive electrode. The electrode separator material is permeable to this evolving oxygen, so that it moves through the starved potassium hydroxide electrolyte-separator system and migrates to the negative plate. Here the oxygen and water at the negative-plate electrolyte interface and react to form hydroxyl ions.

This is the reverse of the reaction occurring at the positive plate. Because of the recombination of the oxy-

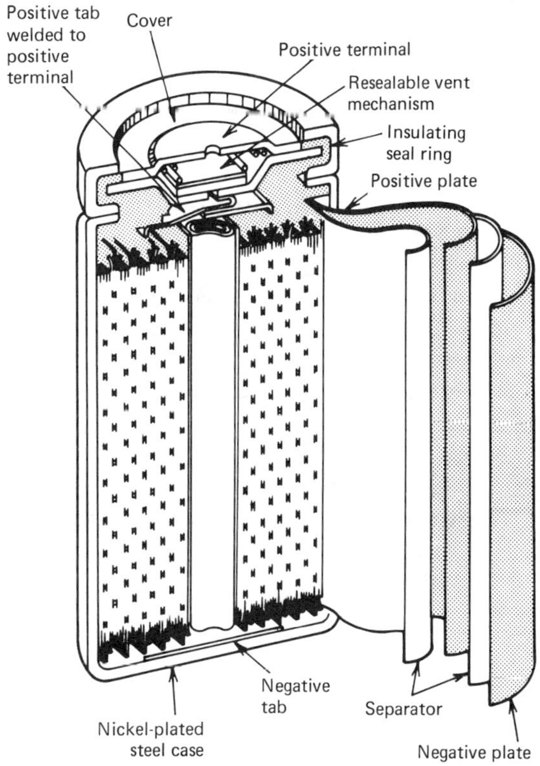

Figure 11.8 Nickel-cadmium cell.[69]

gen evolved, some special sealed nickel-cadmium batteries can be recharged in as little as 15 min.

Significant features of the rechargeable nickel-cadmium battery are:

A. ***Impedance*** Nickel-cadmium cells have a very low internal impedance and a high effective capacitance.

B. ***Cycle Life*** Nickel-cadmium cells have an excellent cycle-life characteristic. Normal cycle life ranges from 300 cycles up to 100 cycles, depending on cell size and construction.

C. ***Discharge*** During use the average voltage of a sealed nickel-cadmium battery will vary from 1.25 to 1.20 V, depending on the discharge rate. At normal discharge rates the characteristic is very nearly flat until the cell approaches complete discharge. Individual cells may be discharge down to 0 V.

Nickel-cadmium cells lose capacity when standing idle after charge. The rate of self-discharge is increased by elevated temperatures. Generally, the capacity loss is greater during the first month of nonuse. After this, the capacity decreases only gradually with age. Cylindrical cells lose their capacity at a more rapid rate than button cells. Cylindrical and button cells are not harmed if not used for long periods of time, but may require two or three cycles to regain full capacity.

Where discharge is carried below 1.0 V/cell a very rapid fall to 0 V occurs. When cells are connected in series one cell may reach zero volts before others because of slight differences in cell capacity. This can result in cells reversing polarity.

While polarity reversal usually causes no immediate damage to nickel-cadmium cells, repetitive reversal should be avoided for maximum possible life. Recharging will correct reverse polarity and restore full capacity. Frequent recharging is the best way to avoid polarity reversal.

D. ***Charging*** Constant-current charging is the recommended charging method. It is obtained by making the supply voltage greater than the battery voltage and by limiting the current flow with series resistance. For full or partially discharged cells, 1.4 times the capacity removed must be replaced.

Due to construction differences, overcharge affects cylindrical and button cells differently. Cylindrical cells can be overcharged at rated currents for extended periods with little effect. However, overcharge of button cells should be avoided, as permanent capacity loss can result.

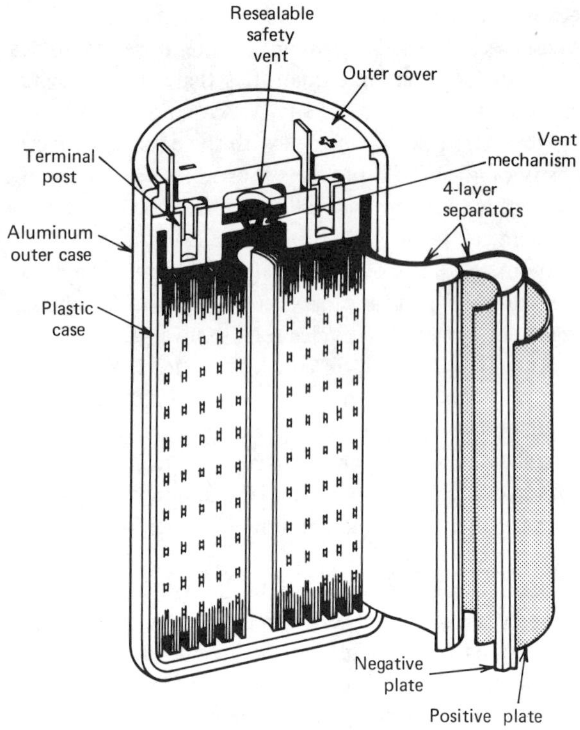

Figure 11.9 Spiral-wound sealed lead–acid battery.[73]

11.3.2 Sealed Lead-Acid[73,74]

A variation of the conventional lead-acid battery, the sealed lead-acid battery uses the same chemical system, but the electrolyte is held in a carrier that prevents spilling if the battery is tilted or used in a horizontal position. The spiral-wound sealed lead-acid battery, Figure 11.9, offers many of the performance advantages of nickel-cadmium rechargeable batteries, such as rechargeability, low internal impedance, and charge/discharge capability in any position.

The construction of the sealed lead-acid cell is similar to that of nickel-cadmium units. Two long, thin plates with separators between them and spiral wound with special construction allow for the recombination of the oxygen generated during overcharge. The plates are built on a formed lead grid loaded with lead dioxide (positive plates). These batteries also have both positive and negative space terminals on top, rather than the familiar axial design of other batteries.

The nominal voltage of the sealed lead-acid battery is 2.0 V, which is higher than that for nickel-cadmium batteries. They are also available in a D size with 2.5 A-hr capacity and a X-size with 5.0 A-hr capacity.

In cyclic applications, the sealed lead acid battery will deliver up to 500 cycles. Although this is less than for the nickel-cadmium cell, it is often compensated for by a lower price.

The sealed lead-acid battery will remain charged for considerably long periods, with a room temperature discharge rate of approximately 6%/month. However, the battery should not be allowed to remain discharged for extended periods of time, or it may not remain capable of being recharged. The battery can be recharged in 16–20 hr, but for the longest life recharge times up to 48 hr are recommended.

Two types of this battery exist. The more recently developed smaller-size batteries can replace rechargeable nickel-cadmium units in some instances. A major use of the smaller sealed lead-acid battery is to provide uninterrupted power to large computer systems. Also, the low impedance of these cells extends their high-rate discharge capabilities, making them ideal for maintaining clean emergency power.

The older type of sealed lead-acid battery uses a gelled electrolyte. Its significant features are:

A. ***Construction*** There is no liquid of any kind to splash around or ooze out; thus the battery remains completely sealed unless it is relieving internal pressure. Therefore, there is no need to check liquid levels or to perform other routine maintenance techniques. Also, the sealed construction does not require special handling precautions or expensive containers.

B. *Size and Weight* A high power-to-weight ratio makes the battery ideal as a self-contained portable power pack for transportable equipment.

C. *Temperature Range* A fully charged battery will operate over a temperature range of from -76°F (-60°C) to $+140^\circ$F ($+60^\circ$C). Batteries are rated at $+68^\circ$F ($+20^\circ$C) with capacity increases above this temperature and decreases below this temperature.

D. *Types* The cell is available in two types—one for float/standby service where maximum life in terms of years is desirable and another for cycle service where the battery is expected to be discharged frequently.

E. *Memory* Some applications require a battery to undergo short repetitive periods of use that consume only a small portion of the total battery capacity. Under these cyclical conditions, sealed nickel-cadmium batteries, for instance, forget their actual capacity and become conditioned to deliver only the small amount of power required, that is, they have a "memory." Therefore, they fail when called upon to deliver full power.

The sealed lead-acid battery does not exhibit this "memory" characteristic. Thus the battery delivers full rated power when needed, regardless of the previous use history.

F. *Discharge* High discharge currents are possible due to the low internal resistance of the sealed lead-acid battery. This allows a small battery to be used in applications wherein the average current drain is high. Also, there is no danger of permanent cell reversal if the battery is totally discharged due to an extended power outage or because the equipment accidently is left "on."

G. *Charging* When completely recharged, the battery is capable of 200-500 charge-discharge cycles. The number of possible cycles increases proportionately to the amount of capacity removed on each cycle.

Reserve electrolyte allows the battery to be easily recharged even after complete total discharge. No complicated two-step charging is required and no discharge-voltage cutoff is required. Also, having reserve electrolyte allows the battery to gas without affecting the life of the battery.

H. *Shelf Life* Gelled-electrolyte batteries lose very little capacity when stored because of the use of special calcium grids. The storage loss ranges from 2 to 3%/month at room temperature, 10 to 12% at $+95^\circ$F, and less than 0.5% at 0°F.

I. *Venting* Self-sealing vents prevent high pressures from occurring during charging and also prevent air from entering the battery during discharging. Excessive pressure, such as that caused by charger failures, will not occur because of the low relief pressure required to open the vent.

11.3.3 Silver-Zinc/Cadmium

Many applications require a lighter, more powerful, power source than the nickel-cadmium battery. Such tasks are usually performed by silver-zinc and silver-cadmium battery packs. The capacity of these power systems ranges from 0.1 to 750 A-hr.

Silver-zinc and silver-cadmium batteries provide substantially more electrical energy for their size and weight than do other rechargeable batteries.

The silver-zinc rechargeable battery can provide as much as six times more energy per unit of weight or volume than other secondary battery types. Its extremely low internal resistance permits discharges at rates as high as thirty times the ampere-hour capacity rating, while a flat voltage characteristic alleviates voltage-regulation problems. Custom cells have been built with capacity ranging from 0.1 to 20,000 A-hr.

The silver-cadmium battery combines the high-energy and space-and-weight-saving characteristics of the silver-zinc battery with the long-life, low-rate characteristics and better resistance to overcharge of the nickel-cadmium battery. The silver-cadmium battery also lives long on the shelf in charged or uncharged conditions, has a level output voltage, and is mechanically rugged. Watt-hour capacity per unit weight and volume is two to three times greater than a comparable nickel-cadmium battery, and it retains its charge better.

11.3.4 Alkaline Rechargeable

Alkaline-manganese-zinc rechargeable dry cells and batteries are low-cost, rugged, portable power sources that combine the major advantages of both primary and secondary batteries. The batteries can be used in any equipment in which primary batteries can be used and, in AA-, C-, and D-size cells, are directly interchangeable with carbon-zinc and alkaline primary cells. Alkaline rechargeable cells and batteries are currently being used as power sources in portable radios, televisions, phonographs, tape recorders, cameras, flashlights, and cordless small appliances.

Although the total number of discharge-charge cycles that can be obtained from alkaline rechargeable cells is less than that obtained from nickel-cadmium cells, in most applications alkaline rechargeable cells offer a substantial savings in cost when compared to nickel-cadmium cells.

The basic alkaline-manganese-zinc rechargeable cell is composed of a zinc negative electrode and a manganese dioxide positive electrode. Potassium hydroxide is used as the electrolyte. The cell features a special internal construction and each cell is hermetically sealed. The required number of identical cells are assembled in series and sealed in a metal case to produce a battery of the desired terminal voltage.

The open-circuit voltage of a rechargeable alkaline dry cell is 1.5 V. Alkaline rechargeable cells have excellent charge retention characteristics when stored at room temperature. The cells also have a very large power reserve that may exceed twice their rated capacities early in their cycle life. In applications that require immediate power this reserve power can be used.

However, total cell life or the ability of the cells to be rechargeable may be sacrificed by "deep" discharging. Deep discharging occurs when the cells are discharged below 0.9 V/cell because an irreversible secondary chemical reaction takes place within the cells that adversely affects the ability of the cells to be recharged. As the cell is discharged to its rated capacity, the terminal voltage actually drops. During the early cycle life the terminal voltage may be 1.1–1.2 V/cell after a normal discharge period. As the cell is repeatedly cycled the terminal voltage will eventually measure between 0.9 and 1.0 V/cell after withdrawing its rated capacity. If the discharge period is shortened, however, many hours of operation can still be obtained before the terminal voltage will drop below 0.9 V/cell.

PART FIVE

Special-Function Components

CHAPTER TWELVE

Function Modules

12.1 OPERATIONAL AMPLIFIERS[75-77]

The operational amplifier is the most widely used analog subassembly. However, the basis for choice, the subtleties of using operational amplifiers (op amps) in circuits for best results, especially in precision measurement and control, and the varieties of possible applications are understood in varying degrees by users.

12.1.1 Application Groups

Operational amplifiers in seven groups are described below to help enable the user to determine and select the best operational amplifier available for a design requirement.

A. ***General Purpose*** General-purpose op-amps are suited for a wide variety of applications. They give moderately good performance over a wide range of parameters at moderate cost. This applications group contains both FET and bipolar-input models with frequency responses from 0.5 to 1.5 MHz and offset voltages as low as 1 mV.

B. ***Low-Temperature Drift*** Low-temperature drift amplifiers are best suited for applications where accuracy must be preserved over a substantial temperature range. These amplifiers are optimized to minimize the initial input offset voltage and input offset voltage change with temperature. Input offset drifts from 0.1 to 10.0 μV/°C are available within this group. Chopper-stabilized op amps represent the best available in overall accuracy and long-term stability.

C. ***Low-Bias Current*** Low-bias current op amps consist of a group of varactor diode and FET input designs. This group includes amplifiers with input bias currents from 0.01 pA to 1.00 nA. Applications with large feedback resistances or large source resistance (long time constants, integrators, current sources, and so forth) and buffer applications will benefit from the use of low-bias current amplifiers.

D. ***Wide Band*** Wide-band op amps have bandwidths greater than 10 MHz. This group also contains fast-settling and high slew-rate amplifiers. These amplifiers reduce the phase errors at high frequencies and accurately reproduce complex waveforms. These amplifiers are well suited for pulse, video, fast-settling, and multiplexing applications.

E. ***High Voltage*** The amplifiers in this group are designed to provide large output voltage swings and to operate on wide ranges of supply voltage. Output voltages greater than ±10 V and up to ±145 V are available in this applications group (up to 290 V, single supply). These amplifiers provide good frequency response and performance in other parameters. Most models have electrically isolated packages and automatic thermal sensing and shutdown. Most units have FET inputs to minimize bias-current errors when the amplifier is used with the large resistances usually found with high-voltage amplifiers.

F. *High Current* These amplifiers provide output currents from ±10 mA to ±2 A (±5-A peak). They are used with small load resistances, coax-cable impedances, and with power-booster applications. Many units have self-contained thermal sensing and shutdown to automatically protect the amplifiers from overheating and damage. Often these units have electrically isolated packages.

G. *Unity-Gain Buffer (Power Booster)* Unity-gain buffer amplifiers have a wide variety of applications. They are used to boost the output-current capability of another amplifier, to buffer an impedance that might load a critical circuit, or to be an input-impedance converter from an input that must not be loaded. These amplifiers may also be used inside the feedback loop of another op amp to form a current-boosted composite amplifier.

12.1.2 Selection

In general, the objective of amplifier selection should be to choose the least expensive device that will meet the physical, electrical, and environmental requirements imposed by the application. This suggests that a general-purpose amplifier will be the best choice in all applications where the desired performance can be met. Where this is not possible, it is usually because of limitations encountered in two areas—bandwidth requirements and/or offset and drift parameters.

A. *Gain Bandwidth* Although all selection criteria must be met simultaneously, determination of the bandwidth requirement is a logical starting point. Where DC information is required and where frequent requirements are relatively modest (full power response below 100 kHz, unity gain of less than 1.5 MHz), other criteria will probably influence the final choice. It is important, however, to choose an amplitude with which an adequate value of loop gain is assured (at frequency of interest) to obtain the desired accuracy.

B. *Offset and Drift* In the majority of op amp applications final selection is determined by the DC offset and drift characteristics. To undertake amplifier selection in these cases it is necessary to translate the requirements as follows, assuming that bandwidth requirements have been established at this point:

1 What input impedance must the circuit present to the signal source?
2 How much drift error can be tolerated?

C. *Current Amplifier* In selecting a current amplifier the most important consideration is current noise, and bias and current drift. To make a proper selection, pick an amplifier with an error current over the operating temperature that is small compared to the signal current. Do not overlook current noise, which may be more important than current drift in many applications.

D. *Steady State* Steady-state applications involve amplifying or otherwise manipulating continuous sinusoidal, complex, or random waveforms. In these applications the significant issues in choosing an amplifier are as follows:

1 Is DC coupling required?
2 What closed-loop gain and bandwidth are required?
3 What loop gain is required or alternatively what gain stability, output impedance, and/or linearity are necessary?
4 What full-power response and/or slew rate are required?

12.2 INSTRUMENTATION AMPLIFIERS

An amplifier circuit optimized for performance as an instrumentation amplifier gain block has high input impedance, low offset and drift, low linearity, stable gain, and low effective output impedance. It is commonly used for applications that capitalize on these advantages. Examples include transducer amplifiers for various types of transducers, such as strain-gauge bridges, load cells, thermistor networks, thermocouples, current shunts, biological probes, weather gauges. Other applications include recorder preamplifiers, multiplexer buffers, servo-error amplifiers, current sensors, signal conditioners in process-control and data-acquisition systems, and in general measurements of small differential signals riding on common-mode voltages.

The small size, low cost, and high performance of these amplifiers offer an attractive approach for data-acquisition applications assigning a fixed-gain amplifier to each transducer and locating the amplifier physically near the transducer. This approach largely eliminates common-mode noise pickup problems, since a high-level signal (rather than a low-level transducer signal) is then retransmitted to the data-gathering station. The result is a higher signal/noise ratio at the output. Using one amplifier per point may well be more economical, as well as offering better performance and flexibility than the approach using low-level multiplexers.

12.3 ISOLATION AMPLIFIERS

The isolation amplifier (or isolator) has an input circuit that is galvanically isolated from the power supply and the output circuit. Isolators are intended for applications requiring safe, accurate measurement of DC and low-frequency voltage or current in the presence of high common-mode voltage (to thousands of volts) with high CMR, line-receiving of signals transmitted at high impedance in noisy environments, and for safety in general-purpose measurements where DC and line-frequency leakage must be maintained at levels well below certain mandated minima. Principal applications are in electrical environments of the kind associated with medical equipment, conventional and nuclear power plants, automatic test equipment, and industrial process-control systems.

12.3.1 Applications

In general, most applications can be broadly categorized into the following four types:

A. Amplifying and measuring low-level signals in the presence of high common-mode voltages.

B. Breaking ground loops and/or eliminating source ground connections. The isolation amplifier provides full floating input, eliminating the need for connections to source ground, and thus allow two-wire hookup to the signal sources.

C. Providing an interface between patient-monitoring equipment and the transducer/devices that may be in physical contact with the patient. Such applications require high isolation voltage levels and very-low leakage currents.

D. Providing isolation protection to electronic instruments/equipment. Large common-mode voltages occasionally cause hazardous electronic faults. Low leakage currents and high isolation voltage capability of isolation amplifiers help protect instruments against damage caused by such faults.

12.3.2 Selection

When one or more of the following conditions/requirements are present in an application, an isolation amplifier will generally be the right choice as a signal conditioning device when:

A. Ohmic isolation between the signal source and the output is a requirement with isolation impedance between the input and the output greater than 10 MΩ.

B. Excellent common-mode noise and voltage rejection is a requirement (CMR greater than 100 dB).

C. It is necessary to process signals in the presence of, or riding on, high common-mode voltages (CMV greater than 10 V).

12.4 ANALOG CIRCUIT FUNCTIONS

Analog circuit act as building blocks with which to perform a variety of instrumentation, computation, and control functions. They provide a broad range of versatility and proven, ready-to-use computational function circuits for developing simple or complex systems. The analog circuit functions include multipliers, dividers, multifunction converters, true RMS/DC converters, logarithmic amplifiers, voltage and window comparators, peak detectors, precision oscillators, and filters. The multifunctional converter also provides multiply, divide, square, square root, exponentiate, roots, sine, cosine, arctangent, vector magnitude RMS/DC, and logarithmic amplifier functions.

The availability of these relatively complex functions as precise, versatile, easy-to-use, low-cost building blocks has broadened the scope of practical analog-circuit systems and greatly simplified analog-circuit designs. The names of most analog-circuit functions are self-explanatory and describe the main functions they perform. These functions are used mostly for processing (handling) and/or conditioning of analog signals and usually, although not always, for simulation of algebraic and/or trigonometrically expressed analog computations.

12.4.1 Multipliers and Dividers

These devices accept analog voltages and multiply, divide, square, and/or square-root them, depending on device properties and connections.

A. ***Multiplication*** For two inputs V_x and V_y a multiplier will provide the output $E_{out} = V_x V_y / E_{ref}$, where E_{ref} is a dimensional constant usually of 10 V nominal value. Multipliers are used for modulation and demodulation, fixed and variable remote gain adjustment, power measurement, and mathematical operations in analog computing, curve fitting, and linearizing. If the inputs may be of either positive or negative polarity, and the output polarity is in a correct relationship for multiplication, the device is called a "four-quadrant" multiplier that reflects the four quadrants of the X,Y plane.

B. ***Squaring*** If $V_x - V_y = V_{in}$ a multiplier's output will be V_{in}^2/E_{ref}. A four-quadrant multiplier used as a squarer will have a positive output, whether or not V_{in} is positive or negative. Squarers are useful in frequency doubling, power measurement of constant loads, and mathematical operations.

C. ***Division*** For a numerator input V_z and a denominator input V_x an analog divider will provide the output $E_{out} = E_{ref}\ (V_z/V_x)$. If V_z may be of either positive or negative polarity the device is described as being a "two-quadrant" divider, and the output will reflect the polarity of V_z. Analog dividers are used to compute ratios, such as efficiency, attenuation, or gain. They are also used for fixed and variable remote gain adjustment, ratiometric measurements, and for mathematical operations in analog computing.

D. ***Square Rooting*** For a numerator input V_{in} and a denominator input E_o, the output fed back to the denominator input, the output of a divider is $E_o = E_{ref}\ (V_{in}/E_o)$; hence $E_o = \sqrt{E_{ref} V_{in}}$. A square-rooter works in one quadrant, but some devices need external diode circuitry to prevent latchup if the input polarity changes even momentarily. Square roots are used in vector and rms computation, to linearize flowmeters, and for mathematical operations in analog computing.

12.4.2 Logarithmic Amplifiers

These devices are complete, self-contained modules that provide output voltage proportional to the logarithmic or antilogarithmic (exponential) of an input quantity. These modules operate on the instantaneous values of the inputs from DC to an upper cutoff frequency below 1 MHz.

Log amplifiers in the log mode are useful for applications requiring compression of wide-range analog input data, linearization of transducers having exponential outputs, and analog computing that ranges from simple translation of natural relationships in log form, for example, computing absorbance as the log ratio of input currents, to the use of terms involving arbitrary exponents and multiterm products and ratios.

12.4.3 RMS Converters

These devices are high-accuracy true-RMS/DC conversion-integrated circuits and modules. Devices of this class compute the instantaneous square of the input signal, average it, and take the square root of the result, to provide a DC voltage that is proportional to the RMS of the input.

Excellent pretrimmed performance, improvable by optional trims, makes these devices ideal for all types of RMS instrumentation where amplitude measurements must be made with high accuracy, independent of wave shape.

Applications include noise measurement, for example, thermal noise, transistor noise, and switch-contact noise. True-RMS measurement is a technique that provides consistent theoretically valid measurements of noise amplitude, standard deviation, from different sources having different properties.

True-RMS devices are also useful for measuring electrical signals derived from mechanical phenomena, such as strain, stress, vibration, shock, expansion, bearing noise, and acoustical noise. The electrical signals produced by these mechanical actions are often noisy, nonperiodic, nonsinusoidal, superimposed on DC levels, and require true-RMS for consistent, valid, accurate measurements. RMS converters are also useful for accurate measurements on low-repetition-rate pulse trains having high crest factors (ratio of peak to RMS), and for measurements of the energy content of SCR waveforms at differing firing angles.

12.5 DIGITAL-TO-ANALOG CONVERTERS (DACs)[78–81]

DACs accept the discrete binary outputs of computers and change them into continuous analog quantities. In general, they convert mathematical results into usable electrical quantities. D/A converters are used to generate and modulate waveforms, stimulate devices under test, drive motors, or display information.

12.5.1 Basic Elements

Most digital-to-analog converters consist of three major blocks—a precision reference, a set of resistors forming a ladder network, and switches that connect or disconnect the resistor ladder and the reference. Some DACs also include an output amplifier that buffers the current from the ladder and interfaces it to the circuits at the DAC's output.

A. The reference that biases the resistor ladder is usually internal to the converter. In some cases, however, it is supplied by a precision voltage or current already available elsewhere in the system. Internal DAC references usually consist of a temperature-compensated diode buffered by an amplifier that maintains the Zener's optimum operating point.

B. The resistor network or ladder is made up of weighted values that assign a particular binary weight to each of the switches. It directs a portion of the energy from the reference to the analog output determined by the state of the computer-controlled switches.

C. The switches connect parts of the ladder to the reference, causing the output to change. Usually each individual switch is controlled by one digit from the input.

12.5.2 Applications

D-to-A converters are useful in any system where both digital and analog signals are present. Some of the more common applications include:

A. Data processing output interface; driver for displays, plotters, and so on.

B. Programmable power supply or function generator in automatic test equipment.

C. Tool interface in numerically controlled machining.

D. Interface for automatic process control to control temperature, flow rates, etc.

E. Digital communications; digital-to-audio interface.

F. Feedback network in A-to-D converters.

12.5.3 Selection

Before specifying a particular converter, consider the following:

A. ***Description of the Digital Input***

1 Is the digital code a natural binary format, or another such as Binary-Code-Decimal (BCD), Gray Code, Two's Complement, or another code?

2 Which logic family must the digital input be compatible with? That is, what are the voltage levels representing logic "1" and logic "0"? Should the DAC be compatible with CMOS, TTL, ECL, or another logic type?

3 How fast will the input change, and how fast must the DAC respond to the input change? This will determine the speed requirements of the internal DAC switches and the settling time of the output circuit.

4 What resolution is required? That is, how many individual digital inputs or bits are needed? This will establish the number of increments into which the analog output can be divided.

B. ***Error Budget*** How much of the error in the system transfer function can be allotted to the D-to-A converter? This figure is usually on the same order of magnitude as 1/resolution or $1/(2^n - 1)$. It determines the constraints placed on errors due to gain, offset linearity, power supply rejection, and parameter temperature drift.

C. ***Environmental Conditions***

1 What is the operating temperature over which the DAC must perform? Most commercially available D-to-A converters are specified to operate over a temperature range of from either 0 to +70°C, -25 to +85°C, or -55 to +125°C.

2 The storage or nonoperating temperature range for the DAC should also be considered. DAC prices can increase dramatically for wide temperature-range models, so specify the narrowest range possible.

3 Is there a need for a hermetically sealed package? Most integrated-circuit converters employ thin-film nichrome or diffused resistors that can degrade significantly in the presence of moisture.

D. ***Power-Supply Constraints*** Conventional DACs require ±15 and +5 VDC while dissipating less than 500 mW of power, although this can vary widely. Low-power CMOS converters can dissipate less than 20 mW, while fast-settling DACs may require in excess of 10 W. In most DAC types there is a direct tradeoff between power dissipation and speed.

E. ***Physical Size*** How much real estate can be devoted to the D-to-A converter? Converters using integrated-circuit technology range in size from 14-pin dual-inline packages (DIPs) to 32-pin triple-wide DIPs. Fast-settling converters can be in modular encapsulated packages of 2 in. X 4 in. X 0.4 in. or larger.

F. ***Analog Output Description*** Most voltage-output DACs have output ranges of ±10, ±5, ±2.5, 0–5, or 0–10 V. Current output DACs usually have ±1 or 0 to 2 mA outputs.

1 What should be the characteristics of the analog output?

2 What is the required dynamic range?

3 Must it be unipolar or bipolar?

4 What must be the settling time and drive capability?

12.5.4 Key Specifications

After taking the above considerations into account, select an appropriate converter after evaluating the manufacturer's key specifications.

A. ***Resolution*** An n-bit converter has a resolution of $1/(2^n - 1)$. Resolution can be expressed as the number of

Table 12.1 Maximum Available Resolution versus Binary Bits[75]

Binary Bits (n)	(2^n)	Equivalent Percent or Fraction of Range of Least-Significant Bit*		Residual $1 - \sum_{1}^{n} (\frac{1}{2^n})$
		Percent	ppm	
1	2	50.0	500 000.	0.5
2	4	25.	250 000.	0.25
3	8	12.5	125 000.	0.125
4	16	6.25	62 500.	0.062 5
5	32	3.125	31 250.	0.031 25
6	64	1.562 5	15 625.	0.015 625
7	128	0.781 25	7 812.5	0.007 812 5
8	256	0.390.625	3 906.25	0.003 906 25
9	512	0.195 313	1 953.13	0.001 953 13
10	1 024	0.097 656	976.56	0.000 976 56
11	2 048	0.048 828	488.28	0.000 488 28
12	4 096	0.024 414	244.14	0.000 244 14
13	8 192	0.012 207	122.07	0.000 122 07
14	16 384	0.006 104	61.04	0.000 061 04
15	32 768	0.003 052	30.52	0.000 030 52
16	65 536	0.001 526	15.26	0.000 015 26
17	131 072	0.000 763	7.63	0.000 007 63
18	262 144	0.000 381	3.81	0.000 003 81
19	524 288	0.000 191	1.91	0.000 001 91
20	1 048 576	0.000 095	0.95	0.000 000 95
21	2 097 152	0.000 048	0.48	0.000 000 48
22	4 194 304	0.000 024	0.24	0.000 000 24
23	8 388 608	0.000 012	0.12	0.000 000 12
24	16 777 216	0.000 006	0.06	0.000 000 06

*May be limited by noise and other uncertainties in actual circuit.

bits or as a percentage of full-scale range, as shown in Table 12.1.

B. *Nonlinearity* In most cases it is desirable to have a linearity error (nonlinearity) less than $\frac{1}{2}$ LSB, where the magnitude of the LSB (least significant bit) is determined by the resolution. 1 LSB = full-scale range/$(2^n - 1)$. Nonlinearity is the best measure of a DAC performance. Unlike gain and offset errors, it cannot be adjusted to zero when the DAC is installed in the circuit board.

C. *Differential Linearity* Ideally, as the digital input proceeds from one state to an adjacent state, the analog output should increment by one LSB. In fact, the output may act less than ideally, with resulting differential nonlinearity.

D. *Gain Error* Most converters require that the gain error be removed by means of external adjustments if the error is to be consistent with the accuracy of the DAC.

E. *Monotonicity* Monotonicity is especially important when the converter is used in a feedback loop. If the converter goes monotonic at a point in its transfer function, then the negative feedback will become positive, causing system instability at that point. D-to-A converter monotonicity is usually guaranteed over a particular temperature range. If it is not, the temperature drift and initial value of the differential nonlinearity must be known to determine the temperature at which the converter becomes monotonic.

F. *Settling Time* Settling time is a function of the magnitude of the step change, the size of the error band, the loading of the output, and a number of other parameters. Current-output DACs usually settle faster than do voltage-output DACs, since the effect of parasitic capacitance is less.

G. *Temperature Drift* Gain, offset, linearity, differential linearity, and most other DAC parameters will vary with temperature. In most cases the polarity of the drift cannot be predicted nor will it remain linear as temperature changes. The parameter drift is usually specified as an average value over a temperature range, although the definition can vary from one manufacturer to another.

12.6 ANALOG-TO-DIGITAL CONVERTERS (ADCs)

12.6.1 Basic Elements

A-to-D converters comprise several essential functions, namely, an analog section, a digital data-generating section, data outputs, and digital controls.

A. *Analog Section* The analog section requires a reference, a high-gain comparator, and either a D-to-A converter (successive approximations) or acontrollable integrator. The reference may be internal or external, fixed or variable, and of a specified polarity/sense in relationship to the analog input. In ratiometric conversion the reference is usually external and variable.

B. *Digital Data-Generating Section* In successive approximation types, this section consists of a discrete or integrated successive-approximations register (SAR), its controls, and inputs from the comparator and clock (which is on board; but in many cases permits external clock pulses, frequency adjustment, and/or control). In integrating types this section consists of the clock-pulse generator, the counter(s), the input from the comparator, and the associated controls.

C. *Data Outputs* Factors to consider include coding, resolution, overrange information, levels, format, validity, and timing.

1 Coding is usually binary, including jumper-connected offset-binary and/or two's complement for bipolar input signals. For some types, BCD is available with sign-magnitude for bipolar inputs. Output coding should

always be checked for digital polarity (positive-true or negative-true) of both magnitude and sign information.

2 The resolution, number of output bits, must be sufficient for the application. In addition, the specifications must be checked to ascertain that not only will all 2^n (binary) output codes be present (no missing codes), but they must all be present at any temperature in the operating range and related to the input with sufficient accuracy.

3 Nonlinear integration can cause the conversion relationship to become nonlinear. Successive approximation types have no way of determining overrange; they simply fill up. However, counter types roll over and put out a carry flag to signal overrange.

4 The data levels available at the converter output must be checked (TTL, low-voltage CMOS, high-voltage CMOS), as must the load-driving capability and fanout, and the supply conditions under which appropriate output levels will be furnished.

5 The available choice of output formats must also be as desired, that is, parallel, serial, byte-serial, and/or pulse train. If the converter is intended to communicate directly with a microprocessor data bus the output should have three-state capability, and parallel outputs must be enabled in bytes of 8 or fewer lines. If the output is serial it is usually NRZ (non-return-to-zero) and should be accompanied by a set of synchronized clock pulses.

6 A status, or busy, or EOC output changes the state to indicate when the data become valid. The exact nature of this transition should be specified as to polarity, timing, levels, and so forth. For serial data the exact relationship between the data and the synchronizing clock should be specified to indicate when each bit becomes valid and for how long.

7 In general, the timing of the whole conversion process must be clearly understood, especially if high speeds are necessary for either conversion or for communication with a processor, or both. The timing diagram on specification sheets are usually accompanied by adequate descriptions of the conversion process and specifications of the critical interface parameters.

D. *Controls* The functions, action (levels or edges), polarity, and timing of all control inputs and outputs should be clearly understood, as well as their loading characteristics and dependence on the supply. In addition to the essential start-conversion command input and a status output, various control commands may be available, such as clock inhibit, high- (low-) byte enable, status-enable, and short cycle (for speeding conversion at the cost of resolution).

12.6.2 Power Supplies

Appropriate power supplies should be made available that consider the logic levels and analog signals to be employed in the system. The appropriate degree of power-supply stability to meet the accuracy specifications should be employed. Any recommended external protection circuitry should be planned for.

In almost all cases, separate analog and digital grounds are required. The ground wiring should be such as to minimize digital interference with high-accuracy analog signals, while ensuring that a connection between grounds can always exist at one point, even if the "mecca" point is inadvertently unplugged from the system.

12.6.3 Application Checklist

The designer will generally require specific information in the following categories before proceeding to the selection process.

A. What is the accurate description of input and output? What is the analog signal range? What is the source or load impedance? What is the digital code needed (e.g., binary, binary offset, two's complement, BCD)?

B. What is the needed data throughput rate?

C. What are the control interface details?

D. What does the system error budget allow for the converter?

E. What are the environmental conditions (temperature range, time, supply voltage, and so forth) over which the converter should operate to the desired accuracy?

F. What is the analog input voltage range and to what resolution must the signal be measured?

G. What is the requirement for linearity error or relative-accuracy error?

H. To what extent must the various sources of error be minimized as environmental temperature changes?

I. How much time can be allowed in the system for each complete conversion?

J. How stable is the system power supply? What errors will result from power-supply terminal voltage variations of this order?

K. Is monotonicity important to this application or can the system tolerate a few missed codes?

L. What is the character of the input signal? Is it noisy, sampled, filtered, rapidly varying, slowly varying? What kind of preprocessing is to be, or can be, done that will affect the choice and cost of the converter.

12.7 VOLTAGE/FREQUENCY CONVERTERS

12.7.1 Voltage-to-Frequency Converters

Voltage-to-frequency converters (VFCs) convert analog voltage or current levels to pulse trains or squarewaves in a logic-compatible form at frequencies that are accurately proportional to the analog quantity. The output continuously tracks the input signal, responding directly to changes in the input signal. External clock synchronization is not required. V-to-F converters find applications in analog-to-digital converters with high-resolution, long-term high-precision integrators, two-wire high-noise immunity digital transmission, and digital voltmeters.

12.7.2 Frequency-to-Voltage Converters

Frequency-to-voltage converters (VFCs) perform the inverse operation of VFCs. They accept a wide variety of periodic waveforms and produce an analog output proportional to frequency. Combining adjustable threshold, gain, and output offset with low-linearity error, F-to-V converters offer economical solutions to a wide variety of applications where it is required to convert frequency to an analog voltage. Examples are motor-speed controllers, power-line frequency monitors, and VCO stabilization circuits. In analog-to-analog data transmission they convert serially transmitted data in the form of pulse streams back to analog voltage.

12.8 SAMPLE-HOLD AMPLIFIERS (SHAs)

The principal application for sample-hold amplifiers is to maintain an analog-to-digital converter's input constant during conversion, at a value representing the analog input at a certain precisely known time. The characteristics of the SHA are crucial to system accuracy and the reliability of the digital data, especially in 12-bit and/or high-throughput-rate applications.

A sample-hold amplifier has two modes of operation that are programmed by a digital control input. In the sample mode or track mode the output follows the input, usually with a gain of +1. When the mode input switches to hold the output of the SHA ideally retains the last value it had when the command to "hold" was given. It should retain that value until the logic input dictates "sample," at which time the output ideally jumps to the input value and follows the input until the next "hold" command is given.

12.9 FUNCTION MODULE TERMS AND DEFINITIONS

Absolute Maximum Differential Voltage The maximum voltage that can be applied between inputs without causing permanent damage to an amplifier.

Absolute-Value Circuit A circuit that produces a unipolar output signal equal to the magnitude or absolute value of a bipolar input signal.

Accuracy A measure of the deviation of the output level from its predicted value under any input combination.

Accuracy, Absolute The difference between the actual analog output and the output that is expected when a given digital code is applied to a converter.

Accuracy, Relative The deviation of the analog value at any code (relative to the full analog range of the device transfer characteristics) from its theoretical value (relative to the same range) after the full-scale range (FSR) has been calibrated.

Aperture Time The time required in a sample-and-hold circuit for the switch to open after the "hold" command has been given.

Common-Mode Rejection (CMR) The ability of a device to reject the effect of voltage applied to both input terminals simultaneously.

Common-Mode Rejection Ratio (CMRR) The ratio of the differential voltage gain of an amplifier to its common-mode voltage gain.

Common-Mode Voltage (CMV) An undesirable signal picked up in a circuit by both wires making up the circuit, with reference to an arbitrary "ground."

Common-Mode Voltage Gain The ratio of the output signal voltage, ideally zero, to the common-mode input signal voltage.

Comparator A device with two stable output states that signals if an input current or voltage has crossed a threshold. The threshold may be set by one or more other currents or voltages, either fixed or variable.

Conversion Speed The measure of how long it takes an A-to-D converter to arrive at the proper output code. It is the time between the edge of the convert command pulse that starts conversion and the rising edge of the end-of-convert signal that indicates that the conversion is complete.

Differential Linearity In A-to-D and D-to-A converters, the measure of linearity from one digital state to the next.

Input Bias Current The DC input current required at each input of an amplifier to provide zero output voltage when the input signal and input offset voltage are zero.

Instrumentation Amplifier A closed-loop differential-input gain block exhibiting high input impedance and high common-mode rejection, whose primary function is to accurately amplify the voltage applied to its inputs.

Isolation Amplifier A device that provides ohmic isolation, breaks ohmic continuity of an electric signal between the input and the output of the device so as to allow an input circuit to be referenced separately and independently of the output circuitry.

Isolation-Mode Rejection (IMR) The measure of an isolation amplifier's ability to reject common-mode input signals while transmitting the differential signals across the isolation barrier.

Least-Significant Bit (LSB) The input bit carrying the lowest numerical weight or the level shift associated with this bit, which is the smallest possible analog step.

Linearity The maximum deviation of the converter's transfer function from a straight line drawn through its end points.

Linearity Error The maximum difference between actual and "best-straight-line" theoretical output for all pairs of values with all other DC errors nulled.

Linearity Tempco The sensitivity of linearity to temperature over the specified range.

Logarithmic Amplifier An amplifier that develops an output voltage that is proportional to the logarithm of the input signal.

Most Significant Bit (MSB) The input bit carrying the highest numerical weight or the level shift associated with this bit.

Nonlinearity The maximum deviation from a best-fit straight line drawn between the end points of a converter transfer function.

Resolution An indication of the number of possible output levels, usually expressed as the number of input bits that the converter will handle.

RMS Converter A circuit that develops a DC output voltage equal in rms value to an input signal of arbitrary waveform.

Self-Modulation The maximum variation in the output caused by interaction between the signal amplitude and the circuits that accomplish the sample-to-hold mode switching.

Settling Time The total time measured from an input change to the time the output reaches its new value within a specified error band.

Slew Rate The maximum attainable large-signal rate of change of output voltage in either acquisition or tracking.

Successive Approximations A high-speed method of comparing an unknown against a group of weighted references.

Switching Time The time it takes for the switch to change from one state to the other ("delay time" plus "rise time" from 10 to 90%) excluding settling time.

CHAPTER THIRTEEN

Delay Lines

The delay line[82,83] is an extremely versatile device whose basic function is to introduce a time delay in transmitting an electrical signal. This function is utilized in many ways.

An ideal delay line would perform this function without distorting or altering the information represented by the electrical signal. Therefore, while the signal is in the process of being delayed, the line is said to be holding or "storing" the information, be it analog or digital.

Since the delay line is no exception to nature's rule that nothing is perfect, electronic amplification is used to compensate for inevitable losses. The combined result is a highly flexible delay or information-storage device that is compact, rugged, fast, economical (both in cost and in power consumption), reliable, readily reproduced, and applicable to a wide variety of useful applications.

This highly desirable delay phenomenon can be achieved either by purely electrical means, as in electromagnetic lines (lumped constant or distributed constant), or by various electromechanical devices grouped under the terms acoustic or ultrasonic delay lines.

13.1 ELECTROMAGNETIC DELAY LINES

Basically, an electromagnetic delay line is a device that uses the inherent characteristics of a transmission line to provide precise pulse-time delays. Adequate delays are obtained for most electronic circuit applications in miniature packages by special electromagnetic designs that yield relatively high values of both inductances and capacitances. The designs generally take the form of strip lines, distributed inductance/capacitance constants, or lumped constants. The last type, which is made up of a large number of single valued inductors and capacitors in a ladder network, provides the best performance characteristics, but is the most expensive. Distributed and strip lines are generally less expensive and should be selected if requirements permit.

To select a delay line, the circuit designer must first determine a set of device requirements, such as line characteristics and the range of practical values. A rule of thumb for selecting delay-line types is given in Table 13.1.

A further guide is given in Figure 13.1, which indicates the approximate areas covered by the three types of delay lines in the T_d versus T_d/T_r plane. The T_d/T_r ratio is a figure of merit for the delay line and has a bearing on the price (higher T_d/T_r ratio, higher price). Temperature coefficients may be very important in some applications; however, most applications can be satisfied with coefficients of 200 ppm/°C, which again will reduce costs.

Delay lines are available with adjustable delay times. In a lumped parameter line, this is done by switching over a number of taps. Distributed delay lines can be constructed for continuous adjustment by providing a

Table 13.1 Delay-Line Characteristics[82]

Characteristics	Value Limits	Select Type
Delay time, T_d	<10 nsec < 6 μsec 0.1 μsec to 200 msec	Strip Line Distributed Lumped
Delay-to-rise time ratio, T_d/T_r	<20 <10 <250	Strip Line Distributed Lumped
Characteristic impedance, Z_O	25 to 1000 ohms 100 to 2000 ohms 50 to 10,000 ohms	Strip Line Distributed Lumped
Temperature coefficient	>100 ppm/°C >150 ppm/°C > 50 ppm/°C	Strip Line Distributed Lumped

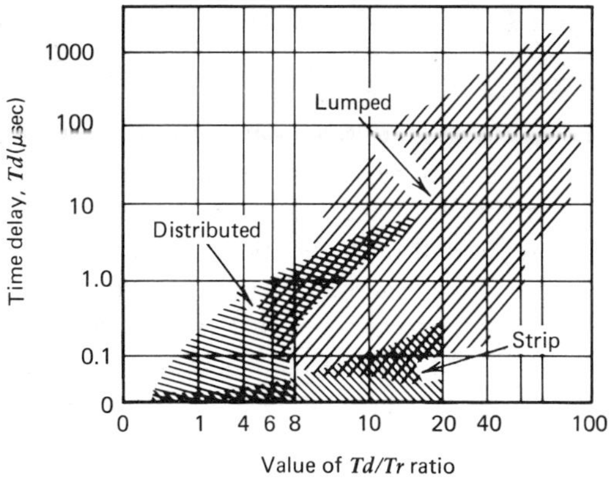

Figure 13.1 Delay line types.[82]

sliding contact over the length of the inductor. Strip lines are generally not adjustable.

13.1.1 Lumped Constant Delay Lines

A lumped parameter delay line is schematized in Figure 13.2, where n is the number of sections used. An input pulse inserted at the left will propagate through the line and appear at the lead terminals T_d microseconds later. For a given pulse delay T_d, the value of T_r goes down with an increasing number of sections n. This relationship closely follows the equation $n = (T_d/T_r)^{1.3}$. The large value of n required for higher values of T_d/T_r is the major factor affecting cost and line attenuation.

13.1.2 Distributed Constant Delay Lines

A distributed line is constructed by winding the inductor coil on a core that has been covered with a metallic coating. The capacitance is then distributed along the inductance, forming an infinite number of "sections." However, the T_d/T_r ratio becomes relatively poor and seldom exceeds 12.

13.1.3 Strip Delay Lines

Strip lines are similar to distributed lines except that a printed inductor is used instead of a coil. The figure of merit for strip lines rarely exceeds 12.

13.1.4 Attentuation

The attenuation of a pulse in a delay line is mostly the result of inductor losses. The attentuation of lines with T_d/T_r less than 100 is generally limited to 6 dB. For T_d/T_r approaching 250, the attenuation may be as high as 20 dB.

13.1.5 Temperature Coefficients

Temperature coefficients of distributed lines are commonly 150–250 ppm. In lumped parameter lines, special components and compensating techniques help achieve coefficients of less than 50 ppm. Strip-line coefficients are highly variable, but generally greater than 100 ppm.

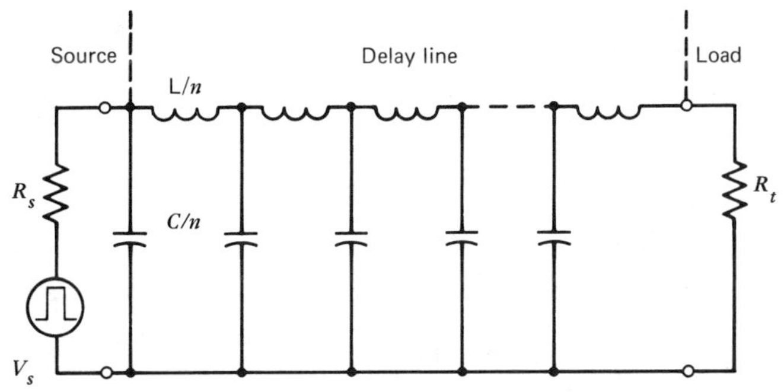

Figure 13.2 Lumped delay line.[82]

13.2 ULTRASONIC DELAY LINES

The ultrasonic delay line converts electrical energy into mechanical energy in the form of acoustic waves, which propagate at a velocity very much slower than electrical energy thereby affording long delay times. The ultrasonic delay line therefore consists essentially of two energy converting transducers (one at the input for conversion from electrical to mechanical; the other at the output for conversion from mechanical to electrical) separated by a transmission medium along which the acoustic waves propagate. The delay time is the time the acoustic wave requires to travel through the medium from input to output transducer.

Ultrasonic delay lines are grouped into categories according to the material used for the transmission medium and the type of transducer used for the electro-acoustic energy conversion. Among the various transmission media are:

A. ***Glass or fused quartz*** blocks in the form of a polygon.

B. ***Glass or fused quartz*** in the form of cylindrical rods.

C. ***Metal blocks*** (aluminum, magnesium, Ni–Fe alloys) in the form of a polygon.

The main types of energy-converting transducers in general use are:

D. ***Magnetostrictive*** whereby the physical dimensions of certain materials change under the influence of a magnetic field.

E. ***Piezoelectric*** the effect whereby certain materials change dimensions when subjected to an applied voltage.

F. ***Optical*** whereby the optical characteristics of certain types of glass change as a result of applied stress.

13.2.1 Acoustic Delay Lines

Wire acoustic delay lines, the most common which employ magnetostrictive transducers, find wide application as digital components. Their unique performance capabilities stem from the fact that mass, the mechanical analog of inductance, can be a much more purely reactive element than any copper-wound electromagnetic inductor. For this reason, acoustic delay lines are able to achieve delay-to-rise time ratios on the order of 20,000 to 1.

Since the velocity of sound in a solid is approximately 3000 m/sec, acoustic delay lines are capable of providing long delay in a small space. Because of the special wire used in their manufacture, wire delay lines approach the stability of a quartz crystal.

In the digital field wire delay lines are used as serial memory elements for bulk storage, input–output buffer storage, and for internal fast-access loops. They are gaining popularity in the telephone industry both as voice and dialing information stores in automatic electronic telephone exchanges. Digital correlators, used in radar and sonar signal processing, and special-purpose computers are often designed using wire delay lines for economical storage.

13.2.2 Principles of Operation

Wire acoustic delay lines, in effect, reduce the velocity of an electrical wave by converting it into a sound wave and then reconverting it back into its electrical form. Figures 13.3 and 13.4 illustrate the principles employed in the two basic types of wire delay lines.

A. ***Longitudinal Mode*** The simplest, Figure 13.3, consists of an input transducer coil with magnetostrictive core attached to one end of a wire sonic waveguide. The other end of the sonic waveguide is attached to an output transducer whose core is polarized by a permanent magnet. At the input transducer, flux change in the coil creates a mechanical stress in the associated core. This stress travels down the delay line as a sound wave. When the wave reaches the output transducer it alters the permeability of the output core. The resulting flux change induces an electrical signal in the output coil. Acoustic absorbers at each end of the delay line prevent signal reflections.

The delay line in Figure 13.3 makes use of longitudinal sound waves, namely, waves in which particle

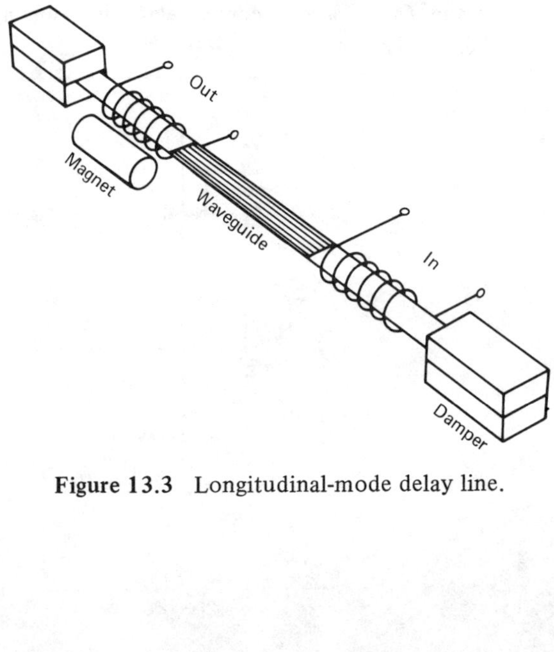

Figure 13.3 Longitudinal-mode delay line.

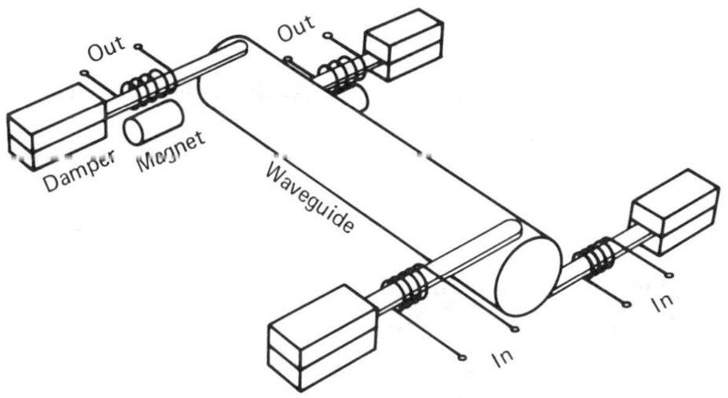

Figure 13.4 Torsional-mode delay line.

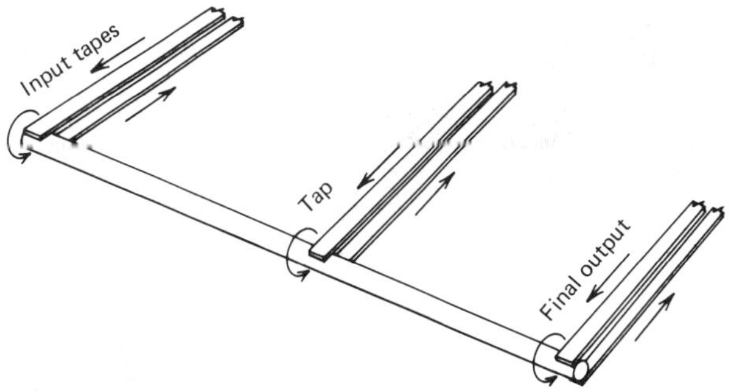

Figure 13.5 Taps on torsional line.

motion is in the same direction as wave motion. It is useful in low to medium storage applications and can readily be tapped.

B. ***Torsional Mode*** The arrangement shown in Figure 13.4 consists of the usual sonic waveguide and transducers. In this case the transducers are joined to waveguide at right angles and are connected so that they produce and respond to a torsion wave in the wire. The torsion wave travels with negligible distortion down the waveguide and requires about 40% less wire for a given delay. This type of delay line is useful for large-capacity storage.

In order to package wire delay lines in a relatively small housing the sonic wave is usually wound into a helix. The ends of the helix are brought out to the transducers, which are located inside and outside of the helix.

13.2.3 Taps

Many logic schemes can be greatly simplified by the use of tapped delay lines. But generally each type of tap limits the performance of the delay line in a manner peculiar to the method employed.

Longitudinal delay lines are well suited for tapping. An additional output transducer, situated at any point between the input and final output transducers, will produce an output at a delay proportional to its distance from the input transducer. Amplitudes will fall off proportionately with distance from the input transducer, but these may be compensated for by suitable termination. A longitudinal line may be tapped as often as every 2 μsec, subject to mechanical limitations, and special techniques are necessary when the first tap is within 5 μsec of the input.

Torsional delay lines can be tapped, but only under special conditions. The most straightforward way of achieving such a tap is to weld an extra set of longitudinal taps at the required point along the torsional delay line medium, see Figure 13.5. These taps are then brought out to a convenient spot in the center of the assembly, and a transducer and damping block attached.

13.3 SPECIFYING DELAY LINES (LUMPED/DISTRIBUTED)

When specifying a delay line bear in mind that there is often more than one way to meet a delay line requirement. Also, needlessly overspecifying a parameter may substantially increase costs. For less critical parameters, it is always best to specify limits instead of a fixed value.

Parameters to be considered when specifying a delay line are as follows. Where appropriate a short description of the parameter is given.

13.3.1 Delay Time

Delay times from 1 nsec to hundreds of nanoseconds are available to meet specific requirements.

13.3.2 Delay-to-Rise Time Ratio

The most critical factor when specifying a delay line is the delay-to-rise time ratio. It should be kept as low as possible. Rise times as fast as 1 nsec are available, but don't overspecify this characteristic. Specifying a slower rise time by only a few nanoseconds can sometimes reduce the cost of the delay line by as much as 50%.

13.3.3 Packaging

Many units are packaged in epoxy cases, see Figure 13.6. Whenever design permits components are mounted on printed circuit boards. Printed-circuit boards simplify assembly of the delay lines and give higher yields and better reliability. They also virtually eliminate loose

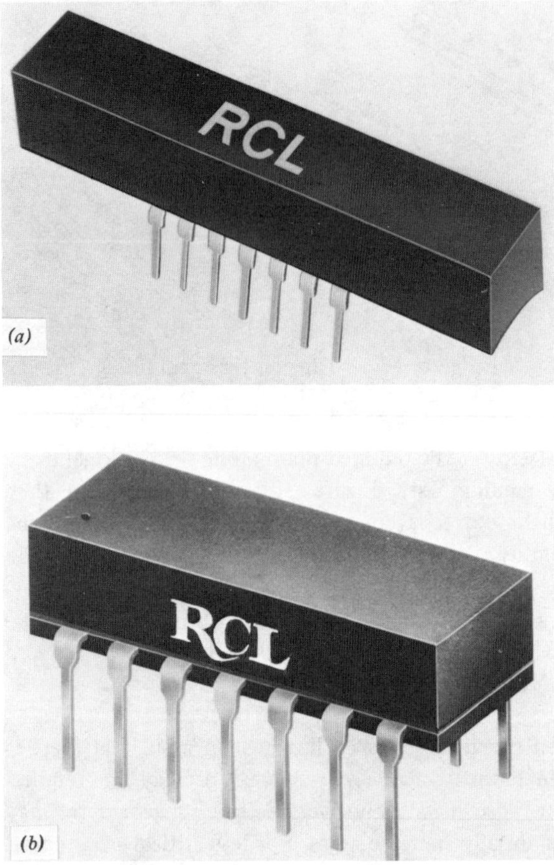

Figure 13.6 Typical delay-line packages. (*a*) SIP package, (*b*) DIP package. (Courtesy of RCL Electronics Div., AMF Inc.)

wires and poor connections during assembly. Also, the manufacturing reproducibility of delay lines is much better when printed-circuit boards are used. Therefore, it is worthwhile when specifying size and configuration of a delay line to try to choose a shape compatible with printed-circuit construction.

13.3.4 Delay Tolerance

Typical delay tolerances are ±2%, although tolerances of ±1% and better can be supplied.

13.3.5 Attenuation

Special core materials are available for applications that require extremely low attenuation. Attenuation is also dependent on circuit design and can often be lowered simply by redesign without the use of special core materials.

13.3.6 Impedance

Impedance values range from 50 to 4000 Ω. Special designs are available for higher impedance values.

13.3.7 Size

Whenever possible delay lines should be compatible in size with standard epoxy cases. As with all electronic packages, significant reductions in size always result in an increased price.

13.3.8 Temperature Coefficient (TC)

Typical temperature coefficients for delay lines are ±50 ppm/°C. If the application demands a lower TC, special circuit designs can give TCs as low as ±20 ppm/°C.

13.3.9 Temperature Range

Standard temperature range is −55 to +125°C. However, if a narrower temperature range can be tolerated economies can sometimes be obtained. Delay lines can also be made to operate over a wider temperature range where necessary.

13.4 DELAY-LINE TERMS AND DEFINITIONS

The definitions given herein are intended to conform to EIA standards on definitions for delay lines. (See Figure 13.7.)

Acoustic Frequency Response The voltage attenuation frequency response measured into a resistive load producing a bandwidth approaching sufficiently close to maximum.

Characteristic Impedance (Z_0) The value of terminating impedance specified by the customer that will provide a minimum of reflection to the input.

Crosstalk (C_s) The amount of input pulse reflected directly into the output pulse.

Delay Time (T_0) The time duration between the 50% point on the leading edge of the input pulse and the 50% point on the leading edge of the output pulse.

Dispersion The variation of delay with frequency.

Feedthrough (leakage) A spurious signal which arrives at the output by electrical coupling.

Frequency Response A representation of the insertion loss or the voltage attenuation as a function of fre-

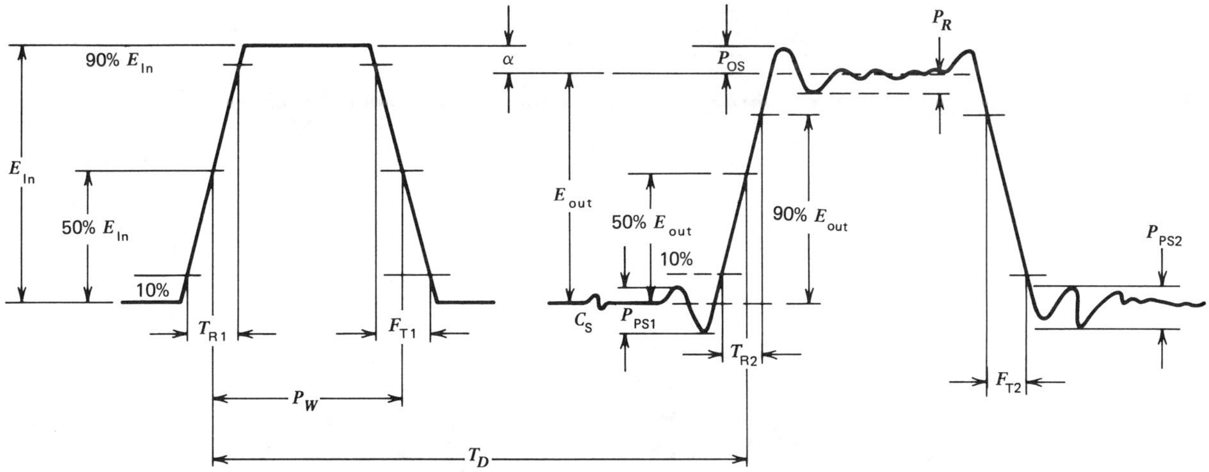

Figure 13.7 Delay-line characteristics.[83]

quency, preferably normalized to a maximum of 0 dB. The particular quantity should be specified, for example, attenuation frequency response, insertion loss frequency response.

Input Fall Time (F_{T1}) The time duration between the 90 and 10% points on the decreasing edge of the input pulse.

Input Rise Time (T_{R1}) The time duration between the 10 and 90% points on the increasing edge of the input pulse.

Input Voltage (E_{in}) The amplitude of the input voltage.

Insertion Loss (Power Insertion Loss) The ratio of the power delivered to that part of the system following the delay line, before the insertion of the delay line, to the power delivered to that same part of the system after insertion of the delay line.

Output Fall Time (F_{T2}) The time duration between the 90 and 10% points on the decreasing edge of the output pulse.

Output Rise Time (T_{R2}) The time duration between the 10 and 90% points on the increasing edge of the output pulse.

Output Voltage (E_{out}) The amplitude of the output voltage.

Postpulse Spurious (P_{PS2}) The output pulse excursions following the main pulse.

Prepulse Spurious (P_{PS1}) The output pulse excursions prior to the main pulse.

Pulse Attenuation (α) The difference in voltage between the input and output pulses.

Pulse Distortion (S) The magnitude of the largest peak amplitude of all spurious responses in either a positive or negative direction relative to the pulse amplitude.

Pulse Overshoot (P_{OS}) The amplitude of the overshoot on the leading edge of the output pulse.

Pulse Top Ripple (P_R) A measure of the deviation from the average amplitude in the output pulse top.

Pulsewidth (P_W) The time duration on the input pulse between the 50% point on the increasing edge and the 50% point on the decreasing edge.

Sonic Delay Line A device that uses electroacoustic transducers and the propogation of an elastic wave through a medium in order to achieve the delay of an electrical signal.

Spurious Level The ratio of a spurious signal to the desired signal, expressed in decibels, measured at a specific frequency over a specified band of frequencies.

Tapped Line A delay line having more than one terminal pair associated with a single sonic delay channel.

Temperature Coefficient of Delay The variation of delay with temperature measured in parts per million per degree centigrade (ppm/°C).

Terminal Impedance The impedance at specified terminals, measured at a specific frequency, under conditions of negligible incident acoustic energy.

Terminal Inductance The imaginary part of the terminal impedance at the specified frequency divided by twice that frequency.

Terminal Resistance The reciprocal of the real part of the effective terminal admittance.

CHAPTER FOURTEEN

Crystal-Controlled Oscillators

14.1 PIEZOELECTRIC EFFECT

The piezoelectric effect (Figure 14.1) occurs when an electric stress is applied to a precut quartz crystal blank in the direction of the X axis and a mechanical stress is produced at right angles to this axis. The mechanical stress along the major Y axis produces electrical charges on both faces of the crystal at right angles to the stress axis. This relationship provides a means of relating mechanical vibrations to electrical stress and is termed the "piezoelectric effect." The polarity of the electrical stress and direction of the mechanical forces are directly related, a change in direction of one causing a corresponding change in the other.

Piezoelectric crystal blanks are cut from natural quartz stones. The angles at which the crystal blank is cut with respect to the natural crystal axes control the temperature coefficient of the crystal unit, as well as its basic mode of vibration. The $Z - Z'$ angle is particularly significant in controlling this characteristic. The $Z - Z'$ angle can be cut for any particular temperature and frequency combination so that the device temperature coefficient is close to zero.

By rotating the cut around one or more major axes a canceling effect or zero temperature coefficient point at a specific temperature and frequency desired is achieved. This means that the frequency of the crystal unit will be very stable throughout the temperature range desired. It is standard practice to control the $Z - Z'$ angle to within plus or minus 1 min of arc of the optimum design angle.

The blank's angle is cut with diamond slurry saws. Double defraction X-ray measurements are used to control the cutting with a high degree of accuracy and repeatability.

14.2 FREQUENCY SOURCES

Frequency sources, also called oscillators, are packaged units that supply a fixed frequency and waveform. The element and frequency-changing, pulse-forming, waveshaping, and/or power-amplifying circuits as may be required to produce the desired output characteristic.

14.2.1 Frequency

Frequency sources have output frequencies from as low as 1 Hz to microwave frequencies, but use crystals in the 1 kHz to 150 MHz range. (See Figure 14.2.) The final choice of crystal frequency is influenced by the frequency stability that must be maintained over a spe-

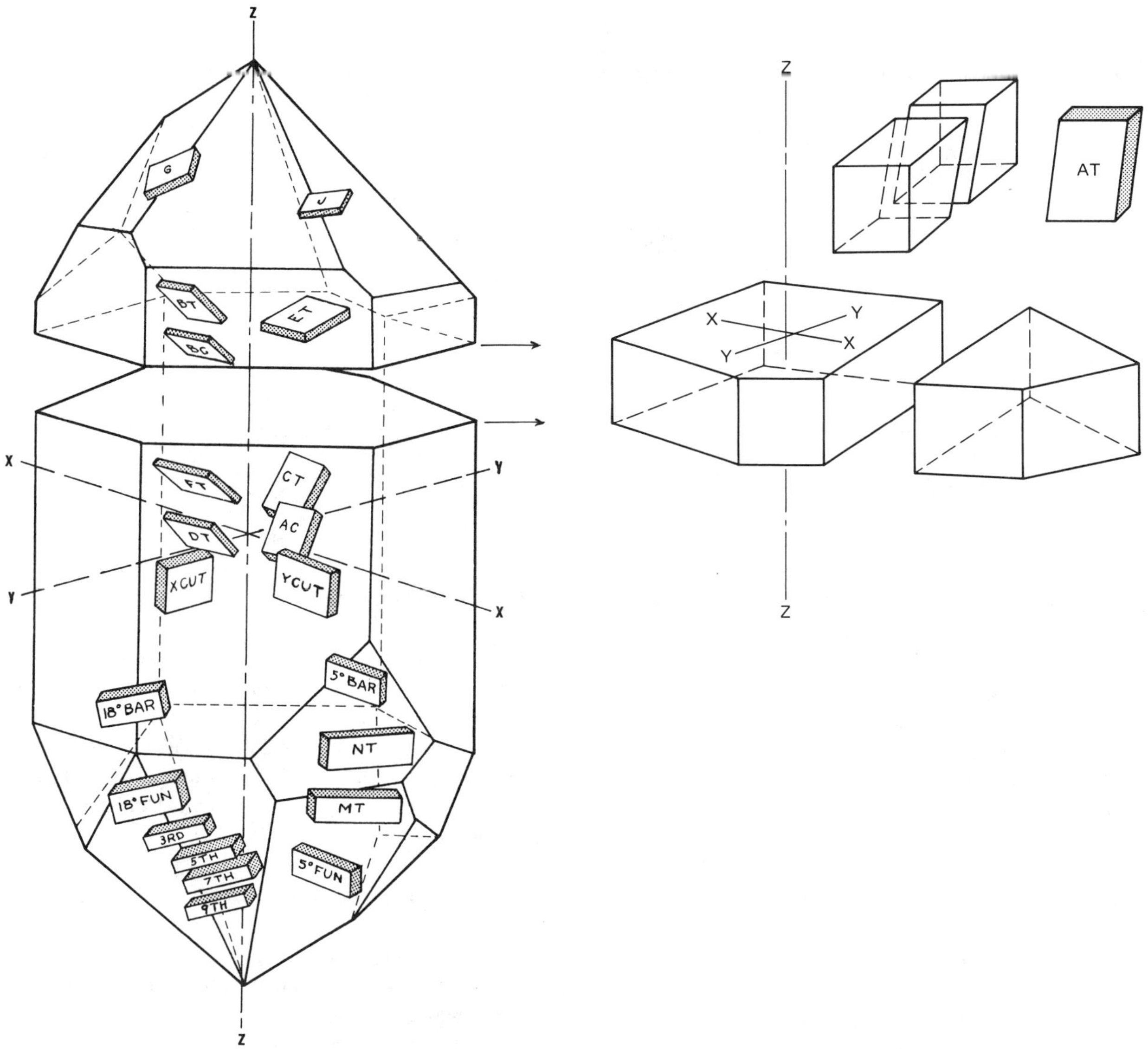

Figure 14.1 Quartz crystal cuts. The illustration at the right displays an *X* section cut from a double terminated quartz crystal (candle) with a *Z* section cut from the *X* section that in turn yields the frequency control industry's most popular cut, the *AT*.

cified period of time and temperature range and the permissible change in frequency as a result of shock and vibration. As a general rule, crystals above 1 MHz are more stable with respect to time, temperature, shock, and vibration.

14.2.2 Temperature

All crystals exhibit a change in frequency with temperature. For the JT crystal the temperature-frequency characteristic can be shifted along the temperature axis. Therefore, the JT crystal is limited to a narrow range over which the frequency can be maintained very precisely.

By choosing a particular temperature and orientation the NT and AT crystals can exhibit much less frequency shift with temperature. The temperature at which the crystal frequency-versus-temperature changes slope is called the turning-point temperature.

14.2.3 Ovens

When the frequency stability over a temperature range must be better than that available from the unstabilized

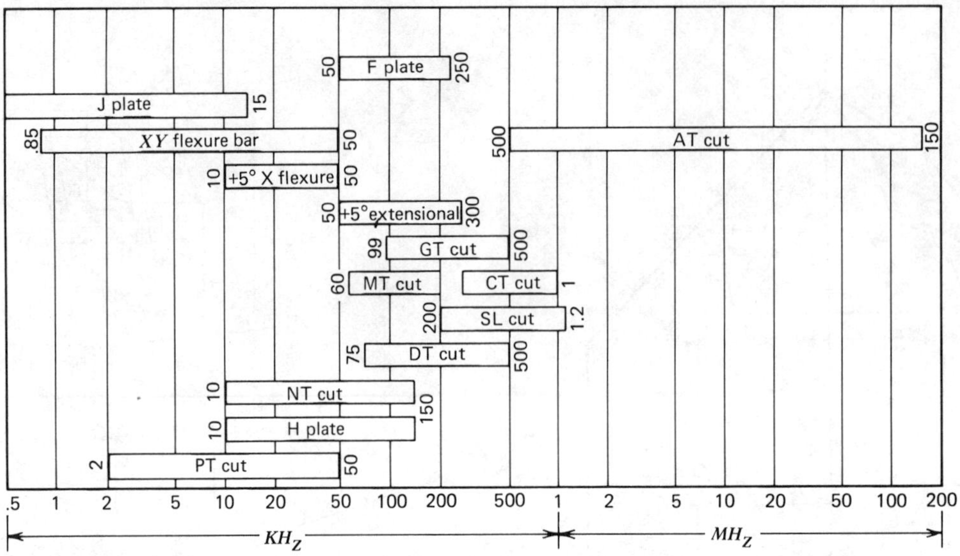

Figure 14.2 Crystal frequencies.

crystal an oven is incorporated in the frequency source. Typically, oven temperatures are controlled by thermostats or proportional circuitry.

Because thermostats are on-off devices, allowance must be made for the peak-power requirements. Proportional oven circuits are continuous-control devices supplying only the power required, proportional to the need, and are generally capable of much closer temperature control.

With either type of temperature control, the oven power required for a given frequency stability is a function of the ambient temperature range. As a guide, wide ranges of temperature require more power, as do better stabilities.

14.3 CONFIGURATIONS[84]

Low-frequency crystals are available in evacuated-glass envelopes, conventional metal holders, and the newer cold-weld metal enclosures (Figure 14.3). High-frequency crystals are manufactured at high-volume rates for both commercial and military users in both metal and glass holders.

Cold-weld metal enclosures, employing compression glass seals and ceramic bases, are available in a variety of conventional and special holder shapes that offer the advantages of heatless, contamination-free, sealed holders. Filter crystals with especially controlled tight tolerances of frequency, capacitances, and inductances, are offered for crystal-filter usage.

14.3.1 Precision Crystals

Precision crystal (A, D, H, J, L) fundamental and overtone types are offered in several types at 1.0, 2.0, 3.0, and 5.0 mHz and special frequencies in the 1.0–5.0-mHz range. These crystals have *Q*s up to several million and aging rates as low as 1×10^{-10} per day, and are specifically designed for frequency-standard and precision-oscillator requirements.

14.3.2 Low-Frequency Crystals

Low-frequency crystals (B, E, F, K, M, N, O, P) from 1000 Hz to 1000 kHz can be furnished in glass-bulb, conventional-shape, soldered holders and in cold-weld holders. They are designed for all types of oscillator and filter applications when low-frequency oscillator or filter resonators are required.

14.3.3 High-Frequency Crystals

High-frequency crystals (A, C, E, I, K, M, N, O) range from 800 kHz fundamental to 200 MHz overtone types, and are offered in conventional metal holders of all sizes, glass-tube envelopes, cold-weld packages, and HC-26/U, HC-27/U, HC-28/U, and HC-29/U "flat-glass" equivalent of military-type metal holders.

14.3.4 Military Crystals[85]

Military-standard crystals (A, B, C, E, O) are commonly designated by "CR-" numbers and manufactured to such

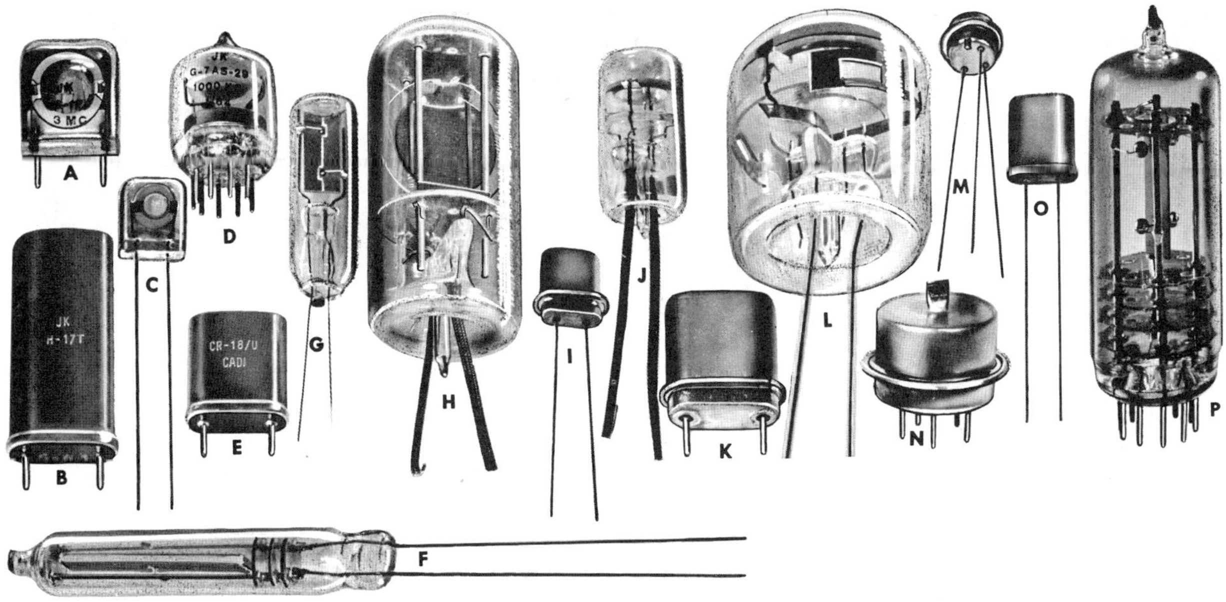

Figure 14.3 Crystal holders. (Reproduced from Ref. 85.)

requirements as MIL-C-3098, MIL-C-12813, MIL-C-17167, and MIL-C-16. (See Table 14.1.)

14.3.5 Cold-Weld Holders

Cold-weld metal crystal holders (I, K, M, N) represent an advance in packaging quartz crystals by virtue of freedom from the effects of heat and contamination inherent with solder techniques used in sealing standard metal holders. In addition, the cold-weld sealing process lends itself to manufacturing finished crystals in transistor cases and other special-shape enclosures.

14.4 CRYSTAL PARAMETERS

14.4.1 Aging

Aging of a quartz crystal is a general term applied to any change in parameters of a crystal unit that takes place over a period of time. This rate of change of frequency is normally exponential in character. The change is at its fastest rate during the first 45 days of dynamic operation.

There are many interrelated factors in aging-leakage of seals, internal contamination, wire fatigue, outgassing of materials, overdriving, and many other small effects. Since the effects of aging cannot be determined by incoming inspection or short dynamic usage, it is imperative when aging is a major problem that the following be adhered to:

A. Operation of a crystal unit should be kept at the lowest ambient temperature possible.

B. Circuit should be designed with drive level kept at absolute minimum.

C. The crystal should be statistically aged at +85°C for 30–45 days prior to its dynamic application.

D. The crystal unit should be dynamically aged in its oscillator circuit prior to final adjustments. This aging can be shortened if static aging has been conducted prior to end equipment application.

14.4.2 Bulk Crystal Losses

Bulk crystal losses are performance characteristics resulting from physical phenomena of quartz and are the sum of acoustic loading, molecular friction, and power transmitted to the mounting or holder. These vary with the mass, mode of resonance, and frequency range of the device.

14.4.3 Phase Jitter

Phase jitter occurs when a crystal is physically subjected to vibration. As the crystal is vibrated a mechanical stress is created on either the *X* or *Y* quartz axis causing a phase change in frequency. If the oscillator circuit is part of a phaselock loop, difficulty can be experienced in an output different from the operating frequency. This phenomenon is dependent on the mechanical frequency of vibration.

Table 14.1 Military Quartz Crystal Units (Reproduced from Ref. 86)

Military Crystal Unit	Military Holder	Frequency Range Nominal KHZ	Operating Temp. Range °C	Frequency Tolerance	C_O PF	Resonance	C_L PF	Modes
CR-1A/AR	HC-11/U	3,600 to 10,000	−55° to +90°	±0.02%	—	P	35.0	F
CR-1D/U	HC-11A/U	3,600 to 10,000	−55° to 90°	±0.005%	—	P	35.0	F
CR-2/U	HC-17/U	200	−40° to +70°	±0.009%	—	P	125.0	F
CR-3/U	HC-17/U	300 to 600	−40° to +70°	±0.02%	—	P	—	F
CR-4/U	HC-17/U	500 to 1,200	−40° to +70°	±0.02%	—	P	—	F
CR-5A/U	HC-17/U	2,000 to 10,000	−55° to +90°	±0.005%	—	P	25.0	F
CR-6A/U	HC-17/U	2,000 to 10,000	−55° to +90°	±0.005%	—	P	13.0	F
CR-7A/U	HC-14/U	3,750 to 10,000	−75°	±0.004%	7.0	P	28.0	F
CR-8A/U	HC-17/U	1,000 to 10,000	−40° to +70°	±0.02%	—	S	—	F
CR-10/U	HC-17/U	5,000	−40° to +70°	±0.005%	—	P	25.0	F
CR-13/U	HC-17/U	445	−40° to 70°	±0.02%	14.0	S	—	F
CR-15B/U	HC-21/U	80 to 200	−40° to +70°	±0.01%	—	P	32.0	F
CR-16B/U	HC-21/U	80 to 200	−40° to +70°	±0.01%	—	S	—	F
CR-18A/U	HC-6/U	800 to 20,000	−55° to 105°	±0.005%	7.0	P	32.0	F
CR-19A/U	HC-6/U	800 to 20,000	−55° to 105°	±0.005%	7.0	S	—	F
CR-24/U	HC-10/U	15,000 to 50,000	−55° to 105°	±0.005%	7.0	S	—	O
CR-25B/U	HC-6/U	200 to 500	−40° to +85°	±0.01%	—	S	—	F
CR-25A/U	HC-6/U	200 to 500	−75° ±1°	±0.002%	—	S	—	F
CR-27A/U	HC-6/U	800 to 20,000	75° ±1°	±0.002%	7.0	P	32.0	F
CR-28A/U	HC-6/U	800 to 2,000	75° ±1°	±0.002%	7.0	S	—	F
CR-29A/U	HC-21/U	80 to 200	75° ±1°	±0.002%	—	P	32.0	F
CR-30A/U	HC-21/U	80 to 200	75° ±1°	±0.002%	—	S	—	F
CR-33A/U	HC-6/U	10,000 to 25,000	−55° to +105°	±0.005%	12.0	P	32.0	O
CR-35A/U	HC-6/U	800 to 20,000	85° ±1°	±0.002%	7.0	S	—	F
CR-36A/U	HC-6/U	800 to 20,000	85° ±1°	±0.002%	7.0	P	32.0	F
CR-37A/U	HC-13/U	90 to 250	−40° to +70°	±0.02%	—	P	20.0	F
CR-38A/U	HC-13/U	16 to 100	−40° to +70°	±0.012%	—	P	20.0	F
CR-39/U	HC-15/U	150 to 330	−55° to 75°	±0.003%	—	S	—	F
CR-40/U	HC-15/U	160 to 330	65° to 75°	±0.003%	—	S	—	F
CR-42A/U	HC-13/U	90 to 250	70° to 80°	±0.003%	—	P	32.0	F
CR-43/U	HC-16/U	80,860	−30° to 75°	±0.01%	45.0	P	45.0	F
CR-45/U	HC-6/U	455	−40° to +70°	±0.02%	5.0	S	—	F
CR-46B/U	HC-6/U	200 to 550	−40° to +85°	±85%	—	P	20.0	F
CR-47A/U	HC-6/U	190-500	75° ±1°	±0.002%	—	P	20.0	F
CR-50A/U	HC-13/U	16 to 100	−40° to +70°	±0.012%	—	S	—	F
CR-51A/U	HC-6/U	10,000 to 61,000	−55° to +105°	±0.005%	7.0	S	—	O
CR-52A/U	HC-6/U	10,000 to 61,000	−55° to +105°	±0.005%	7.0	S	—	O
CR-53A/U	HC-6/U	50,000 to 87,000	−55° to +105°	±0.005%	7.0	S	—	O
CR-54A/U	HC-6/U	50,000 to 125,000	−55° to +105°	±0.005%	7.0	S	—	O
CR-55A/U	HC-18/U	17,000 to 62,000	−55° to +105°	±0.005%	7.0	S	—	O
CR-56A/U	HC-18/U	50,000 to 125,000	−55° to +105°	±0.005%	7.0	S	—	O
CR-57/U	HC-6/U	500	85°C ±1°	±0.001%	7.0	P	32.0	F
CR-58A/U	HC-17/U	800 to 20,000	−55° to +105°	±0.005%	7.0	P	32.0	F
CR-59A/U	HC-18/U	50,000 to 125,000	85° ±1°	±0.002%	7.0	S	—	O
CR-60A/U	HC-18/U	5,000 to 20,000	−55° to +105°	±0.005%	7.0	S	—	F
CR-61/U	HC-18/U	17,000 to 61,000	85° ±1°	±0.002%	7.0	S	—	O
CR-62/U	HC-6/U	800 to 20,000	75° ±1°	±0.001%	7.0	P	32.0	F
CR-63/B/U	HC-6/U	200 to 5555	−40° to +70°	±0.01%	7.0	P	20.0	F
CR-64/U	HC-18/U	2,900 to 20,000	−55° to +105°	±0.005%	7.0	P	30.0	F
CR-65/U	HC-6/U	10,000 to 61,000	75° ±1°	±0.001%	7.0	S	—	O
CR-66/U	HC-6/U	3,000 to 20,000	−40° to +90°	±0.002%	7.0	P	30.0	F
CR-67A/U	HC-18/U	17,000 to 62,000	−55° to +105°	±0.0025%	7.0	S	—	O
CR-68/U	HC-6/U	3,000 to 20,000	75° ±1°	±0.002%	7.0	P	32.0	F
CR-69A/U	HC-18/U	2,900 to 25,000	−40° to +90°	±0.002%	7.0	P	30.0	F
CR-70/U	HC-13/U	4	−40° to +70°	±0.01%	—	S	—	F
CR-71/U	HC-30/U	4,500 to 5,500	+65° to +77°	±0.00008%	4.0	P	32.0	O
CR-72/U	HC-25/Umod	17,000 to 61,000	−55° to +105°	±0.005%	7.0	S	—	O
CR-73/U	HC-29/U	17,000 to 61,000	−55° to +105°	±0.003%	4.0	S	—	O
CR-74/U	HC-26/U	45,000 to 125,000	85° ±1°	±0.001%	4.5	S	—	O
CR-75/U	HC-6/U	50,000 to 125,000	75° ±1°	±0.001%	7.0	S	—	O
CR-76A/U	HC-18/U	16,000 to 61,000	−40° to +90°C	±0.002%	3.5	S	—	O
CR-77/U	HC-25/U	17,000 to 62,000	−40° to +90°	±0.002%	7.0	S	—	O
CR-78A/U	HC-25/U	2,900 to 20,000	−55° to +105°	±0.005%	7.0	P	30.0	F
CR-79/U	HC-25/U	2,900 to 20,000	−55° to 105°	±0.005%	7.0	S	—	F
CR-80/U	HC-18/U	50,000 to 125,000	−40° to +90°	±0.002%	7.0	S	—	O
CR-81/U	HC-25/U	17,000 to 61,000	−55° to +105°	±0.005%	7.0	S	—	O
CR-82/U	HC-25/U	50,000 to 125,000	−55° to +105°	±0.005%	7.0	S	—	O
CR-83/U	HC-25/U	50,000 to 125,000	−40° to 90°	±0.002%	7.0	S	—	O
CR-84/U	HC-25/U	17,000 to 61,000	85° ±1°	±0.002%	7.0	S	—	O
CR-85/U	HC-6/U	800 to 20,000	−40° to +90°	±0.002%	7.0	S	—	F

Table 14.1 *(Continued)*

Military Crystal Unit	Military Holder	Frequency Range Nominal KHZ	Operating Temp. Range °C	Frequency Tolerance	Co PF	Resonance	C_L PF	Modes
CR-86/U	HC-27/U	5,000 to 15,000	−55° to +200°	±0.005%	7.0	S	—	F
CR-87/U	HC-27/U	15,000 to 45,000	−55° to +180°	±0.005%	7.0	S	—	O
CR-88A/U	HC-13/mod	2.75 to 16	0° to +65°	±0.005%	30.0	S	—	F
CR-89/U	HC-32/U	2,120 to 6,200	−55° to +90°	±0.005%	7.0	P	30.0	F
CR-90/U	HC-27/U	3,000 to 5,000	30°	±0.0004%	3.0	P	30.0	F
CR-91/U	HC-11A/U	6,444 to 12,500	−55° to +90°	±0.01%	10.0	P	55.0	F
CR-92/U	HC-11A/U	6,900 to 8,222	−55° to +90°	±0.005%	—	P	50.0	F
CR-95/U	HC-11A/U	2,000 to 9,250	−35° to +75°	±0.005%	—	P	91.0	F
CR-96/U	HC-11A/U	2,000 to 9,250	−35° to +75°	±0.005%	—	P	35.0	F
CR-97/U	HC-18/U	8,000 to 10,000	−40° to +85°	±0.005%	7.0	P	32.0	F
CR-98/U	HC-18/U	50,000 to 134,000	−30° to +71°	±0.0015%	7.0	S	—	O
CR-99A/U	HC-11A/U	44,000 to 69,000	−55° to +105°	±0.005%	7.0	P	32.0	O
CR-100/U	HC-6/U	1,750 to 3,500	−55° to +90°	±0.005%	7.0	P	32.0	F
CR-101/U	HC-35/U	7,000 to 70,000	−55° to +105°	±0.005%	7.0	P	30.0	F
CR-102/U	HC-35/U	50,000 to 90,000	−55° to +105°	±0.0025%	7.0	S	—	O
CR-103/U	HC-35/U	17,000 to 61,000	−55° to +105°	±0.0025%	7.0	S	—	O
CR-104/U	HC-33/U	200 to 580	75° ±1°	±0.002%	—	—	21.5	F
CR-105/U	HC-18/U	48,000 to 70,000	25° ±1°	±0.001%	4.0	S	—	O
CR-106/U	HC-18/U	10,500 to 11,500	−55° to +105°	±0.005%	7.0	P	32.0	F
CR-107/U	HC-18/U	70,000 to 125,000	−55° to +105°	±0.005%	6.0	S	—	O
CR-108/U	HC-26/U	66,000 to 93,000	85° ±1°	±0.005%	5.5	S	—	O
CR-109/U	HC-25/U	10,000 to 11,500	85° ±2°	±0.002%	7.0	P	32.0	F
CR-110A/U	HC-18/U	53,000 to 75,000	−40° to +85°	±0.003%	4.0	S	—	O
CR-111/U	HC-18/U	17,000 to 27,000	40° to 80°	±0.003%	4.0	S	—	O
CR-112/U	HC-18/U	5,000 to 20,000	−55° to +105°	±0.0025%	7.0	S	—	F
CR-113/U	HC-18/U	10,000 to 61,000	75° ±1°	±0.001%	7.0	S	—	O
CR-114/U	HC-25/U	2,900 to 3,850	−55° to +105°	±0.005%	7.0	P	32.0	F
CR-115/U	HC-18/U	24,900 to 34,450	−55° to +105°	±0.005%	4.0	S	—	O
CR-116/U	HC-18/U	50,000 to 125,000	−55° to +105°	±0.0025%	4.5	S	—	O
CR-117/U	HC-25/U	30,000 to 63,000	−40° to +65°	±0.0015%	7.0	S	—	O
CR-118/U	HC-25/U	10,775 to 11,250	−55° to +85°	±0.003%	7.0	P	32.0	F
CR-119/U	HC-6/U	800 to 20,000	−55° to +105°	±0.005%	7.0	P	30.0	F
CR-120/U	HC-27/U	2,000 to 10,000	85° ±5°	±0.0005%	—	P	100.0	F
CR-121/U	HC-27/U	4,000 to 10,000	85° ±5°	±0.0005%	—	P	50.0	F
CR-122/U	HC-18/U	50,000 to 125,000	80° to 90°	±0.001%	7.0	S	—	O
CR-123/U	HC-18/U	50,000 to 125,000	85°	±0.001%	4.5	S	—	O
CR-124/U	HC-18/U	2,900 to 3,800	−55° to +100°	±0.005%	7.0	P	32.0	F
CR-125/U	HC-6/U	1,850 to 3,050	−55° to 105°	±0.005%	7.0	P	32.0	F
CR-126/U	HC-27/U	5,000 to 10,000	85°C	±0.003%	7.0	P	30.0	O
CR-127/U	HC-6/U	10,000 to 61,000	−55° to +90°	±0.0025%	7.0	S	—	O
CR-128/U	HC-18/U	17,000 to 61,000	−55° to +105°	±0.005%	7.0	S	—	O
CR-129/U	HC-18/U	9,000 to 12,000	−55° to +90°	±0.0035%	7.0	P	32.0	F
CR-130/U	HC-6/U	800 to 20,000	−40° to +90°	±0.002%	7.0	P	30.0	F
CR-131/U	HC-6/U	800 to 20,000	70° to 80°	±0.002%	7.0	P	30.0	F
CR-132/U	HC-6/U	800 to 20,000	80° to 90°	±0.002%	7.0	P	30.0	F
CR-133/U	HC-25/U	45,000 to 125,000	−55° to +150°	±0.006%	6.5	S	—	O
CR-134/U	HC-18/U	18,000 to 21,000	+10° to +80°	±0.0025%	6.0	P	32.0	F
CR-135/U	HC-25/U	10,000 to 20,000	+80° to +90°	±0.002%	7.0	P	32.0	F
CR-136/U	HC-18/U	8,000 to 10,000	25°	±0.001%	6.0	P	32.0	F
CR-137/U	HC-18/U	8,500 to 19,000	+20° to +70°	±0.0025%	7.0	P	32.0	F
CR-138/U	HC-6/U	165 to 360	+40° to +85°	±0.012%	—	P	32.0	F
CR-139/U	HC-18/U	20,000 to 22,000	−55° to +105°	±0.005%	12.0	P	30.0	F
CR-140/U	HC-6/U	9,500 to 12,500	−55° to +90°	±0.005%	10.0	P	60.0	F
CR-141/U	HC-18/U	48,000 to 90,000	−30° to +55°	±0.0025%	7.0	S	—	O
CR-142/U	HC-6/U	833.33	75°	±0.001%	7.0	P	32.0	F
CR-143/U	HC-27/U	56,000 to 100,000	75°	±0.00025%	3.0	S	—	O
CR-144/U	HC-36/U	56,000 to 100,000	75°	±0.00025%	7.0	S	—	O
CR-145/U	HC-26/U	66,000 to 93,000	−55° to +105°	±0.0025%	5.5	S	—	O
CR-146/U	HC-33/U	2,300 to 3,500	−55° to +70°	±0.002%	7.0	P	32.0	F
CR-147/U	HC-33/U	200 to 500	0 to +60°	±0.01%	—	P	20.0	F
CR-148/U	HC-6/U	10,000 to 61,000	−55° to +105°	±0.005%	7.0	S	—	O
CR-149/U	HC-18/U	46,000 to 100,000	+25° to +30°	±0.001%	6.0	S	—	O
CR-150/U	HC-27/U	45,000 to 100,000	−55° to +70°	±0.0005%	3.0	S	—	O
CR-151/U	HC-18/U	50,000 to 100,000	25°	±0.002%	6.0	S	—	O
CR-152/U	HC-18/U	16,000 to 61,000	−40° to +90°	±0.002%	3.5	S	—	O
CR-153/U	HC-42-2/U	10,000 to 30,000	0 to 60°	±0.00005%	—	P	20.0	F
CR-155/U	HC-37/U	11,000	−55° to +80°	±0.00025%	7.0	S	—	F
CR-156/U	HC-43/U	35,000	80° to 90°	±0.001%	7.0	S	—	O

14.4.4 Frequency Tolerance

Frequency tolerance is the amount of frequency deviation, plus or minus, from the desired operating frequency at a specific temperature. Accuracy requirements for crystal tolerance are expressed in percentage (±0.005, ±0.0035, ±0.002, ±0.001%).

14.4.5 Temperature Coefficient (TC)

TC is the frequency stability or deviation with temperature change. It is expressed in parts per million change or plus-or-minus percentage change over the operating temperature range. Frequency change with temperature can be either positive or negative and should not be confused with frequency drift.

14.4.6 Drive Level

Drive level is the amount of power dissipation experienced by the crystal in a given operating circuit. It is expressed in milliwatts or microwatts. Excessive drive level will result in possible fracture of the quartz resonator or excessive long-term drift due to aging. An indication of excessive drift level is a constant frequency drift.

14.4.7 Resonant Resistance

Resonant resistance is the value of the crystal's equivalent impedance in the operating resonant circuit. The higher the resistance, the less activity of the quartz and the more drive required to excite the quartz. Resonant resistance is directly affected by the type of crystal cut and operating frequency.

14.4.8 Shunt Capacity (C_0)

Shunt capacity is the capacitance between the terminals of the crystal. It is measured in picofarads and should not be confused with load capacity (C_L) in parallel antiresonant circuit operation.

14.4.9 Quality Factor (Q)

The Q of a crystal unit is the quality factor of the motional arm at resonance. The maximum stability that can be attained by the crystal is related directly to Q. The higher the Q, the smaller the bandwidth. External circuit reactance value changes have less effect on a high-Q crystal and has less warpability (pulling range) than lower-Q devices.

14.4.10 Spurious Response

Spurious response is another mode of vibration generally higher than the desired main mode of oscillation and is generally created by coupling. It is measured as a resistance in ratio to the maximum ESR (equivalent series resistance) of the crystal device. It is specified as 1.5:1 or 2.0:1 ratio and is often referred to as "unwanted modes" or "spurs."

14.4.11 Parallel Resonance

When a crystal is operating at parallel resonance (Figure 14.4), it will look inductive in the circuit. Therefore, a change in circuit reactance values will have the effect of pulling the frequency of the crystal. If the crystal is to be used at parallel resonance, the load capacity that the crystal is looking into should always be specified. This load capacity is the dynamic capacity of the total circuit measured or computed across the crystal terminals. In parallel circuit design the load should be selected to operate the crystal at a stable point on the F_s/F_a reactance curve.

14.4.12 Crystal Equivalent Circuit

The equivalent circuit for a crystal (Figure 14.5) is useful to explain the basic concepts governing the crystal's performance. C_0 represents the static capacitance and is the sum of the capacitance between the electrodes and capacitances added by the wire leads and holder. The R_1-L_1-C_1 branch is known as the motional arm. C_1 represents the motional capacity of the quartz; L_1 is a function of the mass; and R_1 is the sum of the bulk crystal losses.

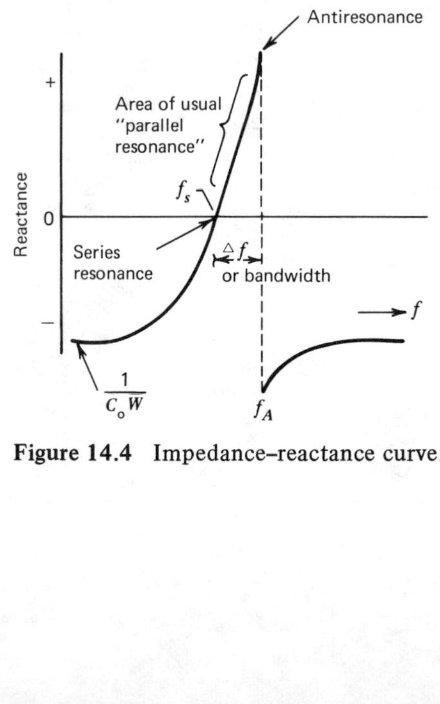

Figure 14.4 Impedance–reactance curve.

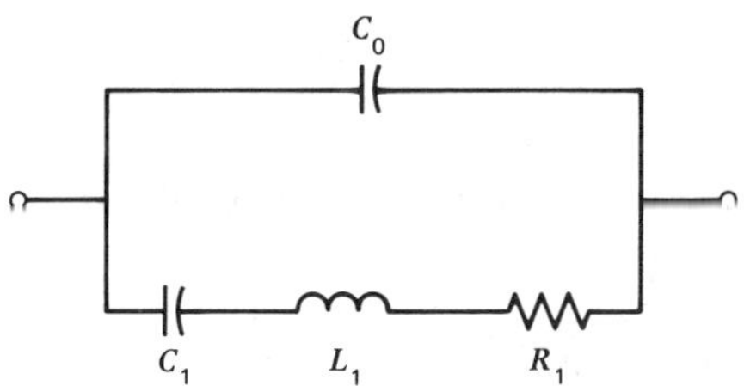

Figure 14.5 Equivalent crystal circuit.

14.5 CLOCK OSCILLATOR[86]

For the moderate-stability crystal-controlled oscillator, where neither temperature compensation nor oven operation are required, there are three primary parameters, namely, output-waveshape, frequency, and accuracy stability.

14.5.1 Output

The vast majority of systems require a crystal-oscillator output that is TTL/DTL compatible, ECL compatible, CMOS compatible or sinusoidal. Any of these outputs can be simply generated by circuits that follow the crystal oscillator stage. The four types are illustrated in Figure 14.6, with a dashed line in each case representing supply-voltage input and the solid line showing the output.

14.5.2 Accuracy/Stability

The most basic element in an oscillator specification is the output frequency. At any given time, however, the oscillator's output frequency will differ from the desired specified frequency, resulting in a frequency error. This error is comprised of the following primary components:

A. ***Initial Accuracy*** Generally defined as the difference between the oscillator output frequency and the specified frequency at +25°C at the time of shipment by the oscillator manufacturer.

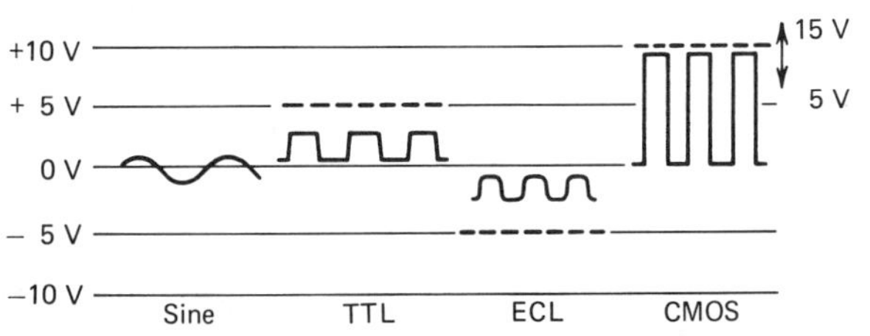

Figure 14.6 Circuit types.[84]

B. ***Temperature Stability*** The angle at which the quartz crystal is cut determines the temperature characteristic of a specific crystal. In a well-designed oscillator the stability versus temperature relationship is determined primarily by the temperature characteristic of the crystal. Therefore, the oscillator manufacturer must select the crystal characteristics that conform with the oscillator circuit in order to insure that the intrinsic stability of the crystal is not degraded.

C. ***Aging*** Aging (long-term stability) refers to the continuous change in crystal operating frequency with time when all other parameters, such as temperature and supply voltage, are held constant. The better the processing of the crystal, the lower the aging rate, that is, the higher the long-term stability.

Improved aging can be achieved by employing a specially processed crystal housed in an evacuated glass or coldweld-sealed holder. Because aging generally introduces a small part of the overall error in moderate-stability clock oscillators, it is often ignored in specifying these devices.

14.6 TEMPERATURE-COMPENSATED CRYSTAL OSCILLATOR (TXCO)

14.6.1 Temperature Stability

The temperature stability of a crystal oscillator can be improved by incorporating in the oscillator circuit components with temperature characteristics approximately equal to and opposite from that of the crystal. The actual technique employed in the more sophisticated TCXOs is based upon the use of a varactor diode in series with the crystal across which a voltage is developed having a characteristic over temperature so as to compensate for the inherent crystal instability.

14.6.2 Aging

In clock oscillators with moderate temperature stability aging is usually of little consequence. However, in highly temperature-stable TCXOs crystal aging becomes a significant factor in the oscillator's overall frequency error. Therefore, TCXOs generally employ specially processed crystals in evaluated glass or coldweld holders. Furthermore, because crystals age at lower rates when current through the crystal is low, an automatic gain-control (AGC) circuit is generally employed in high-grade TCXOs to maintain a low and constant crystal current.

14.6.3 Other Factors

Elements generally not required in simple clock oscillators but included in the proper design of a highly stable TCXO are:

A. ***A frequency-compensation network*** To minimize temperature sensitivity.

B. ***A precision crystal*** Coupled with AGC for minimum aging.

C. ***A tuning adjustment*** Permitting precise setting of frequency.

D. ***Buffering*** Following the oscillator to minimize the effects of external circuit changes.

E. ***An internal regulator*** To minimize the effects of voltage variation.

14.7 OVEN-CONTROLLED CRYSTAL OSCILLATOR

Crystal ovens are of two basic types—those with on-off control and those with proportional control.

14.7.1 On-Off Oven Control

An on-off oven control is one that is turned on when the oven temperature falls below a specified value and turned off when the oven temperature rises above that value. The simplest type is a "snap-action" bimetallic control such as used in a thermostat. However, this type provides only coarse control, and it generates undesired RF interference when the contacts open and close. Therefore, it is rarely used for oscillators.

An improved on-off type oven employs a mercury thermometer to control a transistor switch that turns the oven on and off. Despite the somewhat improved control, this approach suffers from the shortcoming of an on-off system—that of temperature overshoot as the oven is turned on and off.

14.7.2 Proportional Control

A proportional control is an electronic servosystem that continuously supplies power to the oven. It varies the amount of oven power, thereby continuously compensating for ambient temperature changes. Most oven-controlled crystal oscillators use proportionately-controlled designs due to the equipment performance improvement which they offer with little sacrifice in economy.

CHAPTER FIFTEEN

Filters

15.1 GENERAL INFORMATION[87,88]

A filter is a circuit that has a deliberately nonuniform transfer function with respect to frequency. There are many types of filters, but all have in common the fact that their gain or loss varies with the frequency of the signal applied to their input. Most filters are designed to favor, or "pass" signals in some frequency band, and markedly attenuate, or "reject" signals in all others. Some are low-pass, some high-pass, and some bandpass or band-reject. Low-pass filters usually pass frequencies down to dc and up to their so-called "cutoff" frequency. High-pass filters usually attenuate frequencies from DC to cutoff, and pass all others up to some very high frequency at which circuit "strays" limit transmission.

Bandpass filters pass signals that fall between two cutoff frequencies, which may be very different (wide bandpass) or very close in frequency (narrow bandpass). Band-reject filters attenuate signals that fall between two cutoff frequencies, which may, as in the bandpass case, be widely spaced or close together. Figure 15.1 shows typical transfer functions for the types discussed above.

Note that all physically realizable circuits have some filtering effect, as no circuit has infinite bandwidth. In some, the filtering effect is negligible for all signals and noise components of interest; in others, such as integrators, the filtering is very pronounced and inherent in the functioning of the circuit.

15.1.1 Filter Characteristics

A. *The Number of "Poles" of a Filter* Equal to the number of reactive elements that contribute to its transfer characteristics (i.e., ignoring strays or "compensating" or "trimming" reactances), the term "pole" comes from the pole-zero method of network analysis and synthesis. Generally speaking, the sharper the cutoff, the steeper the rolloff, and the higher the stopband attenuation required, the larger is the number of poles required.

B. *Cutoff Frequency* (f_{co}) It has been the practice to say that this frequency is the one at which the transfer characteristics show 3 dB of attenuation with respect to the level of the passband (in a low-pass, bandpass, or high-pass filter). But it is more meaningful to refer attenuation to the signal level at the input. In any filter displaying significant passband ripple it is the practice of both filter and system designers to specify the cutoff frequency as that frequency at which the attenuation increases (and continues to increase) beyond the ripple amplitude.

C. *Gain/Loss in the Passband* Preferably referenced, to the input signal level, the tolerance on passband gain or loss should be stated, as well, because this tolerance can be just as important an error factor as the behavior near cutoff. Some filters have passband ripple; these variations

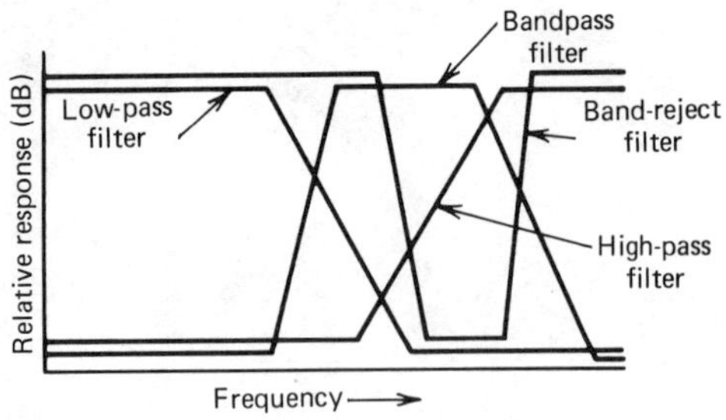

Figure 15.1 Typical filter-transfer functions.[87]

should be separately specified, and not lumped into the gain/loss tolerance–unless, of course, they are trivial compared with that tolerance.

D. ***Phase Shift in a Filter*** Can be as important as the attenuation. (The complete transfer function must include both attenuation and phase response curves, by the way.) In general, phase shift of a filter at any given frequency is directly proportional to the number of poles, and is of the order of 45°/pole at f_{co}.

E. ***Transient Response and Phase Distortion*** The relative phase of the various frequency components is as important as their relative amplitude, in determining the signal waveform. Unless the delay-versus-frequency characteristic is constant in the passband and in the lower-attenuation parts of the transition region, the distortion caused by the filter to the signal to be processed could destroy the accuracy of the system.

15.1.2 Filter Technology[89]

Active and passive filters each have their positive and negative attributes, and the area in which it is difficult to choose between them is really very small. Frequency is the system parameter dictating which filter technology to use. Figure 15.2 compares active filters with digital and passive approaches as a function of frequency. Active filters are most useful in the 1-Hz–1-MHz region. Likewise, digital filters are currently limited to below about 1 kHz, and passive filters are most useful above 10 MHz. Below approximately 100 Hz, active filters are superior because of their smaller size, weight, and lower cost. Above 100 kHz or so, active filters would require op amps with greater bandwidths than are now available, and, therefore, except for very simple applications, the advantage swings to passive filters. Between 100 Hz and 100 kHz is the range requiring thought.

Generally, it turns out to be best to choose passive filters for critical applications and active filters for noncritical work. A critical application can be defined as one in which the theoretical limits of system performance are determined by the filter, or when the filter must meet some unusual specification–wide temperature range, extremely low noise, very wide dynamic range, and so forth.

For a large number of noncritical applications, the active filter is usually the better choice. It is relatively easy to design, it can be made quite compact, and its inductorless design is amenable to a high degree of automation and integration. Finally, broadcast radios notwithstanding, active filters are much easier to tune over large bandwidths than are passive filters.

A. ***Passive Filters*** Passive filters are networks of discrete and/or distributed resistance, capacitance, and inductance. Active filters are circuits containing amplifiers, resistors, and capacitors. (Inductors can be used in active filters, but almost never are.)

Passive filters have very low noise, require no power supplies, and are capable of very wide dynamic range without saturation. But they can be bulky (at low frequencies) and difficult to make adjustable over wide ranges, especially if the resolution of adjustment must be high and precise. They are generally more temperature dependent and more susceptible to magnetic pickup. Near saturation, they tend to be nonlinear, introducing distortion. At low frequencies, they become prohibitively large and expansive, and/or degraded in performance. Passive filters cannot provide power gain–indeed, they always exhibit "insertion loss," and it is difficult to design them to have very high-input or very low-output impedance. They are very sensitive to impedance matching, at both input and output. They are also less flexible than active filters (e.g., it is difficult to simply switch them from low pass to high pass). Finally, they cannot be conveniently programmed by remote digital signals.

B. ***Active Filters*** Active filters are capable of remarkably wide ranges of convenient adjustment including digital programming. Unless carefully designed, they can contribute noise and dc offset to the signal and they do require power supply energy. If properly designed, they have excellent dynamic range, extremely wide frequency ranges, high adjustment resolution, excellent stability and linearity, and significantly higher performance than passive filters, particularly at low frequencies. Their input impedance may be made very high, their output impedance may be made very low, and they can be designed to provide large power gain. Their performance can be made essentially independent of either source impedance or load impedance. They can be configured for simple

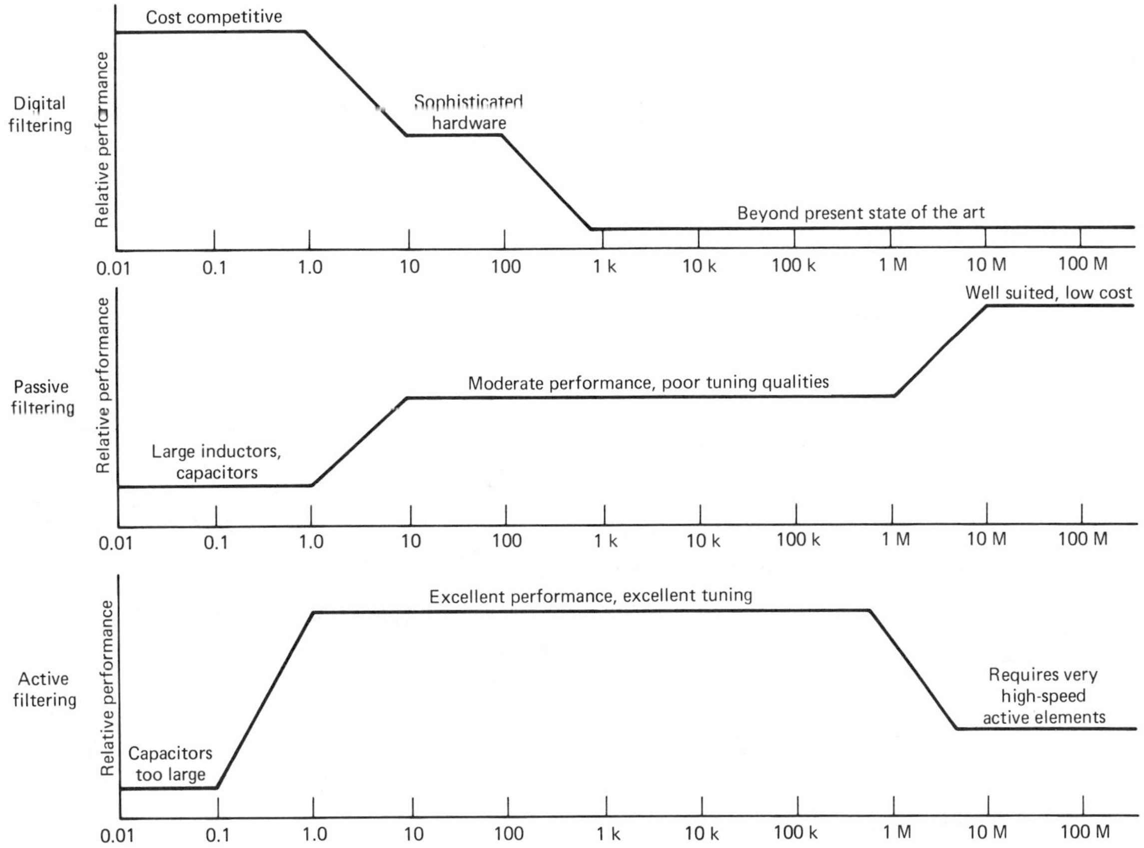

Figure 15.2 Filter technology versus frequency.[89]

switching from low-pass to high-pass, and can be combined, for bandpass or band-reject behavior. They can be made essentially immune to temperature changes, power-supply variation, or external magnetic fields. Finally, they can be cascaded without significant interaction.

Active filters function like simple frequency-selective control systems. Thus, you can create any desired filter characteristic by interconnecting integrators, inverters, summing amplifiers and lossy integrators. Each of three of the more common active-filter types (state-variable, Sallen-Key, and biquad) optimizes a specific aspect of performance.

The performance of an active filter is limited principally by its op amp. As a rough rule of thumb, the gain-bandwidth product of the op amp must be about 10 times greater than the product of Q and the center frequency of the filter. Thus any increase in the op amp's bandwidth translates directly into improved filter performance.

Just as important as a high gain-bandwidth product is a high slew rate. Slew rates are frequently expressed in terms of the op amp's full-power bandwidth.

There are many kinds of active filters, but only one, the biquad, demonstrates a particularly useful characteristic: constant bandwidth. The bandpass output of the state-variable filter, for instance, has a Q that remains constant as the center frequency varies. This constant Q, in turn, implies that the bandwidth decreases at lower frequencies and increases at higher ones. In contrast, the biquad filter, while appearing very much like the state-variable filter, has a bandwidth that remains fixed regardless of center frequency.

Some guidelines for the design of simple active filters are:

1 For a bandpass filter where the bandwidth is more than 100% of the center frequency, use a cascade of low-pass and high-pass filters rather than one bandpass filter.
2 When designing for very high-Q (less than 50) circuits, watch out in single-amplifier designs for oscillation arising from the inherent phase shift of the op amp. Cascading two lower Q filters achieves the same very high-Q response.

Table 15.1 Active Filter Design[89]

VARIABLE	FILTER TYPE			
	STATE VARIABLE	SALLEN-KEY		BIQUAD
		UNITY GAIN	EQUAL R, C	
$f_0 = \frac{\omega_0}{2\pi}$	ASSUME C_F $R_F = \frac{1}{2\pi C_F f_o}$	ASSUME R_F $\left(\frac{C_1}{C_2}\right)^{1/2} = \frac{1}{2\pi R_2 f_o}$	ASSUME R_F $C_F = \frac{1}{2\pi R_F f_o}$	ASSUME C_{FS} $R_F = \frac{1}{2\pi R_F F_o}$
$Q = \frac{K}{d}$	ASSUME R $R_d = \frac{R}{Q}$	ASSUME C_1 $C_2 = \frac{4Q^2}{C_1}$	$K = 3 - \frac{1}{Q}$	ASSUME R_F $R_Q = R_F Q$
GAIN	ASSUME R $R_g = \frac{R}{GAIN}$	GAIN FIXED AT +1	GAIN FIXED AT $3 - \frac{1}{Q}$	ASSUME R $R_g = \frac{R}{GAIN}$

3 The choice of capacitor type deserves consideration. Teflon, polystyrene, and mica always prove excellent choices in filter applications; they have low loss, good stability with temperature and time, and come in 5 and 2% tolerances. Avoid using ceramic or electrolytic capacitors because of loss, drift and polarity sensitivity.

4 Metal-film resistors tend to be very inexpensive in moderate quantities and exhibit excellent tolerance and stability. Use carbon-composition resistors only for noncritical application.

5 The frequency response of the amplifier in a single-op amp filter should be 5 to 10 times the gain-f_c product.

6 In a 2-op-amp filter, frequency response should be 5 to 10 times the square root of the gain-f_c product.

These last two guidelines are intended to be conservative and not absolute. With them, you can determine how well a particular op amp is suited for a given filter application.

Applying Table 15.1 can aid in designing an active filter. First, choose the appropriate row by deciding which variable (f_o, Q or gain) is most important to your design. Then select the column according to the desired filter type. Pick a standard value for the assumed component and then calculate the values of the dependent passive components.

15.1.3 Filter Configurations[87,90–92]

For most applications filters are supplied as single-ended or unbalanced networks, that is, the input and output are common on one side of the line. Unbalanced filters are simpler and less expensive than their balanced counterparts. Many applications, however, require balanced circuits. Phantom circuits and lines that must be protected from stray longitudinal currents are typical.

Filters may be connected in parallel when provision is made in the design to avoid shorting or loading in the pass band of any parallel filter. The impedance outside the pass band may present either a falling or rising characteristic. The falling characteristic would present a heavy load or even a short to the line, while the rising characteristic is quite suitable for paralleling, since it presents minimum or no loading to the passbands of the other paralleled filters. Loading becomes less of a problem as the spacing between the edges of adjacent passbands is increased.

To obtain rising impedance on the paralleled side, low-pass and bandpass filters are internally terminated with a series coil. The impedance of the filter is raised. In the extreme, the ability of parallel narrow-band and/or high-impedance filters is limited by the practicability of manufacturing high inductance coils.

Filters may also be joined in series. In this case, the side connected in series must present a falling impedance characteristic.

Low pass filters allow DC and low-frequency electrical energy to pass with little or no attenuation, but reject high-frequency electrical energy. High-frequency energy is reduced by approximately 6 dB/octave, per element.

The design principle is based on arranging various combinations of inductors and capacitors into filter devices whose networks are called L, Pi, and T type. Block construction is utilized, that is, the inductors and capacitors are mounted side by side in their housing. The advantage of this construction approach is that the filter can handle more RF power. As frequency increases, the impedance of the inductor increases thereby causing the temperature of the inductor to rise. Since the components are side by side and in intimate contact with the housing, the flow of heat generated by the RF power is not restricted and can be dissipated very readily by the housing. In this type of construction, the coaxial monolithic capacitor is attached 360° to the inner and outer conductor. Theoretical analysis and experience indicates that a coaxial arrangement with a circumferential attachment results in optimum attentuation.

A. *L Network Filters* Most effective when a high impedance noise source is to be suppressed. A capacitive input filter configuration is usually used. When a noise source has low impedance, an inductive input circuit is used. The series inductor and shunt capacitor present a high and low impedance source, respectively, to high frequency noise.

B. *Pi Network Filters* Generally used as line filters where the source and load impedances are not known. Pi network filters provide sharp frequency cut off and high suppression of the unwanted noise signals.

C. *T Network Filters* Effectively used in switching circuits. This type of network will cause no deterioration to switch contacts while providing attenuation to high frequencies, thereby increasing the life and reliability of the system.

Most filter configurations are named after their discoverers, or after the mathematical principles applied in their design. Here are the most common designations, briefly characterized:

A. *Butterworth Filters* Achieve maximally flat passbands, with moderately sharp cutoffs, but have moderately nonlinear phase response. They have excellent uniformity of stopband attenuation, and are readily synthesized with almost any cutoff frequency from DC to RF, in either active or passive form.

B. *Chebyshev (Variously Spelled) Filters* Very similar to the Butterworth filters, but achieve faster rolloff near cutoff, at the expense of significant passband ripple. They also have excellent uniformity of attenuation in the stopband. They are "efficient" in their use of reactances (like the Butterworth) and are also easily realizable in both active and passive forms.

C. *Elliptic Filters* Have characteristics similar to the Chebyshev designs, but have large variations in attenuation in the stopband, as well as significant passband ripple.

D. *Bessel Filters* Characterized by excellent phase linearity at the expense of sharpness of cutoff and passband flatness. They are efficient and stable, are easily realized in active or passive form, and have excellent uniformity of stopband attenuation.

E. *"Iterative-RC" Filters* Very linear in phase over the entire passband, but have poorer bandpass flatness, and gentler cutoff characteristics than the Bessel designs. They are otherwise very similar to the Bessel configuration.

15.1.4 Specifying Filters[93]

It is desirable to evaluate filter requirements in conjunction with overall circuit performance. Frequently an inexpensive circuit modification results in simplification of filter requirements and a corresponding reduction in overall cost.

The effect of particular specifications on cost is shown by the following comments.

A. *Insertion Loss* For low-pass and high-pass filters values of 1 dB or less usually present no problem. For bandpass filters, insertion loss varies inversely with bandwidth and Q of the components. In the latter case, filter cost can be lowered and size reduced by being lenient with insertion loss specifications.

B. *Attenuation in the Passband* Allow for the maximum possible variation in order to achieve greatest economy. A filter constructed for 3-dB variation will be less expensive than one that must be flat within 1 dB. In some cases, 1-dB flatness may not be realizable.

C. *Attenuation Outside the Passband* Steep-sloped attenuation curves and high levels of attenuation tend to increase cost. The necessity for more expensive components, additional filter sections and internal shielding results in greater expenses. Filters with more than 80-dB attenuation may be impractical.

D. *Impedance* Source and load impedances should be accurately specified, but avoid tying down the impedance of the filter. Exact filter impedance is difficult to maintain and the specification of this parameter often limits the filter engineer in meeting other requirements.

E. *Special Requirements* These include phase or delay specifications, variations with voltage or temperature, of special services. Such requirements add to cost due to the need for more expensive components and/or additional test time.

F. *Environmental Specifications* Temperature specifications exceeding +85°C, severe shock, and vibration requirements or unusual service conditions may require additional engineering time, more expensive materials plus extra rugged construction.

G. *Adjustment Tolerances* Very narrow passbands, filters with very steep attenuation slopes and tight tolerances require extremely careful adjustment and control in manufacture. Wherever possible, it is advisable to specify tolerances as maximums or minimums.

H. *Size* Extreme miniaturization of units is more expensive than conventional construction. Coils must be wound with much finer wire and the units require special handling in assembly.

I. *Mechanical Construction* Tight mechanical tolerances and special constructions necessitating the use of special tools and jugs increases filter cost, especially on small-quantity orders.

J. *Checklist* When requiring special filters, detailed data on electrical performance required should be supplied.

The following checklist is a guide to the minimum information normally required.

1 **Low-Pass and High-Pass Filters.**
 a Cutoff frequency and passband attenuation.
 b Attenuation outside passband (at least one frequency).
 c Input and output impedance.
 d Operating voltage level.
 e Special requirements, such as phase shift, transient response, stability, direct current through filter, environmental conditions.

2 **Bandpass Filters.**
 a Center frequency.
 b Passband frequencies and attenuation.
 c Attenuation outisde passband (at least one frequency above and below passband).
 d Input and output impedance.
 e Operating voltage level.
 f Special requirements, such as phase shift, transient response, stability, environmental conditions.

3 **Band Elimination Filters.**
 a Center Frequency.
 b Rejection frequencies and attenuation.
 c Passband frequencies and attenuation.
 d Input and output impedance.
 e Operating voltage level.
 f Special requirements, stability, environmental conditions.

4 **Discriminators.**
 a Crossover frequency.
 b Frequency range over which deviation is to be linear.
 c Sketch of circuit in which discriminator is to be used or source and load impedances.
 d Operating voltage level.
 e Output sensitivity.
 f Special requirements, stability, environmental conditions.

5 **Mechanical Information.**
 a Critical dimensions.
 b Type of mounting (studs or inserts).
 c Type of terminals.
 d Weight limitations.
 e Finish, paint, color, stamping.

15.2 WAVEBAND FILTERS[90,94,95]

The constantly increasing complexity and quantity of electronic systems has placed great emphasis on the efficient utilization of available frequency space. Precision electrical waveband filters are key factors in separating communication channels and eliminating extraneous signals. They therefore find extensive application in equipment for telephone, telegraph, telemetering, multiplexing, single sideband communication, carrier elimination and restoration, facsimile, tone systems, paging systems, radar, sonar, and geographic surveying.

Filters are electrical networks that discriminate between different frequencies. They pass a desired range of frequencies while attenuating all others (see Figure 15.3). The elements in a filter, such as coils, capacitors, resistors and crystals, are formed into four principal groupings, namely, low-pass, high-pass, bandpass, and band elimination, each chosen to satisfy specified transmission requirements.

15.2.1 Parameters (see Figure 15.4)

A. ***Attenuation*** An ideal filter would pass all frequencies in a given band without reduction in amplitude and totally suppress all other frequencies. In practice, losses in the filter elements are a major factor in making ideal response unrealizable.

B. ***Component Losses*** A filter is a precision network constructed of close tolerances, stable, high Q components. Components of poor quality cause unstable and unreliable operation, increased insertion loss and inferior attenuation response. The quality factor Q is used to measure losses in reactive elements.

$$Q = \frac{X}{R}$$

where X = reactive impedance (Ω);
 R = equivalent effective series resistance (Ω).

Coils used for filters have lower Q than do capacitors and, therefore, contribute most to insertion loss.

Practical filters are manufactured of coils with Q values up to approximately 200. The improvement in electrical performance realizable with coils of significantly high Q does not generally justify the accompanying increase in size, cost, and possible instability.

Capacitors used in filters range in Q from about 200 for Mylar film to 1000 and higher for polystyrene film and silver mica.

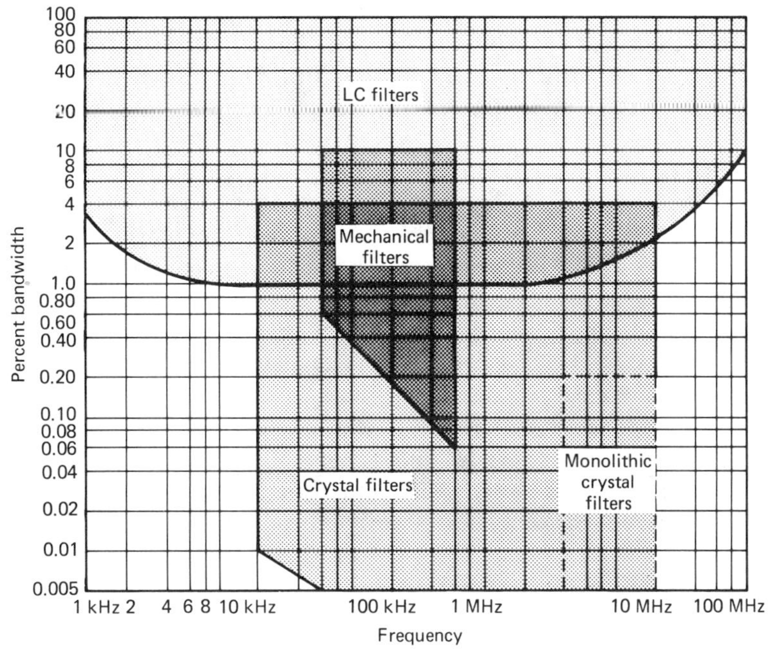

Figure 15.3 Frequency versus percent bandwidth.[95]

A coil with a Q of 100 in combination with a capacitor having a Q of 200 forms a resonant circuit with an effective Q of only 67. The same coil combined with a capacitor with a Q of 1000 would produce a resonant circuit Q of 91. It is apparent that the dissipation characteristics of all elements must be considered in arriving at a sound filter design.

C. *Insertion Loss* Inserting a filter into a circuit results in a power loss, known as insertion loss. The magnitude of the insertion loss depends on the component losses, number of components, and bandwidth.

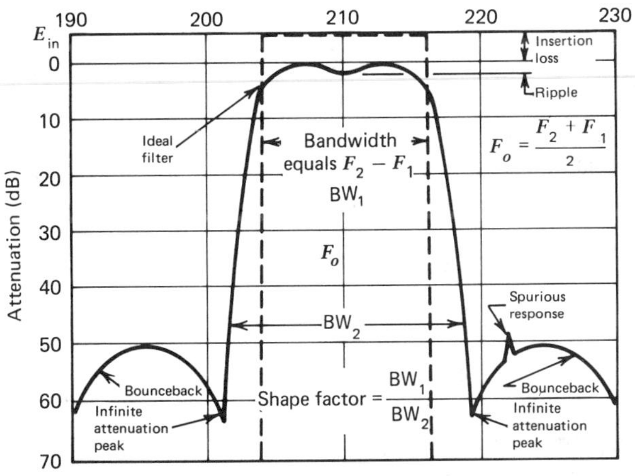

Figure 15.4 Typical bandpass filter response curve.

A very narrow passband results in high insertion loss; widening the passband with the same number of sections reduces the insertion loss. Low-pass and high-pass filters have very little insertion loss (generally 1 dB or less), because of their broad passbands.

Each filter section contributes insertion loss. As filters become increasingly complex through the addition of more section, the insertion loss increases proportionately.

The loss of voltage due to insertion loss of the filter into a circuit can be overcome by increasing the circuit load impedance and modifying the filter to match both the source and new load.

D. *Response to Pulses* In many types of communication circuits the information to be passed is in the form of "bursts" or pulses. Normally, the period of the signal transmitted during the "burst," or "on" time, of transmission, is small compared with the duration of the burst and the repetition rate of the burst is much smaller than the signal frequency.

Purely resistive networks will pass this type of signal without distortion as a delay in time. A filter, like other reactive networks, has both a phase characteristic that is not perfectly linear and a restricted bandwidth. Filters, therefore, delay this burst and may introduce ringing and overshoot. Distortion due to overshoot and ringing can make the transmission unintelligible, especially when the ringing time exceeds the "off" time of the signal.

This type of distortion is minimized when the phase shift versus frequency characteristic of the filter approaches linearity in the passband and when the attenuation-versus-frequency curve is Gaussian, rather than rectangular in shape.

The length of time a burst or pulse of signal information is delayed in passing through a filter is known as envelope delay. The envelope, or shape, of the burst is the locus of signal amplitude peaks within the burst. The slope of the curve showing phase shift versus frequency over the filter passband is equal to the envelope delay attributed to the filter.

In applications where several channels are operated in parallel, it is desirable that filters which perform the same function in each channel have the same bandwidth and envelope delay characteristics. In this way, equal delay compensation can be provided for each channel.

E. ***Impedance*** As a network composed of reactive elements, filters have a characteristic impedance that varies with frequency over the passband. The filter designer can frequently limit the variation by incorporating compensating networks. If the filter specification is too restrictive with regard to impedance variation, however, the filter design may become needlessly complicated. On the other hand, excessive impedance mismatch will cause ripple and deteriorated attenuation response. A practical design is one that provides impedance matching between the source and load sufficiently accurate to assure compliance with other requirements of the specification.

At higher frequencies stray capacity at the input and output can detune the filter and modify the apparent source and load impedance. These stray capacities when known, can be incorporated into the filter design.

F. ***Mechanical*** Filters are constructed to meet best industrial practice and military specification MIL-F-18327 whenever required.

Units are supplied milded or in steel, brass or plastic cases. Metal cases are furnished hermetically sealed and with threaded studs or tapped inserts for mounting and individual solder lug terminals or multiterminal headers. The cases are plated, sprayed with primer and enamel paints, and, finally, stamped or silk screened with identifying information.

Encapsulated or molded units are widely used for printed circuit applications. Straightpin-type terminals, spaced at multiples of 0.1 in., are provided. Mounting studs or inserts can be incorporated, if desired.

G. ***Core Materials and Stability*** The most widely used coil form for filter application is the toroid. This shape provides excellent shielding against stray magnetic fields and high inductance in a small volume (see Section 9.3).

Toroidal cores compressed of molybdenum permalloy powder are suitable for frequencies from below 100 cps to several thousand cycles. By varying the size of the grains and the mixture of grains and binder, the core manufacturer is able to change the effective air gap in the core and thus provide cores of differing permeability. Molybdenum permalloy powder cores have good temperature stability, which can be further improved by special cycling techniques. It is possible to obtain inductance stability of better than 0.5% over a temperature range of -100 to $+220°$F, although typical results are closer to 1%. At low RF frequencies and above low-permeability toroidal cores compressed of iron powder are used. These cores provide Qs and temperature stability comparable to or better than those of molybdenum permalloy.

Filters may be constructed with ferromagnetic pot cores, rather than toroidal cores. By completely enclosing the coil in a magnetic shield, the stray field around pot cores is extremely small. These cores are adjusted by varying the air gap, and when adjusted for minimum core losses, Q's of several hundred may be obtained. The temperature stability depends primarily on the size and stability of the air gap and approaches that of toroidal cores.

At low audiofrequencies coils for filters are constructed with bobbin windings and μ-metal laminations. High inductance and satisfactory Q are obtainable with this construction.

Coils used for filters have a positive temperature coefficient. When resonated with capacitors, the resonant frequency can be stabilized by selecting negative temperature-coefficient capacitors, such as polystyrene film or ceramics.

Filters are often required to maintain their specified characteristics over a wide range of input voltage levels. Consequently, each inductance value within the filter must be essentially independent of voltage level changes. This can be accomplished if (1) core materials and coil design are selected to limit flux density to a range where minimum change in permeability takes place, and (2) the equivalent air gap of the core is sufficiently large so as to minimize the effect of change in core material permeability with flux density variations.

Low-frequency filters are usually more sensitive to voltage level changes than are high-frequency filters. Cores for low frequencies have a relatively smaller air gap and operate at higher flux densities where variations in permeability are large.

It is important that the range of operating levels be stated to ensure proper adjustment of the filter.

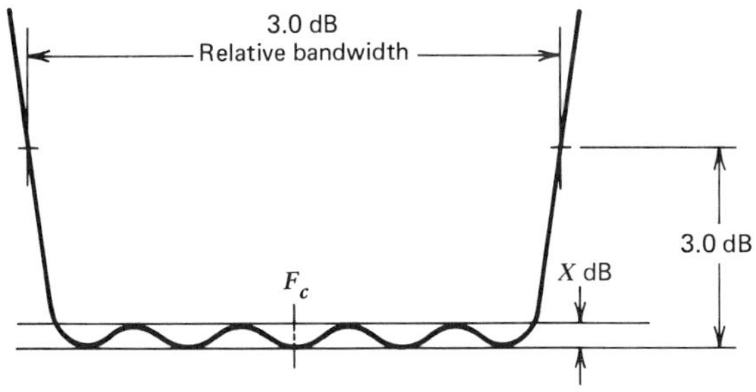

Figure 15.5 Response curve of a theoretical bandpass filter with infinite Q.[96]

15.2.2 Passband Relationships[96]

In many cases it is necessary to know more about the passband of a filter than its insertion loss at the center frequency and its 3-dB bandwidth. Figure 15.5 illustrates the response of a "lossless" (infinite Q) filter of X-dB Chebyschev design. All the attenuation is due to reflection and not to dissipation. This theoretical filter design would provide a flat passband X-dB ripple.

In practice, however, finite values of Q result in dissipative losses; therefore, Figure 15.6 is a more realistic representation of a typical filter's response. The dissipative losses are greater at the band edges than at the center frequency. The passband of the filter becomes rounded at these edges.

Since both reflective and dissipative losses are present in each filter, the ripple caused by the reflective losses becomes superimposed on the rounded passband created by the dissipative losses. As a consequence, it is more meaningful to specify a "relative" bandwidth, as shown in Figure 15.6, than a ripple bandwidth.

15.2.3 LC Filters[97]

LC filters cover the frequency spectrum from DC to 100 MHz for low pass and from 100 Hz to 100 MHz

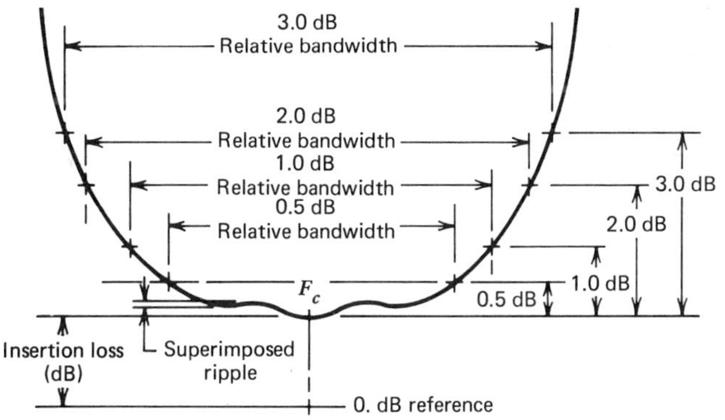

Figure 15.6 Actual filter-response curve with finite Q values.[96]

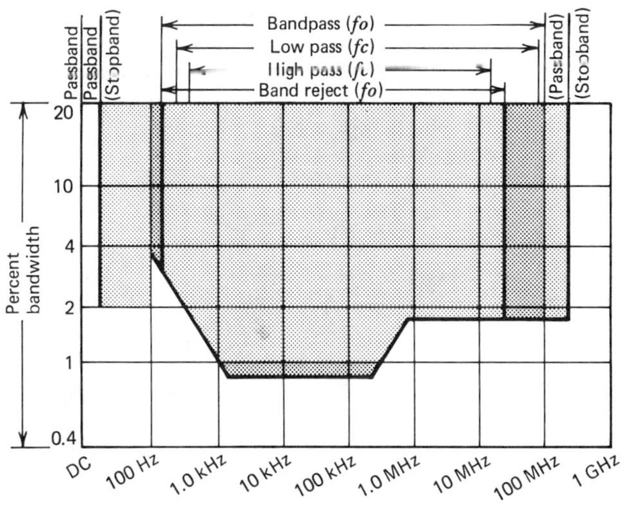

Figure 15.7 LC filter practical design limits.[96]

for bandpass (see Figure 15.7). Components of each filter design are closely matched and temperature-compensated for stability with varying ambient temperature and age. Optimized designs permit the smallest possible volume packaging.

A. Typical Filter Design Types.

1. Bessel.
2. Butterworth.
3. Cauer (elliptic function).
4. Chebyshev.
5. Gaussian.
6. Transitional (Gaussian to 6 dB) (Gaussian to 12 dB).
7. General parameter.
8. Image parameter.
9. Legendre.

B. Design Parameters (see Figure 15.8).

1. Maximum tolerable flat loss.
2. Maximum tolerable ripple.
3. Passband cutoff frequency for LP or HP.
4. Passband bandwidth, $fc_2 - fc_1$, for BP or BR.
5. Stopband frequency for LP or HP.
6. Stopband bandwidth, $fs_2 - fs_1$, for BP or BR.
7. Center frequency, f_0, for BP or BR.
8. Operating and storage temperatures.
9. Input and output impedance.
10. Environmental: shock, vibration, humidity.
11. Volume: length, width, height.

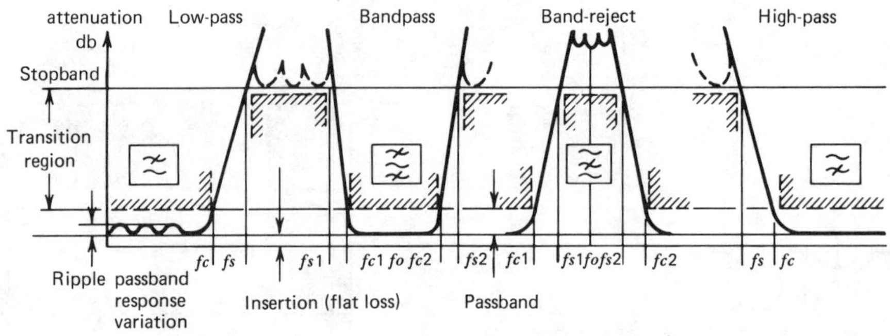

f_c - Cut off frequency (for low pass and high pass)
f_s - Stop band frequency (for low pass and high pass)
f_o - Center frequency (for band pass and band reject)

Figure 15.8 LC filter parameters.[97]

C. Design Considerations.

1 Phase response, linearity.
2 Envelope, group delay.
3 Return loss.
4 Harmonic distortion.
5 Longitudinal balance.
6 Transient response.
7 Shielding requirements.

D. Types of construction.

1 Open circuit board.
2 Plated brass.
3 Plated cold roll steel.
4 Plated Co-netic steel.
5 Molded plastic.

15.2.4 Crystal Filters[85,91]

The crystal filter is a practical, economical means for achieving better utilization of the existent frequency spectrum. The attraction of crystal filters is that they provide high selectivity with excellent temperature stability and a wide variety of frequency characteristics. With their use, a system design is simplified by the elimination of many components that have reliabilities much less than that of quartz crystals. A simple aspect is that the crystal element may have a Q in the hundreds of thousands, an inductance in the thousands of henries, a stability of one part in ten to the seventh per day with no electromagnetic coupling to any coil in the system.

Crystal units are well suited for use in filters because of their high Q and stable frequency characteristics. The most extensive use of crystal units in filters has been in bandpass filters with a lesser demand for band-reject filters and a still smaller demand for low- and high-pass filters. Combination of crystal units and high Q capacitors yield narrowband filters having attenuation characteristics unobtainable by combinations of inductors in addition to crystal units and capacitors yield wider passbands. However, all designs are necessarily limited by the ratio of the static capacity to the motional capacitance of the crystal unit. Most crystal bandpass filters make use of the symmetric lattice network or one of its many derivations. Ladder structures are sometimes used for very narrowband filters eliminating the need for transformers required in the lattice derivations in unbalanced circuits.

Narrow-band crystal filters can be designed with bandwidths as narrow as 0.005% of the center frequency and up to 0.7% of the center frequency. The lower limit is determined primarily by the Q of the filter element while the higher limit is established by the zero-pole spacing of the particular crystal element being used.

Bandwidths for the wideband category extend from 0.6% of center frequency to 10% of center frequency. The lower limit is determined primarily by the Q of the inductances used in conjunction with the crystals. Theoretically, the upper limit depends on the zero-pole spacing of the crystal element. However, in practice it is usually determined by limitations in coil design. The wider bandwidth filters require extremely high inductances with little distributed capacity. In some instances, primarily in the frequency range where AT cut crystals are used, the upper limit is set by spurious modes in the crystals. The shape factor of both narrow-band and wide-band filters depends on the number of sections being used in the particular filter (see Section 14.1).

15.2.5 Mechanical Filters[94]

Mechanical filters provide high adjacent channel rejection by steep-skirted selectivity while achieving a flat-topped frequency response. Electrically and mechanically stable, the filters tolerate extreme temperature changes with minor frequency shift and give continuous service without aging or breakdown. The filters are sealed in molded-phenolic cases or hermetically sealed metal cases. Values of frequency stability with temperature vary from 2 to 10 ppm/°C between -40 and +85°C for various filter types. In general, they may be operated over wider temperature ranges with slight degradation of characteristics.

Applications for the wide range of filters include single side-band, high-performance transmitting and receiving equipment, multiplexing equipment, missile guidance systems, frequency synthesizers, Doppler radar, data transmission systems, precision navigation equipment, spectrum analyzers, FM communications receivers, CB transceivers, and others.

The mechanical filter is a mechanically resonant device that receives an electrical signal, converts this signal into mechanical vibrations, rejects unwanted frequencies within the mechanical structure, and then converts the mechanical vibration back into electrical energy. The filter consists of three basic elements: (1) transducers that convert electrical signals into mechanical vibrations, (2) high Q mechanically resonant metal discs, and (3) disc coupling wires. Figure 15.9 shows an electrical signal applied to the input coil. This produces an alternating magnetic field that passes through the magnetostrictive rod attached to the first disc. Since a magnetostrictive material changes its dimensions when subjected to a magnetic field, the rod, when biased properly, will vibrate at the frequency of the impressed signal. The rod, in turn, drives the first disc which, by means of the short coupling wires connecting the discs, drives the next disc, and so on, until the signal reaches the output transducer. Strains developed in the output magnetostrictive rod produce an alternating magnetic field, which induces a voltage across the output coil.

The transducers not only convert energy from one form to another, but also reflect the source and load resistances into the mechanical circuit. The reflected impedances provide a termination for the filter.

The multielement filter in the electrical analog (Figure 15.9) illustrates how the center frequency of the filter is determined by the metal discs. Each disc is represented by a parallel resonant circuit. Increasing the number of discs increases the skirt. Selectivity of the filter skirt selectivity is specified as shape factor, which is the ratio of 60-dB bandwidth. Shape factors are as low as 1.2 to 1.

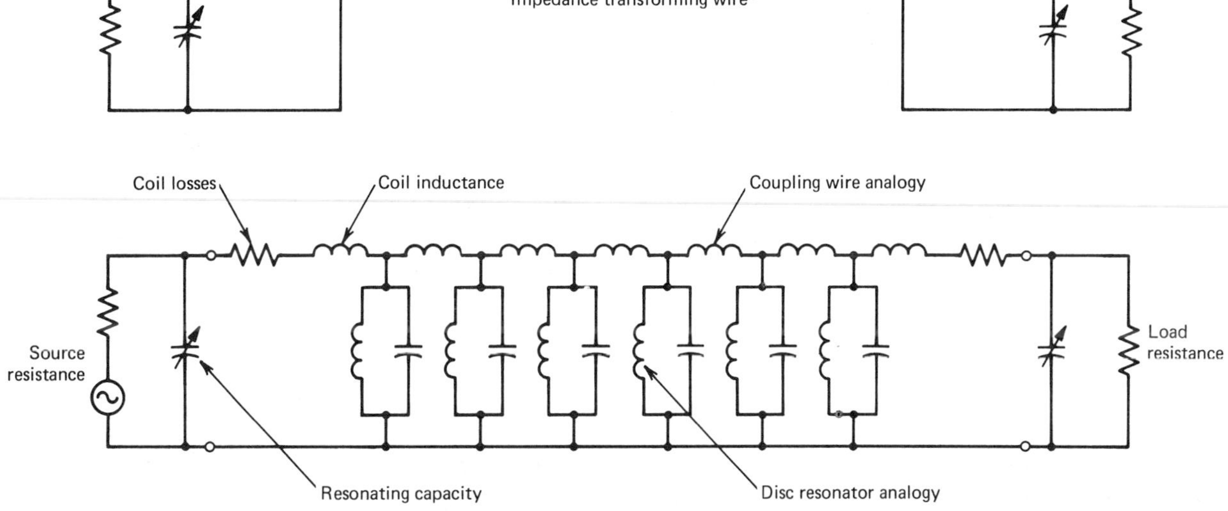

Figure 15.9 Mechanical filter analogy.[94]

15.3 RFI/EMI FILTERS[98–100]

Any piece of equipment that uses or controls electricity is a potential source of RFI (radiofrequency interference) or EMI (electromagnetic interference). EMI/RFI problems increase exponentially with the increased average density of sources.

EMI/RFI problems can be reduced by taking design precautions or by shielding the devices with commercially available coating. In some cases, however, these undesirable emissions are unavoidable by-products of an electronic process, because they travel on the same wires that bring power or information into or out of the unit, so that the only solution in these cases is to provide a method of halting the flow of these electronic parasites into or out of the device in question. At the same time, the normal flow of information and/or power must not be disturbed.

EMI/RFI filters are designed to do just that. When used on power lines, they will allow the relatively low power frequencies (50–400 cps) to pass unaffected. Frequencies considerably higher are attenuated.

Most digital equipment malfunctions that result from interference happen because of common mode noise (conducted RFI that is measurable from either or both lines-to-ground) or line-to-ground noise in the frequency range of 1–20 MHz. Differential mode noise (measurable conducted RFI between 2 AC lines) or line-to-line noise consists of lower frequencies, this latter energy is not as easily radiated or conducted over the power system. Consequently, this line-to-line noise is less of a problem, unless it is generated close to susceptible equipment.

AC power line filters are simple LC networks. They must fit in limited space and, if properly designed, they should not make operation hazardous by permitting excessive leakage currents to flow through their capacitors from the AC line to ground. And, they must not fail when exposed to line transients as high as 1500-V peak. During lightning storms, transient voltage this high is a common occurrence on power lines.

The balun construction has an additional user advantage, because the inductors are essentially unaffected by 50/60-Hz line currents and core saturation does not occur during startup overloads. This means little power is lost in the filter.

The location and installation of the filter is critical. For most effective RFI attenuation, the filter should be located near the line-cord entrance point. Ideally, the filter's location should be considered during the primary stages of the equipment's mechanical design. Filters are available with integral AC line-cord connector, either in U.S. or European types.

15.3.1 Conducted Radiofrequency Interference (RFI)

RFI energy conducted via the power connection into digital electronic equipment may cause malfunctions if the circuitry is susceptible to it. This interference can occur in two different modes.

A. *Common Mode Interference* Present as a voltage from all the power connections to ground.

B. *Differential Mode Interference* Present as a voltage between the power connections.

Most susceptibility problems are caused by common mode interference. This form of RFI energy is readily distributed within power-distribution systems. It is also induced on the power cord by radiated RFI energy and is radiated from such wiring into the adjoining space.

Differential mode interference is troublesome if the offending source is in close proximity to the susceptible equipment. This form of RFI is quickly attenuated in the building wiring. It is neither received by, nor radiated from, the power cord.

Noisy equipment such as switching regulators will generate RFI over a very wide frequency spectrum. This RFI energy will have both common mode and differential mode components.

RFI filters can be thought of as "mismatching networks" at high frequencies. The greater the mismatch is to the terminating impedance, the more effective is the filter in attenuating RF energies (see Figure 15.10).

An RFI filter for use in a commercial/industrial environment will have to work between either of the following sets of termination impedances:

Condition	Power Line Z at High Frequencies	Equipment Z at High Frequencies
A	Low	High
B	Low	Low

Condition A pertains to equipments containing linear power supplies. Condition B pertains to equipments containing switching power supplies, shunt regulators, or synchronous motors.

Since the power line is considered a low impedance, the filter should present to it a high impedance, that is, a series inductance. If the equipment presents a high impedance, the filter must present a low impedance, that is, a shunt capacitor. If the equipment like the power line is a low impedance at high frequencies, the filter must present a high impedance, requiring a series inductor.

Once the basic filter design has been chosen, any filter with the proper current rating, the desired mechanical features (Figure 15.11), and the right price can be

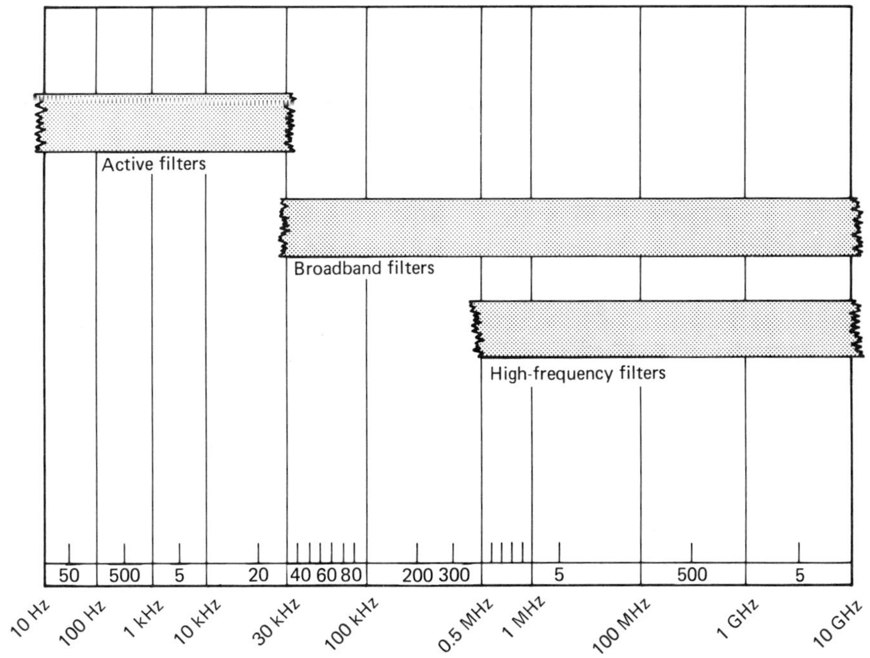

Figure 15.10 EMI filter spectrum.[92]

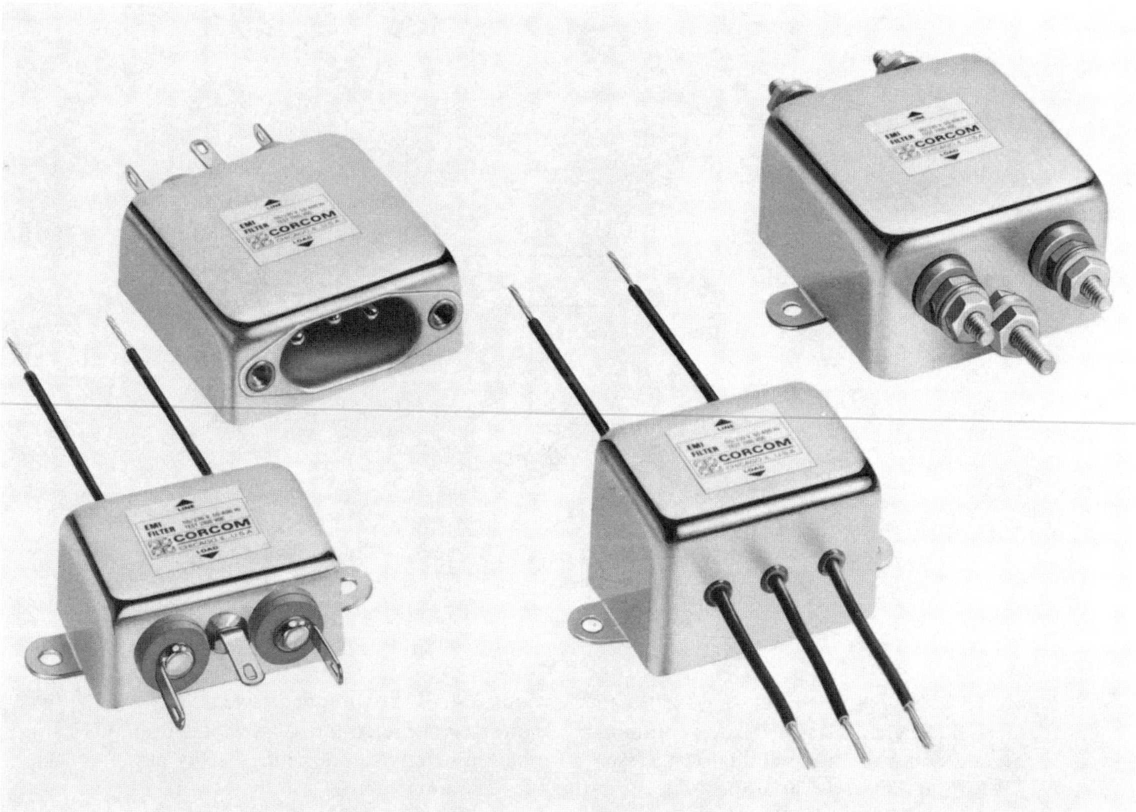

Figure 15.11 Typical EMI filter packages. (Courtesy of Corcom, Inc.)

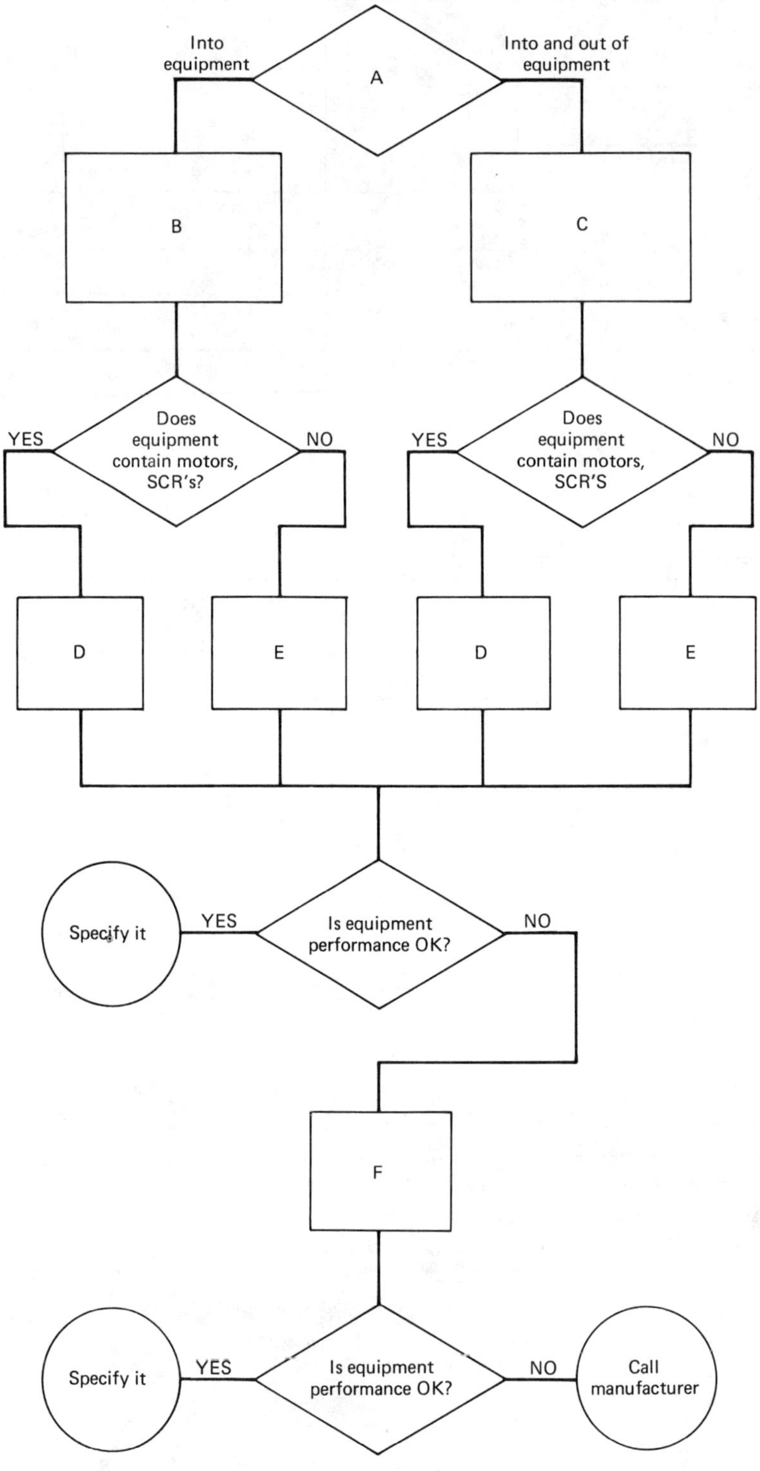

Figure 15.12 Filter selection.[98] A—Is the primary purpose of the filter to attenuate interference? B—Select a filter with good common mode interference rejection. C—Select a filter with good common mode and good differential mode interference rejection. D—Try any $N = 3T$ filter network of proper current rating. E—Try any $N = 2L$ filter network of proper current rating. F—Try same filter type, same network but with less inductance or with higher current rating.

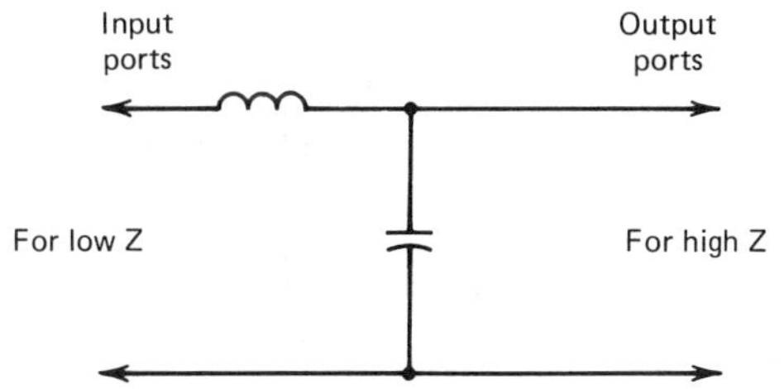

Figure 15.13 L network.[99]

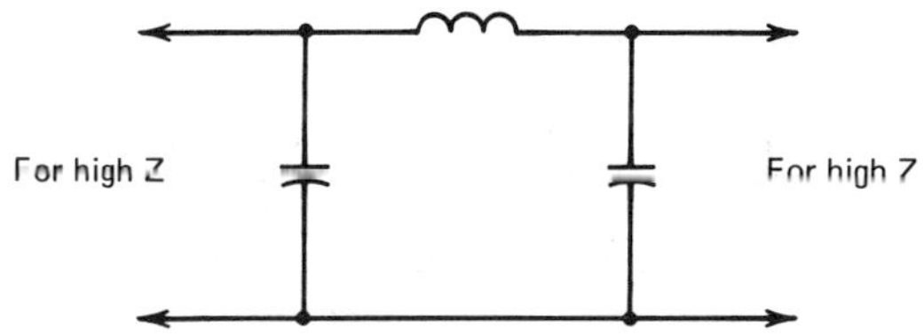

Figure 15.15 Pi network.[99]

tried. The flow chart in Figure 15.12 should prove helpful in choosing a filter for most applications. If the first filter selected fails to protect the equipment, or if too much RFI is conducted to the power line, another filter with the same circuit configuration, but with less inductance or higher current-carrying capacity, should be tried. If the second attempt fails, then the filter manufacturer should be contacted for assistance.

15.3.2 Filter Networks

The filter can be thought of as a two-terminal network. This network, like all such networks, presents input and output impedances that vary with, amongst other factors, frequency. This variance is the key to EMI/RFI filters. If they are to attenuate the high or unwanted frequencies, they must present an impedance mismatch at high frequencies at both ports. As a mismatching network, the filter prevents the undesired noise from being effectively transferred to the output, which is the desired effect. On the other hand, it is equally important that power or useful information not be attenuated.

Accordingly, filters do not mismatch or attenuate these desirable frequencies. The power line itself always presents low impedance to RF (radiofrequencies). The device being powered may present either high or low impedance to RF. Typically, linear power supplies fall into the former category, while switching power supplies, shunt regulators, and synchronous motors may be included in the latter.

A proper value of inductance presents a high impedance to RF. Thus, a port that is to face the power line or a low impedance device should present a series inductor. A capacitor of the correct size presents a low impedance to RF, and it follows that a port that faces a device exhibiting a high impedance to RF should present a shunt capacitance.

An L network (Figure 15.13) consists of a series inductor followed by a shunt capacitor. This is suited for low-impedance input and high-impedance output. The T network of Figure 15.14 is for low Z in and low Z out, because series inductors face both ports. The pi network of Figure 15.15, presenting shunt capacitance at both ports, is for high impedance RF at both ends. Finally, the double L network of Figure 15.16 is like two ordinary L networks. The series inductor at the input makes this device suited for low-input impedances. The shunt capacitor matches the network to high values of equipment impedances. The two L sections provide more vigorous filtering than normally possible with one.

Since these circuits are mismatch networks, it does not matter which way the signal is traveling. A mismatch is presented to RF, whether it is traveling from the equipment to the power line or from the power line to the equipment. Thus the same filter can be used to protect the device from environmental noise or the environment from the RFI/EMI generated by a device.

Effective EMI/RFI filtering can be achieved with capacitors alone, if certain types of capacitors are used. When special filter capacitors are used, they are constructed to have essentially zero lead length. These filter capacitors have positive foils connected directly to the feedthrough (signal carrying) line. The grounded foil is fastened directly to the metal case or ground.

While still not ideal capacitors, these types of devices do more closely approach the ideal capacitor.

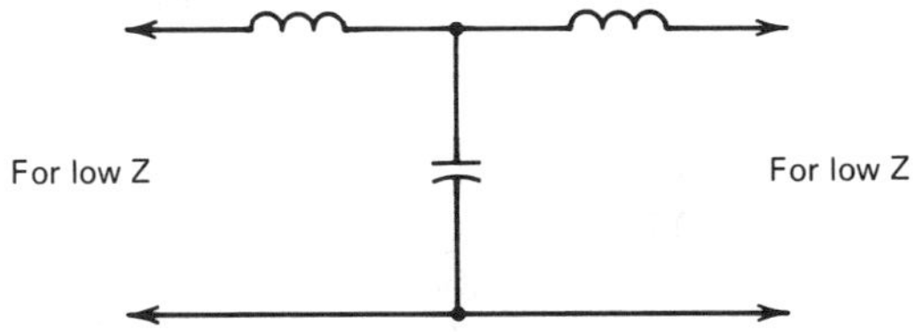

Figure 15.14 T network.[99]

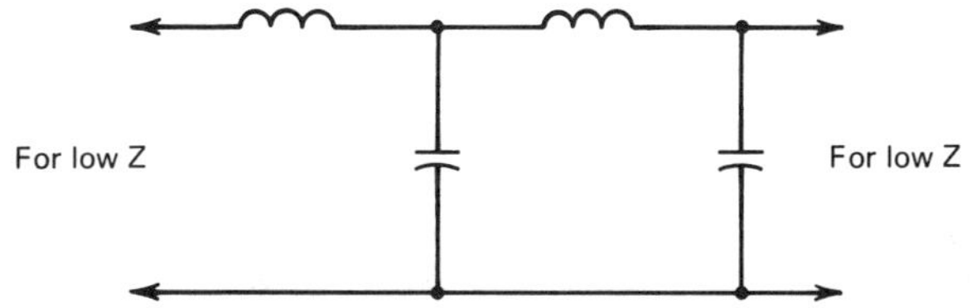

Figure 15.16 Double-L network.[99]

15.3.3 Specifying RFI Filters

Specifying power line interference filters present some unusual problems. It is not generally practical to define the filter in such a way that any filter manufacturer can produce a component that will perform in the equipment as the prototype did. This is because the filter performance is a function of the impedances to which it is connected, and these impedances are not generally known at frequencies other than 50 or 60 Hz.

The filter performance is also a function of its internal impedances, that is, its capacitive and inductive components. These can be specified, but again it is their high-frequency characteristics that are critical, and these are not easily determined. The only practical solution is to evaluate the filter in the equipment and then to specify it so that incoming inspection may verify that the product uniformity is acceptable. The following items should be included in a filter specification:

A. Filter circuit configuration.

B. Mechanical dimensions.

C. Operating voltages currents and frequencies.

D. Operating temperatures and humidity ranges.

E. Applicable specifications, that is, IEC, VDE, UL, CSA, and so on.

F. Hipot, insulation resistance, leakage current, and overload current requirements.

G. Incoming inspection test, that is, inductance and capacitance measurements to be made between terminals and terminals to case, or MIL-STD-220 insertion loss. (Although these insertion loss data cannot be used in evaluating filter effectiveness, it is a useful incoming inspection procedure for measuring product uniformity.)

H. Type approval (evaluation of the filter in the equipment is required in every case before engineering approval can be granted).

The last requirement means that all component suppliers must have their filters evaluated in the equipment. This is necessary, because even if the filters have the same 1-kHz inductance and capacitance values, they may not all perform equally in the same equipment.

15.3.4 Specifications

There are many national and international specifications on interference control and safety. CISPR Publication 9-1967 catalogs all national limits on radio interference. Germany, Switzerland, and Canada have specifications limiting conducted and radiated RFI. CISPR 1 is a specification of RFI measuring apparatus. In the United States, the FCC is expected to issue rules governing the RFI emitted by electronic data processing equipment.

IEC 161 sets standards for capacitors used to attenuate RFI. This is a model for German, British, Swedish, Swiss, and other international specifications. The safety requirements limiting line-to-ground leakage currents vary from country to country, but everywhere requirements are becoming more restrictive than in the past. ANSI C101.1-1973 proposes a maximum of 500 μA for use in the United States.

Many of these specifications are available, some in English, from the American National Standards Institute. (See Table 15.2.)

A. *ANSI (American National Standards Institute)* 1430 Broadway, New York, N.Y. 10018. Publishes specifications for use in the United States, distributes American and international standards and specifications.

B. *CISPR (Comité Internationale Spéciale des Perturbations Radioeléctriques)* 1 rue de Varembe, Geneva, Switzerland. A special committee of the IEC dealing with specifications on radio interference.

C. *CSA (Canadian Standards Association)* 78 Rexdale Blvd., Rexdale, Ontario, Canada. Writes safety and radio frequency interference specifications for Canada. Conducts tests to ensure compliance.

D. *IEC (International Electrotechnical Commission)* 1 rue de Varembe, Geneva, Switzerland. Writes technical specifications recommended for use as standards by all countries.

E. *SEV (Schweizerischer Elektro-Technischer Verein)* 8008 Zurich, Seefeldstrasse 301, Switzerland. Writes safety and audio frequency interference specifications for Switzerland. Conducts tests to ensure compliance.

F. *UL (Underwriter's Laboratories)* 207 East Ohio Street, Chicago, Illinois 60611. Writes safety specifications for the United States. Conducts tests to ensure compliance.

G. *VDE (Verband Deutscher Elektrotechniker)* VDE Prufstelle, D-6050 Offenbach (Main), Merianstrasse 28, West Germany. Writes safety and radio frequency interference specifications for West Germany. Conducts tests to ensure compliance.

Table 15.2 National and International Interference Filter Requirements and Specifications[98]

Country	Leakage current equipment requirement			Hi-pot test		Insulation resistance		High voltage pulse test	Current overload test	Approving organization name	City	Applicable specifications
	Max I ma	At volts	Hz	volts	Hz	Min ohms x 10[6]	At volts					
Canada	5.0	120	60	1500 1000 +2 V_O	0 60[5]	—	—	—	—	CSA	Ontario	C22.2 #8
France	—	—	—	—	—	—	—	—	—	LCIE	Paris	—
Germany	3.5 0.5	250 250	50[1] 50[2]	2250 or 1500 4.3 x V_O +1075 2000	0[3] 50[3] 50[4] 50[5]	12,000	100	—	75 x I rated for 2.5 cycles	VDE	Offenbach	0560-1 0560-7 0550-1 0550-6 0565-1[6] 0565-2[6] 0565-3[6]
Sweden	3.5 0.5	250 250	50[1] 50[2]	1500	50	6000	100	—	—	SEMCO	Stockholm	IEC 161
Switzerland	0.5 0.5	250 250	50[1] 50[2]	2000 or 1000 +4 V_O	50 50	3000	100	Yes, by 5.12	—	S.E.V.	Zurich	1016.1959 1003.1959
United Kingdom	3.5 0.5	250 250	50[1] 50[2]	2250 or 1500 4.3 x V_O +1075 2000	0[3] 50[3] 50[4] 50[5]	12,000	100	—	25 x I rated for 1 min	B.S.I.	London	BS 2135 BS 613
U. S.	5.0	125	60	1075	60	—	—	—	—	UL	Various	UL 478

1 — Class I equipment (ground wire)
2 — Class II equipment (double insulated)
3 — Line-to-ground
4 — Line-to-line
5 — Line-to-case
6 — Proposed
V_O — Line voltage

15.4 FILTER TERMS AND DEFINITIONS

Attenuation The ratio of output voltage (E_2) at the reference frequency to the output voltage (E_3) at the test frequency.

$$\text{Attenuation in decibels} = 20 \log_{10} \frac{E_2}{E_3}$$

Bandpass A type of filter. Frequencies falling within the specified frequency band are passed with minimum attenuation, all other signals being attenuated.

Bandstop A type of filter. Frequencies falling within the specified frequency band are attenuated, all other signals being passed.

Bandwidth Frequency difference between points occurring at the same relative attenuation level.

Center Frequency (F_o) The reference frequency of bandpass filters or the frequency of maximum rejection for band-reject filters.

Conducted Interference Any undesired signal conducted in either direction over power lines or interconnecting cables, that may interfere with the normal intended use of equipment/system.

Cutoff Frequency That frequency marking the edge of the passband.

Electromagnetic Interference Either conducted or radiated electrical impositions, including transients, generally within the frequency spectrum of 30 Hz to 10 kHz, that may cause the malfunction of the intended operation of equipment.

Envelope Delay The time delay due to transmission through a network of the envelope of a complex signal. This can be expressed as the variation of phase shift with frequency.

Filter An important parameter of bandpass and bandstop filters:

$$\text{Bandpass: } Q = \frac{f_o}{\text{3-dB bandwidth}}$$

$$\text{Bandstop: } Q = \frac{f_o}{\text{Ultimate attenuation bandwidth}}$$

Flatness of Response The variation between the peak and valley of the response curve in the passband.

Geometric Mean Frequency (f_o) The square root of $f_1 \times f_2$, where f_1 and f_2 are lower and upper frequencies, respectively, at which a particular signal attenuation occurs, usually but not necessarily taken as 3-dB relative attenuation. An important parameter of bandpass and bandstop filters.

Highpass A type of filter. All frequencies below the specified f_c are attenuated; all above are passed.

Input Impedance (Z_{in}) The impedance looking into the filter's input terminals with the filter properly loaded at the output terminals.

Insertion Loss The ratio of power delivered to the load before insertion of the filter, to the power delivered to the load after insertion of the filter.

LC Filter, Bandpass A frequency-selective network made of discrete inductive and capacitive components which passes signals falling within the specified frequency band, all other signals being attenuated.

LC Filter, Bandstop A frequency-selective network made of discrete inductive and capacitive components that attenuates signals falling within the specified frequency band, all other signals being passed.

LC Filter, High Pass A frequency-selective network made of discrete inductive and capacitive components that passes signals falling above the specified cutoff frequency, all other signals being attenuated.

LC Filter, Low Pass A frequency-selective network made of discrete inductive and capacitive components that passes signals falling below the specified cutoff frequency, all other signals being attenuated.

Longitudinal Balance A measure of output when a signal is applied to both input terminals simultaneously and to ground.

Low Pass A type of filter. All frequencies below the specified f_c are passed; all above are attenuated.

Output Impedance (Z_{out}) The impedance looking into the filter's output terminals with the proper resistor across the input terminals.

Overshoot The maximum output amplitude compared with the steady-state amplitude, that is,

$$\% \text{ Overshoot} = \frac{(\text{maximum output})}{(\text{steady-state output})} - 1 \times 100$$

Passband Range of frequencies over which signals will suffer less than a specified amount of relative attenuation (usually 3 dB).

Passband Ripple The difference from peak to valley of the amplitude response in the passband measured in decibels.

Percent Bandwidth The ratio of the difference between frequencies of specified attenuation, such as the cutoff frequencies F_L and F_H, to the center frequency (F_o).

$$\% \text{ BW} = \frac{F_h - F_L}{F_o} \times 100$$

Radiated Interference An undesired signal that radiates from energized equipment/systems or from the supplying power lines.

Reference Frequency That frequency at which the insertion loss is measured and to which all attenuation measurements are referred. In bandpass filters, the reference frequency may be the center of the passband or the frequency at which maximum output occurs. In low-pass and high-pass filters, the reference is a frequency well within the flat portion of the passband.

Relative Attenuation Attenuation measured with the point of minimum attenuation taken as 0 dB, or

Relative attenuation = attenuation - insertion loss

Ringing Time The duration of the oscillatory transient.

Rolloff Rate The ultimate slope of the transition region of the transfer characteristic, usually expressed in decibels per octave of frequency change (an octave is 2:1 ratio of frequencies) and occasionally expressed in decibels per decade of frequency change (10:1 frequency ratio).

Rolloff Slope See Shape Factor.

Shape Factor The ratio of the bandwidth at a high level of attenuation to that of the passband.

$$\text{Bandpass: } S = \frac{\text{Ultimate attenuation bandwidth}}{\text{3-dB bandwidth}}$$

$$\text{Low pass: } S = \frac{\text{Ultimate attenuation frequency}}{f_c}$$

$$\text{Bandstop: } S = \frac{\text{3-dB bandwidth}}{\text{Ultimate attenuation bandwidth}}$$

$$\text{Highpass: } S = \frac{f_c}{\text{Ultimate attenuation frequency}}$$

Source Impedance (R_s) and Load Impedance (R_L) The impedance of the generator or driving source and of the load are almost always resistive, whereas a filter's input and output impedance is generally reactive, except for active filters.

Stopband That band of frequencies that the filter discriminates against.

Susceptibility The characteristic of the equipment/system that responds to the interfering signal—either conducted or radiated—and malfunctions.

Transfer Function The ratio of output to an input signal (or, more properly, the ratio of the output spectrum to the input spectrum), expressed in terms of both magnitude ratio and phase shift.

Ultimate Relative Attenuation Required attenuation of signals whose frequencies are far removed from the passband, measured in decibels, with the point of minimum passband attenuation taken as 0 dB.

PART SIX
Solid State Devices

CHAPTER SIXTEEN

Diodes and Rectifiers

16.1 SEMICONDUCTORS[101]

Solid-state devices are small but versatile units that can perform an amazing variety of control functions in electronic equipment. Like other electron devices, they have the ability to control almost instantly the movement of charges of electricity. They are used as rectifiers, detectors, amplifiers, oscillators, electronic switches, mixers, and modulators.

In addition, solid-state devices have many important advantages over other types of electron devices. They are very small and light in weight (some are less than an inch long and weigh just a fraction of an ounce). They have no filaments or heaters, and therefore require no heating power or warmup time. They consume very little power. They are solid in construction, extremely rugged, free from microphonics, and can be made impervious to many severe environmental conditions. The circuits required for their operation are usually simple.

16.1.1 Semiconductor Materials

Unlike other electronic devices, which depend for their functioning on the flow of electric charges through a vacuum or a gas, solid-state devices make use of the flow of current in a solid. In general, all materials may be classified in three major categories—conductors, semiconductors, and insulators—depending on their ability to conduct an electric current. As the name indicates, a semiconductor's material has poorer conductivity than a conductor, but better conductivity than an insulator. The material most often used in semiconductor devices is silicon.

A. ***Resistivity*** The ability of a material to conduct current (conductivity) is directly proportional to the number of free (loosely held) electrons in the material. Good conductors, such as silver, copper, and aluminum, have large numbers of free electrons. Insulators such as glass, rubber, and mica have very few loosely held electrons.

Semiconductor materials lie in the range between these two extremes. (See Figure 16.1.)

Pure silicon has a considerably higher resistivity, in the order of 60,000 Ω-cm. As used in semiconductor devices, however, these materials contain carefully controlled amounts of certain impurities that reduce their resistivity to about 2 Ω-cm at room temperature (this resistivity decreases rapidly as the temperature rises).

B. ***Impurities*** Impurity elements which are added to silicon crystals to provide excess electrons include arsenic and antimony. When these elements are introduced, the resulting material is called *n* type because the excess free electrons have a negative charge. Semiconductor material containing "holes" or positive charges is called *p*-type material. *P*-type materials are formed by the addition of aluminum, gallium, or indium.

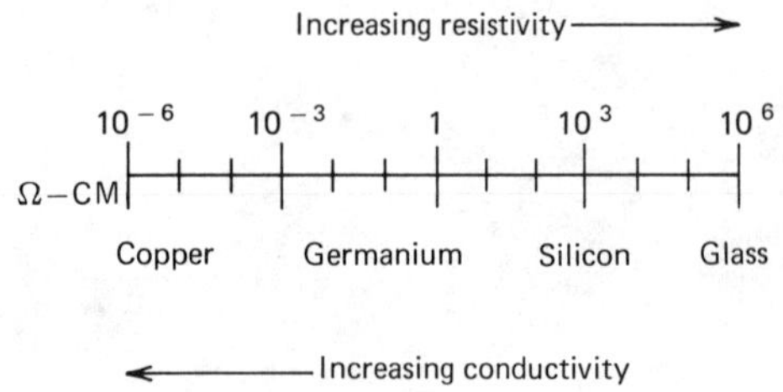

Figure 16.1 Resistivity of typical conductor, semiconductors, and insulator.[101]

Although the difference in the chemical composition of *n*-type and *p*-type materials is slight, the differences in the electrical characteristics of the two types are substantial, and are very important in the operation of semiconductor devices.

16.1.2 *p–n* Junctions

When *n*- and *p*-type materials are joined together, as shown in Figure 16.2, an unusual but very important phenomenon occurs at the interface where the two materials meet (called the *p–n* junction). An interaction takes place between the two types of material at the junction as a result of the holes in one material and the excess electrons in the other.

16.1.3 *n–p–n* and *p–n–p* Structures

A *p–n* junction biased in the reverse direction is equivalent to a high-resistance element (low current for a given applied voltage), while a junction biased in the forward direction is equivalent to a low-resistance element (high current for a given applied voltage). Because the power developed by a given current is greater in a high-resistance element than in a low-resistance element, power gain can be obtained in a structure containing two such resistance elements if the current flow is not materially reduced. A device containing two p–n junctions biased in opposite directions can operate in this fashion. Such a two-junction device is shown in Figure 16.3.

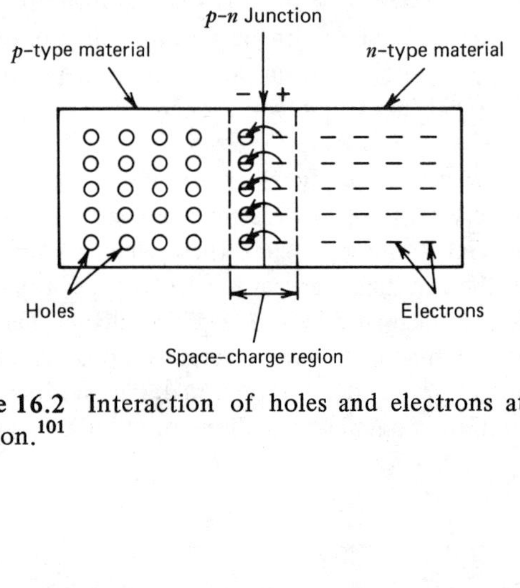

Figure 16.2 Interaction of holes and electrons at p–n junction.[101]

The operation of *p–n–p* devices is similar to that shown for the *n–p–n* device, except that the bias-voltage polarities are reversed, and electron-current flow is in the opposite direction.

16.1.4 Specifications and Standards[102]

There are two major military specifications used for the procurement of standard solid-state devices by the military. These specifications are MIL-S-19500, which covers devices such as discrete transistors, thyristors, and diodes, and MIL-M-35810, which covers microcircuits, both hybrid and monolithic.

These standards also cover a wide range of requirements for the manufacturer on such things as:

A. The procedure and requirements for a manufacturer to become certified to manufacture MIL-spec parts.

B. The requirements for qualifying parts.

C. Product-assurance provisions in such areas as quality control, inspection procedures, personnel training, cleanliness, failure analysis, and documentation.

D. Test methods and procedures.

E. Marking and identification of product.

F. Preservation and packing.

Both MIL-M-38510 and MIL-S-19500 attempt to make available to the designer of military equipment a list of standard, qualified, general-purpose parts which are acceptable to the military. Although MIL-S-19500 and MIL-M-38510 do not cover every solid-state device available on the market, and do not attempt to do so, enough devices are available to build most military equipment. Use of these devices makes the job of spare parts inventory far simpler for the military and the job of specification negotiations far easier for the equipment manufacturer.

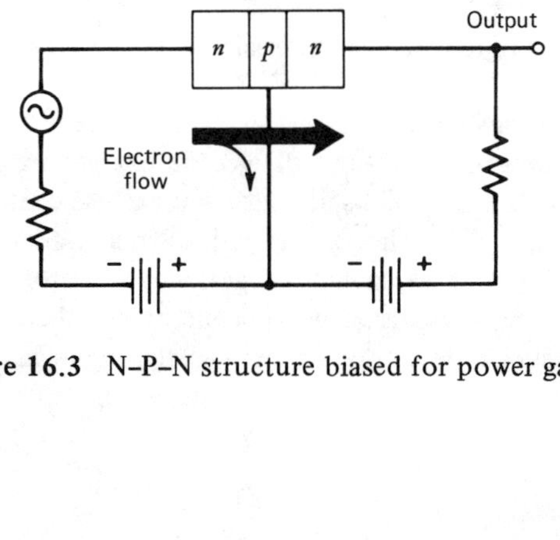

Figure 16.3 N–P–N structure biased for power gain.[101]

Many solid-state devices, both discrete types and integrated circuits, are not covered by military specifications, either because they are too new or are not used in sufficient quantities. Many of these devices offer the most recent technological advances or have special performance characteristics that offer advantages to the designer of high-reliability equipment.

Slash-series devices are supplied in several reliability levels (identified by a slash (/) number following the basic device type number) that approximate the Class A, B, or C reliability requirements defined by MIL-STD-883 for integrated circuits or the JAN, JANTX, JANTXV, or JANS requirements defined by MIL-S-19500 for discrete types.

Devices designated as JAN types receive lot screening only and are the least expensive. Devices designated as TX receive some 100-percent screening (primarily burn-in) and a tight lot-sampling plan. Not all detailed specifications include TX requirements. Devices designated as TXV are tested the same as TX devices; however, they receive an additional visual inspection prior to sealing the package. Only a few detailed specifications include TXV testing.

The type designation for semiconductor devices furnished under MIL-S-19500 are formulated as follows:

JAN	XN	YYYY	ZZ
Prefix (as applicable)	Component designation	Identification number	Suffix letter(s)

A. *Prefix* The four levels of product assurance from the lowest level to the highest level are JAN, JANTX, JANTXV, and JANS. A higher assurance level product is substitutable for a lower assurance level product. The JAN, TANTX, JANTXV, or JANS prefix is used as applicable.

B. ***Component Designation*** Semiconductor devices are identified by the prefix *XN*. *X* is usually a number one less than the number of active element terminations.

C. *Identification Number* It is recommended that each type of semiconductor device intended for standardization be assigned an identification, serially, by the Joint Electron Device Engineering Council, a council sponsored by the Electronic Industries Association and The National Electrical Manufacturers Association. This assignment provides the component designation, the identification number and, if applicable, a suffix letter.

D. *Suffix Letter* Suffix letter(s) may be incorporated in a type number to symbolically describe matching of

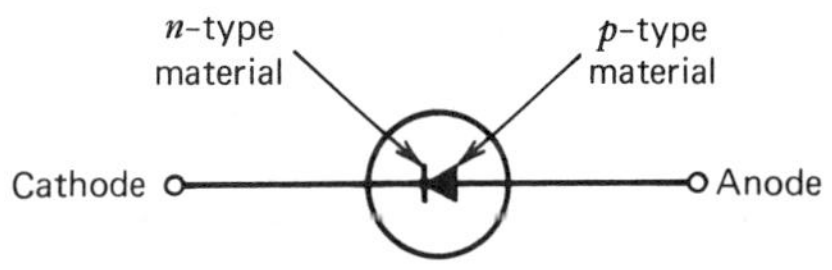

Figure 16.4 Schematic symbol for a semiconductor diode.[101]

separate devices (*M*), reverse polarity (*R*), or a letter to indicate a modified version of the device that can be substituted for the basic numbered unit and devices having previous suffix letters.

16.2 DIODES—GENERAL[103–105]

The simplest type of solid-state device is the diode, which is represented by the symbol shown in Figure 16.4. Structurally, the diode is basically a *p–n* junction.

The *n*-type material, which serves as the negative electrode, is referred to as the cathode, and the *p*-type material, which serves as the positive electrode, is referred to as the anode. The arrow symbol used for the anode represents the direction of "conventional current flow"; electron current flows in a direction opposite to the arrow.

Because the junction diode conducts current more easily in one direction than in the other, it is an effective rectifying device. If an ac signal is applied, electron current flows freely during the positive half-cycle, but little or no current flows during the negative half-cycle.

One of the most widely used types of solid-state diode is the silicon rectifier. These devices are available in a wide range of current capabilities, ranging from tenths of an ampere to several hundred amperes or more, and are capable of operation at voltages as high as 1000 V or more. Parallel and series arrangements of silicon rectifiers permit even further extension of current and voltage limits.

The circuit symbol, commonly used junction schematic, and electrical characteristic for the basic diode/rectifier are shown in Figure 16.5a–c, respectively.

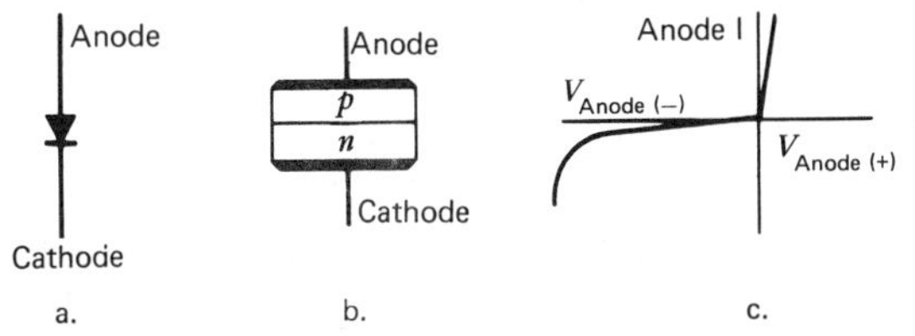

Figure 16.5 Diode/rectifier. (*a*) Circuit symbol, (*b*) junction schematic, (*c*) electrical characteristic.[103]

Figure 16.6 Diode/rectifier case outlines. (Reproduced from Ref. 118.)

Diodes and rectifiers are supplied in a wide variety of mechanical configurations, as shown in Figure 16.6.

16.3 TUNNEL DIODES AND RECTIFIERS

A tunnel diode is a small *p–n* junction device having a very high concentration of impurities in the *p*-type and *n*-type semiconductor materials. This high-impurity density makes the junction depletion region (or space-charge region) so narrow that electrical charges can transfer cross the junction by a quantum-mechanical action called "tunneling." This tunneling effect provides a negative-resistance region on the characteristic curve of the device that makes it possible to achieve amplification, pulse generation, and RF-energy generation.

The circuit symbol and commonly used junction schematic for a tunnel diode are shown in Figure 16.7*a*, *b*, respectively; the electrical characteristic is shown in Figure 16.8.

16.3.1 Construction[106]

The structure of a tunnel diode is extremely simple, as shown in Figure 16.9. A small "dot" of highly conductive *n*-type (or *p*-type) material is alloyed to a pellet of

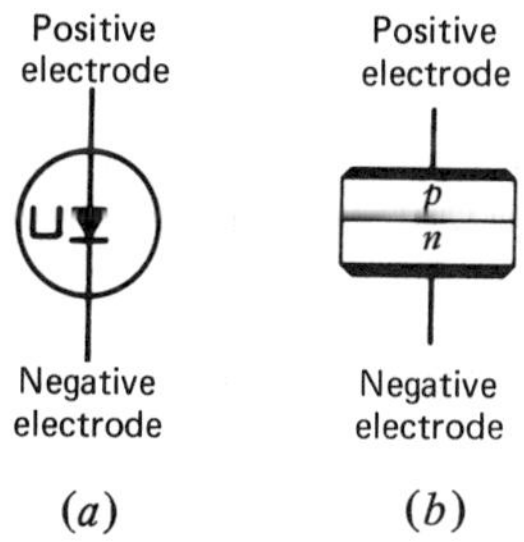

Figure 16.7 Tunnel diode. (*a*) Circuit symbol, (*b*) junction schematic.[103]

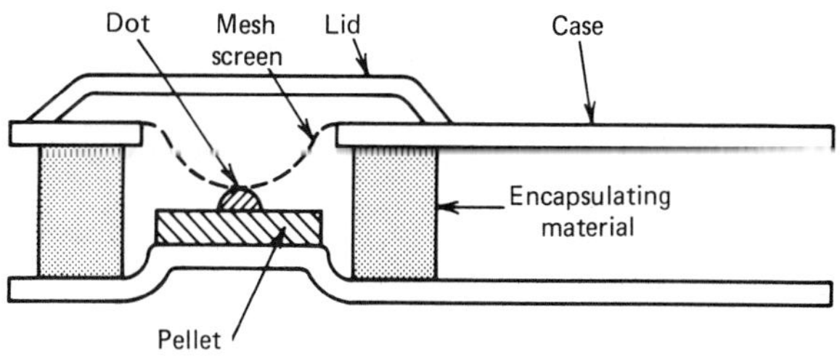

Figure 16.9 Structure of a tunnel diode.[101]

highly conductive *p*-type (or *n*-type) material to form the semiconductor junction. The pellet (approximately 0.025-in.^{2}) is then soldered into a low-inductance, low-capacitance case. A very fine mesh screen is added to make the connection to the "dot." The device is then encapsulated, and a lid is welded over the cavity.

The three types of materials sometimes used in tunnel diodes are gallium antimonide (GaSb), germanium (Ge), and gallium arsenide (GaAs). The GaSb diode is usually used in the first stage of a multistage amplifier because it provides the lowest noise figure, 0.5–1 dB lower than the Ge diode.

Germanium devices offer high speed, low noise, and low rise times (as low as 40 psec). Gallium arsenide diodes have a voltage swing almost twice that of germanium devices and, as a result, can provide power outputs almost four times as high. Because of their power-handling capability, gallium arsenide tunnel diodes are being used in an increasing number of applications and appear to be particularly useful as microwave oscillators.

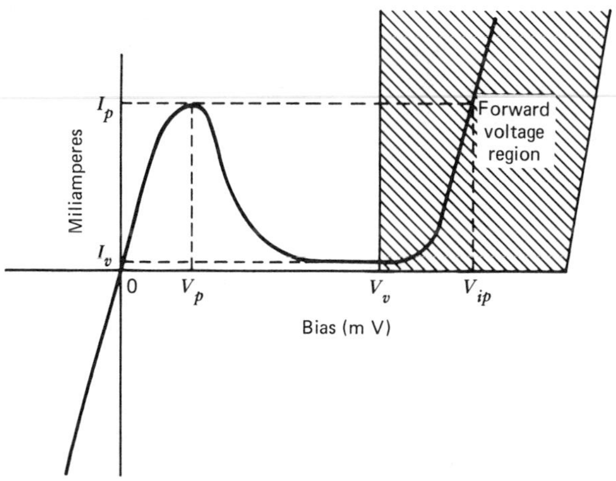

Figure 16.8 Typical current-voltage characteristic for a tunnel diode.[101]

16.3.2 Characteristics

Conventional diodes do not conduct current under conditions of reverse bias until the breakdown voltage is reached. Under forward bias, they begin to conduct at approximately 300 mV. However, the tunnel diode is highly conductive for all reverse biases. Similarly, under conditions of small forward bias, the tunnel diode current rises rapidly to a sharp maximum peak I_p. At intermediate values of forward bias, the tunnel diodes exhibits a negative-resistance characteristic and the current drops to a deep minimum valley point I_v. At higher values of forward bias, the tunnel diode exhibits the diode characteristic associated with conventional semiconductor current flow.

16.3.3 High-Current Tunnel Diodes

High-current tunnel diodes are basically the same as conventional tunnel diodes, except that they have a larger junction area to permit the flow of higher currents. In addition, they use a different package (high-current devices generally use a rectifier package such as the DO-4 or DO-8), and have a much smaller value of series resistance (generally in the order of 0.010 Ω or less).

High-current tunnel diodes are used as low-voltage inverters in circuits having low-impedance DC power sources. They can also be used for efficient inversion of the output of solar cells, thermoelectric generators, or thermionic converters, and as overload detectors in DC and AC power supplies, pulse generators, high-speed switches, and oscillators.

16.3.4 Tunnel Rectifiers

In addition to its negative-resistance properties, the tunnel diode has an efficient rectification characteristic that can be used in many rectifier applications. When a tunnel diode is used in a circuit in such a way that this rectification property is emphasized rather than its

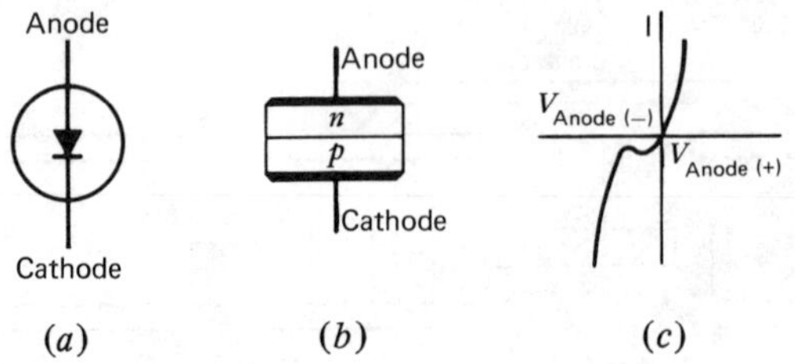

Figure 16.10 Tunnel rectifier. (*a*) Circuit symbol, (*b*) junction schematic, (*c*) electrical characteristic.[103]

negative-resistance characteristic, it is called a tunnel rectifier.

In conventional rectifiers, current flow is substantial in the forward direction, but extremely small in the reverse direction (for signal voltages less than the breakdown voltage for the device). In tunnel rectifiers, however, substantial reverse current flows at very low voltages, while forward current is relatively small. Consequently, tunnel rectifiers can provide rectification at smaller signal voltages than conventional rectifiers, although their polarity requirements are opposite. (For this reason, tunnel rectifiers are sometimes called "back diodes.")

The circuit symbol, commonly used junction schematic, and electrical characteristic for the tunnel rectifier are shown in Figure 16.10*a–c*, respectively.

16.4 VARACTOR DIODES

A varactor or variable-reactance diode is a microwave-frequency *p-n* junction semiconductor device in which the depletion-layer capacitance bears a nonlinear relation to the junction voltage, as shown in Figure 16.11*a*. When biased in the reverse direction, a varactor diode can be represented by a voltage-sensitive capacitance $C(v)$ in series with a resistance R_s, as shown in Figure 16.11*b*. This nonlinear capacitance and low series resistance permit the device to perform frequency-multiplication, oscillation, and switching functions. Very low noise levels are possible in circuits using varactor diodes because the dominant current across the junction is reactive and shot-noise components are absent.

16.5 VOLTAGE-REFERENCE (ZENER) DIODES[107,108]

Voltage-reference or Zener diodes are silicon rectifiers in which the reverse current remains small until the breakdown voltage is reached and then increases rapidly with little further increase in voltage. The breakdown voltage is a function of the diode material and construction, and can be varied from one volt to several hundred volts for various current and power ratings, depending on the junction area and the method of cooling. A stabilized supply can deliver a constant output (voltage or current) unaffected by temperature, output load, or input voltage, within given limits. The stability provided by voltage-reference diodes makes them useful as stabilizing devices and as reference sources capable of supplying extremely constant current loads. The circuit symbol and electrical characteristic for Zener diodes are shown in Figures 16.12 and 16.13, respectively.

16.6 COMPENSATING DIODES[107]

Excellent stabilization of collector current for variations in both supply voltage and temperature can be obtained using a compensating diode operating in the forward direction in the bias network of amplifier or oscillator circuits.

The use of a compensating diode also reduces the variation in transistor idling current as a result of supply-voltage variations. Because the diode current changes in proportion with the supply voltage, the bias voltage to the transistor changes in the same proportion and idling-current changes are minimized.

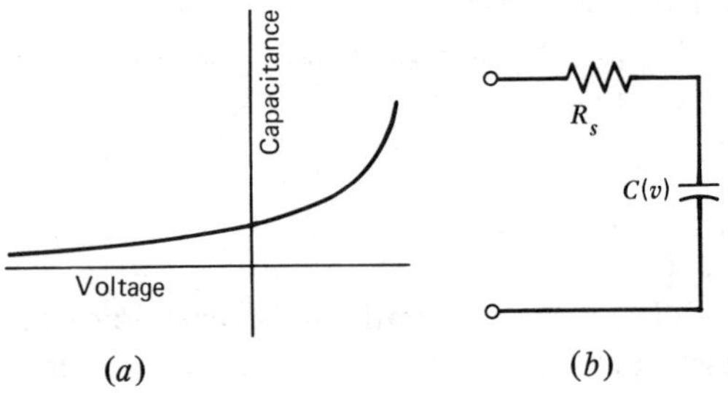

Figure 16.11 Varactor diode. (*a*) Electrical characteristic, (*b*) equivalent circuit.

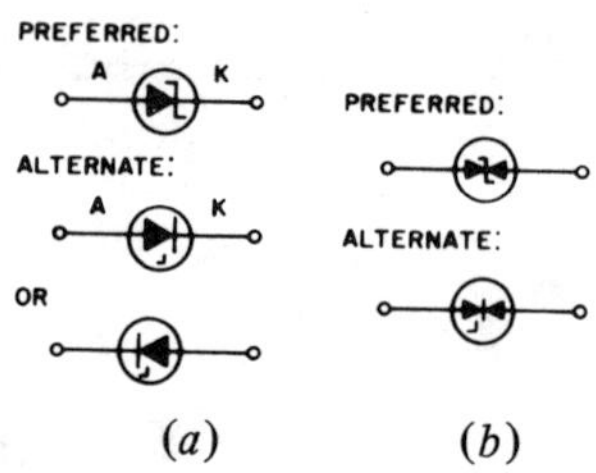

Figure 16.12 Zener and avalanche diode circuit symbols.[107] (*a*) Unidirectional, (*b*) bidirectional.

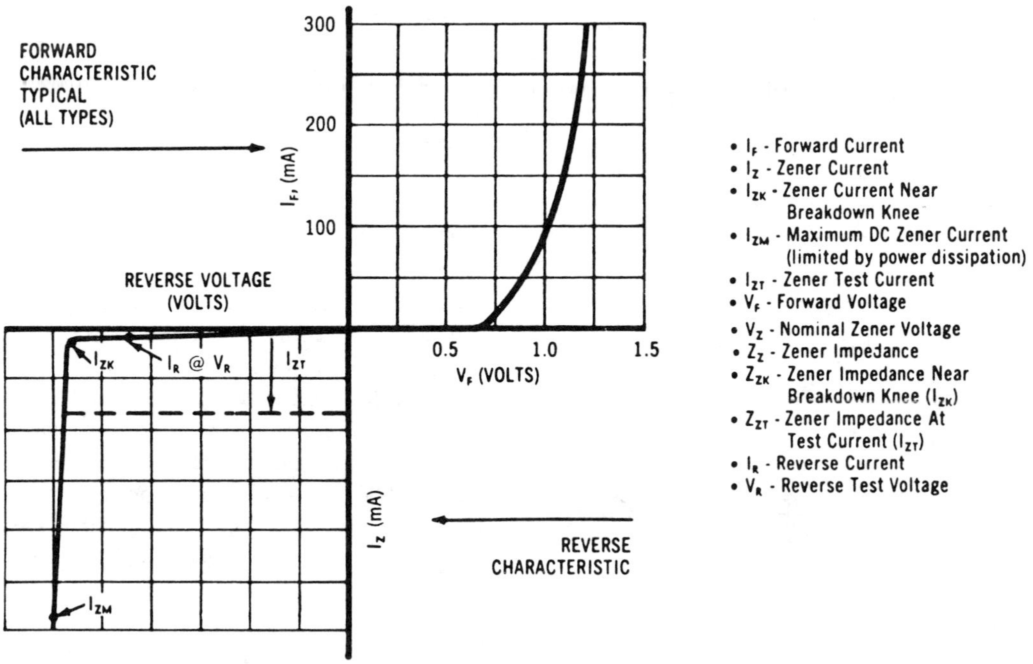

Figure 16.13 Silicon zener diode electrical characteristic.[107]

Temperature compensated reference diodes are made possible by taking advantage of the differing thermal characteristics of forward and reverse-biased silicon PN junctions. A forward-biased junction has a negative temperature coefficient of approximately 2.0 mV/°C. Reverse biased junctions above 5.0 V have a positive temperature coefficient, and therefore it is possible by judicious selection of combinations of forward- and reverse-biased junctions to obtain a device that shows a very low temperature coefficient due to cancellation. Because of the differing impedance versus temperature characteristics of the junctions involved, optimum temperature stability is obtained by operating in the Zener current range at which the temperature coefficient is a minimum.

16.7 *p-i-n* DIODES[109]

A *p-i-n* diode is a silicon semiconductor consisting of a layer of intrinsic (high-resistivity) material of finite area and thickness which is contained between highly doped *p*- and *n*-type material. When the diode is forward-biased, charge is injected into the intrinsic region. The density of charge in the intrinsic region and its geometry determines the conductance of the device, while the lifetime determines the approximate low frequency limit of useful application.

Typical packages in which *p-i-n* diodes are available are illustrated in Figure 16.14. With the exception of the Mini-strip, Microstrip Post and the Stripline packages, all the other packages introduce additional reactive elements in the form of package inductance L_p and package capacitance C_p. Typical values for different packages are also given in Figure 16.14.

In some applications these parasitic elements can be either "tuned out" by additional external reactances or actually utilized by forming a resonant circuit around the diode. But the bandwidth of such structures is limited. When these elements cannot be tuned out, performance over a broad bandwidth will generally suffer.

16.8 MICROWAVE DIODES[110]

Components used for control of microwave signals fall into one of three categories–diode, ferrite, or electromechanical. The characteristics of the three types are sufficiently dissimilar that most applications dictate a single approach. The important characteristics of the three types are compared in Table 16.1.

Most of the microwave signal control functions in modern systems are achieved with diodes. The basic technique is simple. One or more diodes are incorporated into a transmission line as series or shunt (or

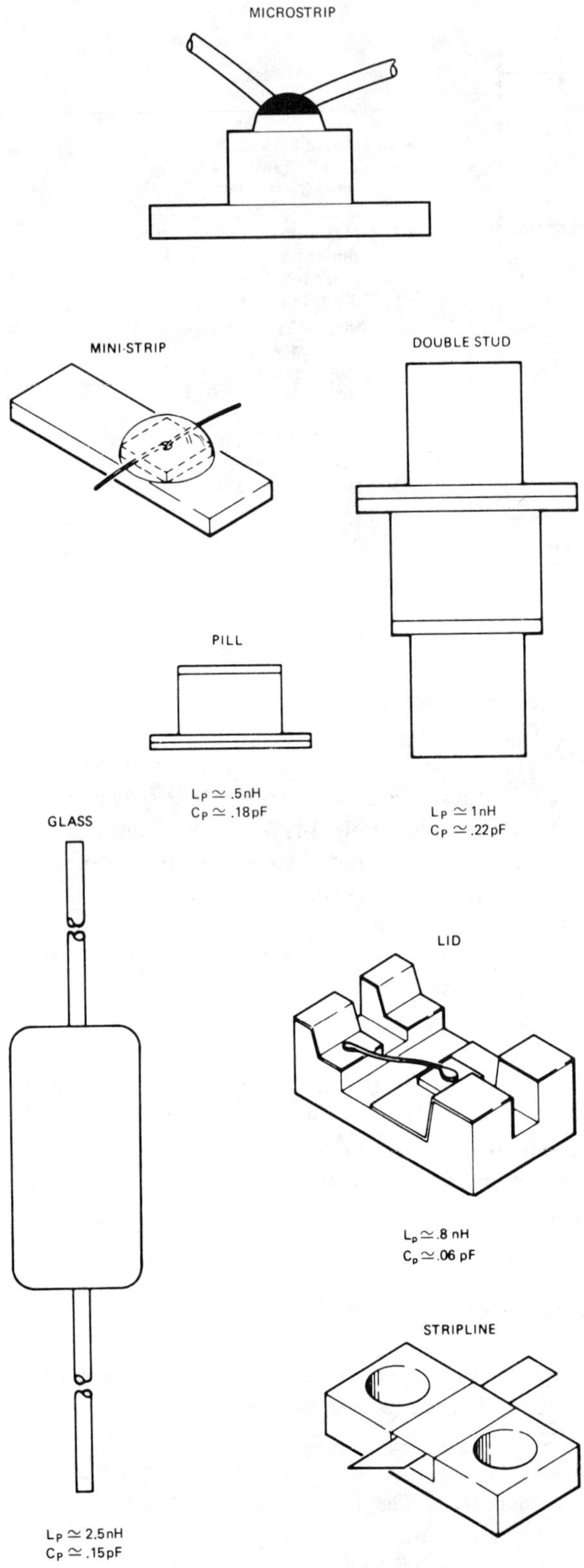

Figure 16.14 PIN diode package outlines.[109]

both) elements. The RF resistance of the diodes can be varied by the application of steady state or time-varying bias, and the microwave signal is controlled (that is, passed, rejected, attenuated, or modulated) according to the bias applied.

16.9 SILICON RECTIFIERS

Silicon rectifiers, like other semiconductor diodes, are essentially cells containing a simple *p–n* junction. As a result, they have low resistance to current flow in one (forward) direction, but high resistance to current flow in the opposite (reverse) direction. They can be operated at ambient temperatures up to 200°C and at current levels as high as 40 A, with voltage levels as high as 1000 V. In addition, they can be used in parallel or series arrangements to provide high current or voltage capabilities.

Because of their high forward-to-reverse current ratios, silicon rectifiers can achieve rectification efficiencies in the order of 99%. When properly used, they have excellent life characteristics that are not affected by aging, moisture, or temperature. They are very small and lightweight and can be made impervious to shock and other severe environmental conditions.

16.9.1 Thermal Considerations

Although rectifiers can operate at high temperatures, they are sensitive to sudden temperature changes because of the extremely small crystals used in their structure. The thermal capacity of a silicon rectifier is quite low, and the junction temperature rises rapidly during high-current operation. Sudden rises in junction temperature caused by either high currents or excessive ambient temperature conditions can cause failure. Consequently, temperature effects are very important in the consideration of silicon rectifier characteristics.

16.9.2 Reverse Characteristics

When a reverse-bias voltage is applied to a silicon rectifier, a limited amount of reverse current begins to flow. This reverse current flow increases slightly as the bias voltage increases, but then tends to remain constant even though the voltage continues to increase significantly. However, an increase in operating temperature multiplies the reverse current considerably for a given reverse bias.

At a specific reverse voltage (which varies for different types of diodes), a very sharp increase in reverse current occurs. This voltage is called the breakdown or

Table 16.1 A Comparison of Diode, Ferrite, and Electromechanical Switches[110]

Characteristic	Diode Switches	Ferrite Switches	Electromechanical Switches
Switching speed	Nanoseconds	Microseconds (low field) Milliseconds (high field)	Milliseconds
Insertion loss	Higher	Low	Low
Isolation	Typical to 80 dB	Typical 20 dB (low field) Typical 60 dB (high field)	Typical 90 dB
Can be used for analog applications	Yes	Yes	No
Lifetime (cycles)	Unlimited	Unlimited	Several million cycles for best quality switches
Size and weight	Least	Greater	Greater
Ability to withstand severe environment	Best	Temperature sensitive	Temperature, vibration, and shock sensitive
Bandwidth	Octaves to multidecades	Octaves or less	DC to 26.5 GHz

avalanche (or Zener) voltage. In many applications, rectifiers can operate safely at the avalanche point. If the reverse voltage is increased beyond this point, however, or if the ambient temperature is raised sufficiently, "thermal runaway" results and the diode may be destroyed.

16.9.3 Forward Characteristics

A silicon rectifier usually requires a forward voltage of 0.4–0.7 V to overcome the potential barrier at the *p–n* junction. A slight rise in voltage beyond this point increases the forward current sharply. Because of the small mass of the silicon rectifier, the forward voltage drop must be carefully controlled so that the specified maximum value for the device is not exceeded. Otherwise, the diode may be seriously damaged or destroyed.

In certain applications, close control of ambient temperature is required for satisfactory operation. Close control is not usually required, however, in power circuits.

16.9.4 Ratings

Three current ratings are usually given for silicon rectifiers: the maximum average forward current, the peak recurrent forward current, and the maximum surge current. The first of these currents refers to the maximum average value of current permitted to flow in the forward direction for a specified ambient or case temperature. Typical average current outputs range from 0.5 A to as high as 40 A for single silicon diodes.

The peak recurrent forward current is the maximum repetitive instantaneous forward current permitted under stated conditions. The maximum surge current is the maximum nonrepetitive peak current of a single forward cycle. The surge, or fault, current is permitted for only a very short time interval (about 8 msec). Surge currents generally occur when the equipment is first turned on or when unusual voltage transients are introduced in the AC supply line. Protection against excessive currents of this type can be provided in various ways.

Because these maximum current ratings are all affected by thermal variations, ambient-temperature conditions must be considered in the application of silicon rectifiers.

16.9.5 Controlled Avalanche Rectifiers[111]

Reverse biases, when applied to silicon rectifiers, can, if large enough, cause the rectifier to avalanche. The term "avalanche" refers to that critical reverse voltage at which further increases in voltage result in rapidly increasing reverse current flow. This avalanche effect is commonly used in voltage regulator diodes, which are frequently referred to as Zener, or breakdown diodes. Conventional power rectifiers are normally not designed for operation near this critical avalanche voltage level and will usually be destroyed if it is exceeded even momentarily.

Reverse biases sufficient to cause avalanche may result from transient overvoltages, or voltage spikes, in supply line power, or may be caused by line switching, feedback from line-connected equipment, arcing in the distribution system, lightning, or other causes.

However, the most common cause is the inductive "kick," a voltage many times higher than the impressed voltage, produced by the collapsing magnetic field in an inductive component when current through it is abruptly cut off. This voltage will rise until it causes sufficient current to flow to dissipate the stored energy, and may cause arcing in the switch, avalanching of the rectifier, or other component breakdown. Unless a safe channel is provided for the dissipation of this energy, destruction of one or more components may result. The inductive kicks generated by power transformers, filter chokes, motors, and magnetic devices such as solenoids are frequently sufficient to damage conventional silicon rectifiers.

Silicon rectifiers capable of absorbing large amounts of transient reverse avalanche energy without being destroyed are commonly called Controlled Avalanche Rectifiers, and generally have a minimum and maximum reverse breakdown voltage specified. Special manufacturing processes are used to produce a silicon die that avalanches uniformly across the entire area of the junction, rather than at discrete points as in conventional devices. This process considerably increases the energy levels which the rectifier can absorb.

A controlled avalanche rectifier is not a replacement for transient protection devices in all situations, but it does provide adequate protection wherever avalanche energy levels are within the rectifier specifications. Among such applications are permanent magnet motors, solenoid actuators, and transformer power supplies. The advantages of using a controlled-avalanche rectifier are:

A. Greater inherent tolerance to overvoltage transients.

B. Protection of other circuit elements that might be damaged with the use of high PIV-rated devices.

C. Higher circuit reliability.

D. Potential cost savings in most applications through the use of lower-rated devices.

E. Elimination of transient suppressors in many applications.

F. Compatibility with inductive circuit and load elements.

G. Greater device reliability resulting from improved internal design and tighter testing levels.

16.9.6 Fast-Recovery Rectifiers[112]

A silicon rectifier cannot make an instantaneous switch from forward conduction to reverse blocking. For a momentary period, prior to full blocking, the rectifier is a short circuit, conducting current freely in the wrong direction, whereas the electrical charge that has been built up by forward conduction is "swept out."

In typical "slow" recovery silicon rectifiers, the time required for the rectifier to recover and perform its blocking function may be from 5 to 50 μsec. In some types of applications, this may be too slow for efficient rectification. Furthermore, it results in a reverse spike of considerable energy that may be highly undesirable in certain conditions and applications.

The most obvious applications for fast recovery rectifiers are those involving high frequencies where slow recovery devices produce low rectification efficiencies, and overheat due to internal losses. Generally speaking, for rectifier purposes high frequencies may be defined as the band from 10 to 200 kHz.

In addition, waveforms such as sawtooth or squarewaves have high-frequency components, even though the repetition rate is low, and often require fast recovery rectifiers for satisfactory operation.

Actual rectification efficiencies are not easily determined analytically. The best practical method of determination is to insert the rectifier in the actual circuit and observe its performance.

There is a group of applications for fast-recovery rectifiers that is less immediately obvious. The reverse recovery "spike" itself has a high-frequency content and is therefore a source of noise and radio frequency interference, RFI. Since RFI is related to spike energy, it can be reduced quite effectively through the use of fast recovery rectifiers, which can result in a reduction in the amount of supplementary RFI filtering required.

The spike-reducing characteristics of fast-recovery rectifiers may also improve the ripple characteristics of DC power supplies (particularly with squarewave inputs), while maintaining minimum values on filter components. This can mean savings in parts costs as well as significant contributions to size and weight reduction.

16.9.7 Applications

Silicon rectifiers are used in a continually broadening range of applications. Originally developed for use in such equipment as DC-to-DC converters, battery chargers, mobile power supplies, transmitters, and electroplating devices, silicon rectifiers are also used in power supplies for radio and television receivers and phonograph amplifiers, as well as in such applications as in-line-type modulators, hold-off and charging diodes, pulse-forming networks, and brushless alternators. They are also being used in many aircraft applications because of their small size, light weight, and high efficiency.

16.10 DIODE AND RECTIFIER TERMS AND DEFINITIONS

Backward Diode A device in which quantum-mechanical tunneling leads to a current-voltage characteristic with a reverse current greater than the forward current, for equal and opposite applied voltages.

Current Regulator Diode A diode that limits current to an essentially constant value over a specified voltage range.

Detector Diode A device that converts RF energy into dc or video output.

Gunn Diode A microwave diode that exhibits negative resistance arising from the bulk negative differential conductivity occurring in several compound semiconductors, such as gallium arsenide, and that operates at a frequency determined by the transit time of charge bunches formed by this negative differential conductivity.

Impatt Diode (Impact, Avalanche, and transit time diode) A semiconductor microwave diode that, when its junction is biased into avalanche, exhibits a negative resistance over a frequency range determined by the transit time of charge carriers through the depletion region.

LSA Diode (Limited Space-Charge Accumulation Diode) A microwave diode similar to the Gunn diode except that it achieves higher output power at frequencies, determined by the microwave cavity, that are several times greater than the transit-time frequency by avoiding the formation of charge bunches or domains.

Matched Pair A pair of diodes identical in outline dimensions and with matched electrical characteristics. The two diodes may both be forward polarity, or one forward and one reverse polarity, or both reverse polarity.

Microwave Diode A two-terminal device that is responsive in the microwave region of the electromagnetic spectrum, commonly regarded as extending from 1 to 300 GHz.

Mixer Diode A microwave diode that combines RF signals at two frequencies to generate an RF signal at a third frequency.

Semiconductor Rectifier Diode A device having an asymmetrical voltage-current characteristic, used for rectification.

Semiconductor Signal Diode A device having an asymmetrical voltage-current characteristic and used for signal detection.

Trapatt diode (trapped plasma avalanche transit time diode) A microwave diode that, when its junction is biased into avalanche, exhibits a negative resistance at frequencies below the transit time frequency range of the diode due to generation and dissipation of trapped electron-hole plasma resulting from the intimate interaction between the diode and a multiresonant microwave cavity.

Tuning Diode A varactor diode used for RF tuning including such functions as automatic frequency control (AFC) and automatic fine tuning (AFT).

Tunnel Diode A device in which quantum-mechanical tunneling leads to a region of negative slope in the forward direction of the current-voltage characteristic.

Varactor Diode A two-terminal semiconductor device that uses its property of capacitance variance with applied voltage.

Voltage-Reference Diode A diode that is normally biased to operate in the breakdown region of its voltage-current characteristic and that develops across its terminals a reference voltage of specified accuracy, when biased to operate throughout a specified current and temperature range.

Voltage-Regulator Diode A diode that is normally biased to operate in the breakdown region of its voltage-current characteristic and that develops across its terminals an essentially constant voltage throughout a specified current range.

CHAPTER SEVENTEEN

Transistors

17.1 TRANSISTOR, GENERAL[101]

When a second junction is added to a semiconductor diode to provide power or voltage application, the resulting device is called a transistor. The three regions of the device are called the emitter, the base, and the collector. In normal operation the emitter-to-base junction is biased in the forward direction, and the collector-to-base junction biased in the reverse direction.

Different symbols are used for *n-p-n* and *p-n-p* transistors to show the difference in the direction of current flow in the two types of devices. In the *n-p-n* transistor shown in Table 17.1 the electrons flow from the emitter to the collector. In the *p-n-p* transistor shown in Table 17.1 the electrons flow from the collector to the emitter. In other words, the direction of DC electron current is always opposite to that of the arrow on the emitter lead. (As in the case of semiconductor diodes, the arrow indicates the direction of "conventional current flow" in the circuit.)

The first two letters of the *n-p-n* and *p-n-p* designations indicate the respective polarities of the voltages applied to the emitter and the collector in normal operation. In an *n-p-n* transistor the emitter is made negative with respect to both the collector and the base, and the collector is made positive with respect to both the emitter and the base. In a *p-n-p* transistor the emitter is made positive with respect to both the collector and the base, and the collector is made negative with respect to both the emitter and the base.

17.1.1 Bipolar Transistors

All bipolar transistors consist of three layers of semiconductor material, usually silicon. The structure forms two back-to-back *p-n* junctions. The input (emitter-base) junction serves as the source or injector of current carriers; the output (base-collector) junction collects the injected current carriers.

17.1.2 Field-Effect Transistors (FETs)

The field-effect transistor is another type of solid-state device that is becoming increasingly popular in electronic circuits. These transistors derive their name from the fact that the current flow in them is controlled by variation of an electric field established by the application of a voltage to a control electrode, referred to as the "gate."

In contrast, current flow in bipolar transistors is controlled by variation of the current injected into the base terminal. Moreover, the performance of bipolar transistors depends on the interaction of two types of charge carriers: holes and electrons. Field-effect transistors, however, are unipolar devices. As a result their operation is basically a function of only one type of charge carrier,

Table 17.1 Transistors (Reproduced from Ref. 103)

NAME OF DEVICE	CIRCUIT SYMBOL	COMMONLY USED JUNCTION SCHEMATIC	ELECTRICAL CHARACTERISTICS		MAJOR APPLICATIONS
n-p-n Transistor	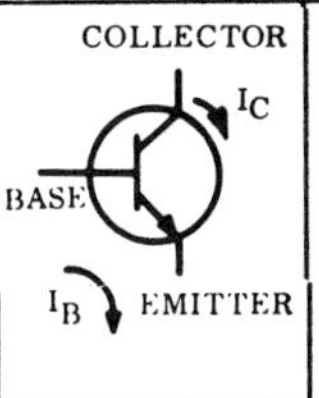	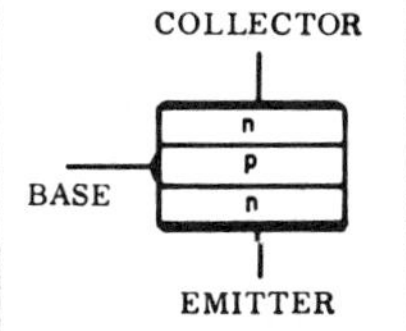	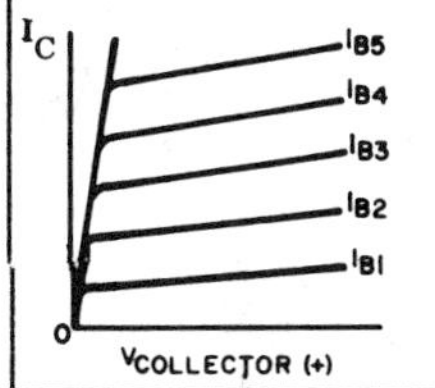	Constant collector current for given base drive	Amplification Switching Oscillation
p-n-p Transistor	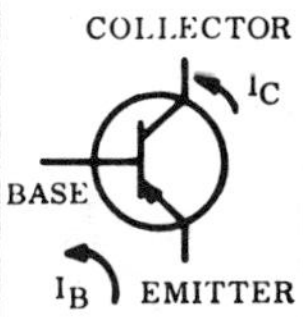	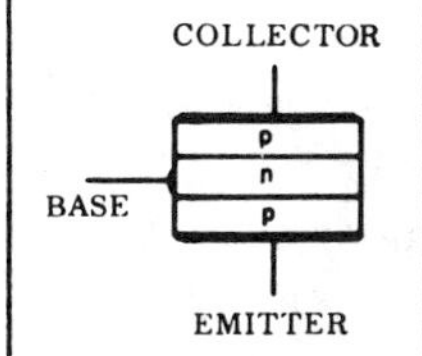	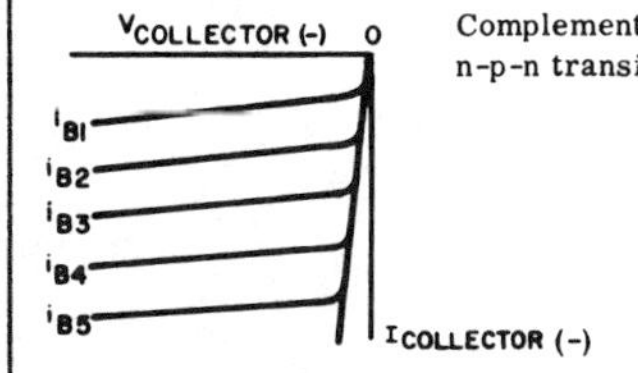	Complement to n-p-n transistor	Amplification Switching Oscillation
Unijunction Transistor (UJT)	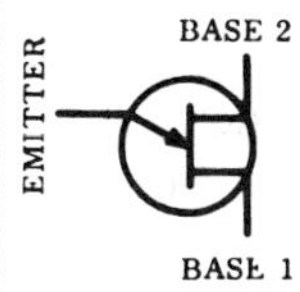	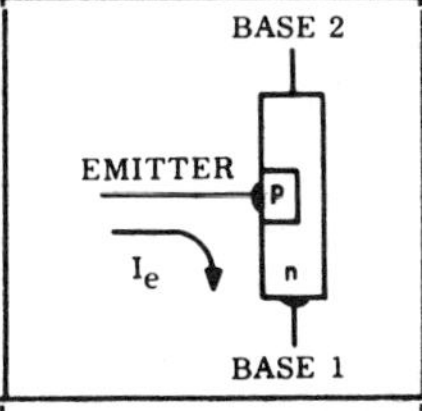	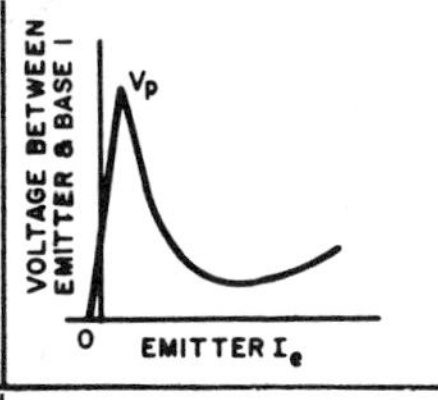	Unijunction emitter blocks until its voltage reaches V_p; then conducts	Interval timing Oscillation Level Detector SCR Trigger
Complementary Unijunction Transistor (CUJT)	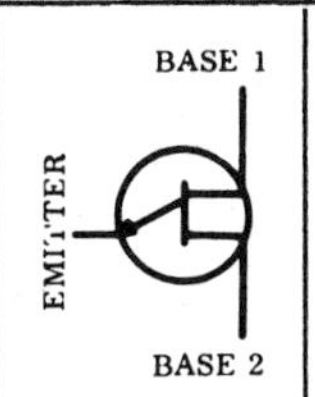	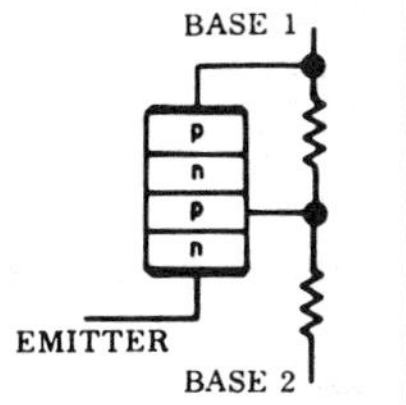	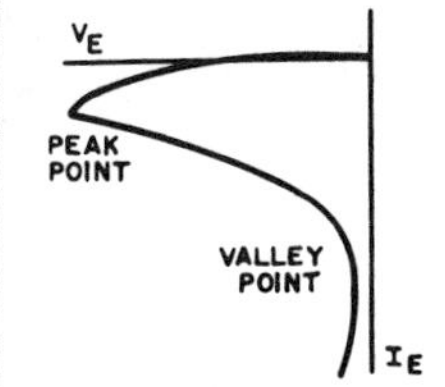	Functional complement to UJT	High stability timers Oscillators and level detectors
Programmable Unijunction Transistor (PUT)	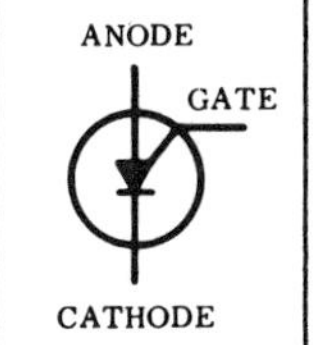	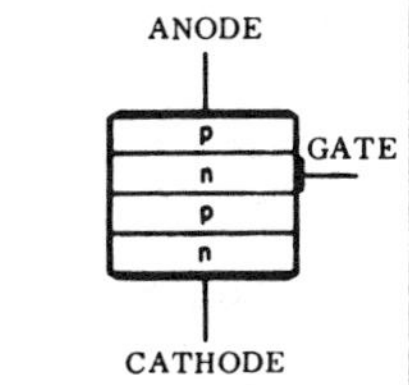	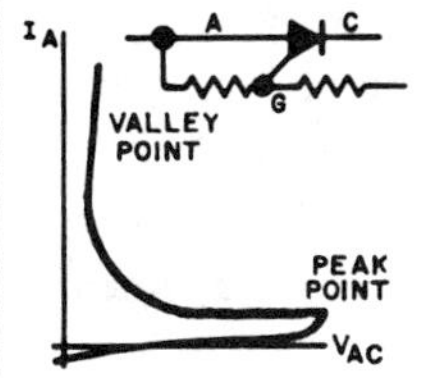	Programmed by two resistors for V_p, I_p, I_V. Function equivalent to normal UJT.	Low cost timers and oscillators Long period timers SCR trigger Level detector
Photo Transistor	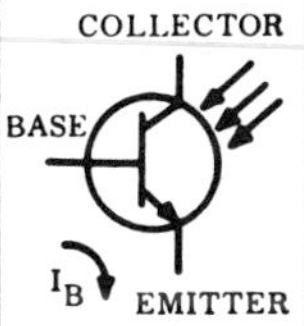	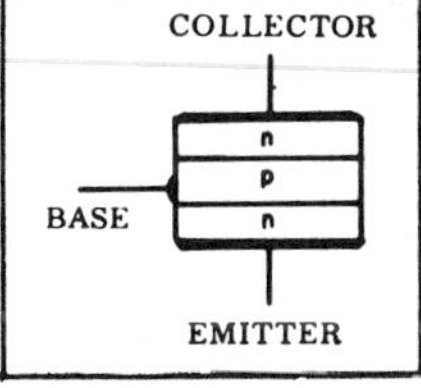	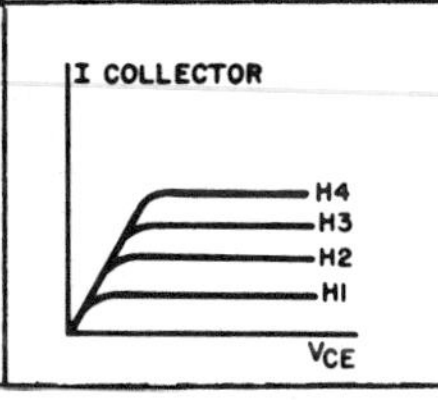	Incident light acts as base current of the photo transistor	Tape readers Card readers Position sensor Tachometers

i.e., holes in p-channel devices and electrons in n-channel devices.

Discrete-device field-effect transistors are classified on the basis of their control-gate construction, as either junction-gate types or metal-oxide-semiconductor (MOS) types. Although both types operate on the basic principle that current conduction is controlled by variation of an electric field, the significant difference in their gate construction results in unique characteristics and advantages for each type.

17.2 DESIGN AND FABRICATION[113,114]

The performance of transistors in electronic equipment depends on many factors besides the basic characteristics of the semiconductor material. The two most important factors are the design and fabrication of the transistor structure and the general circuit configuration used.

The ultimate aim of all transistor fabrication techniques is the construction of two parallel *p–n* junctions with controlled spacing between the junctions and controlled impurity levels on both sides of each junction. A variety of structures have been developed during the course of transistor evolution. (See Section 16.1.)

17.2.1 Alloy Transistors

To form *p–n–p* germanium transistors indium containing aluminum or gallium alloys into *n*-type germanium. The germanium alloy produces *p–n–p* transistors with excellent saturation voltages, good gain, and excellent safe operating area (SOA). The basic *p–n–p* germanium–alloy structure is shown in Figure 17.1.

A major disadvantage of the alloy transistor is its susceptibility to "punchthrough." Once base resistivity has been optimized the voltage at which punchthrough occurs can be increased, but only by increasing the physical base width with its attendant detrimental increase in diffusion capacitance and base store.

A second disadvantage of the simple alloy process is its lack of precise control. Because alloying depth is somewhat irregular, alloy transistors of the same type often vary considerably in values. Furthermore, the irregular alloying depth prevents the design of transistors with extremely narrow base widths as needed for high-speed switching. In general, the switching rate of alloy units is limited to about 500 kHz.

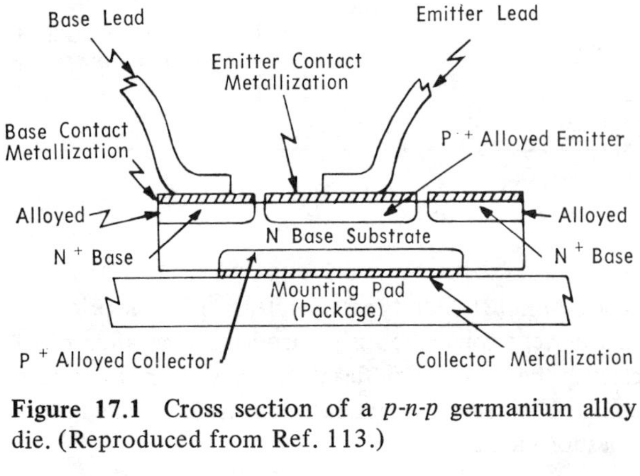

Figure 17.1 Cross section of a *p-n-p* germanium alloy die. (Reproduced from Ref. 113.)

17.2.2 Diffused-Base Alloy Transistors

A further improvement of the alloy transistor results from a graded distribution of impurities in the transistor base region. This yields devices with much higher voltage ratings than those of other types of alloy transistors. The "built-in" electric field associated with gradation also reduces the transit time of carriers through the base region.

The penalties paid for these improvements are a reduction of β (the common-emitter forward current-transfer ratio) and emitter avalanche breakdown voltage, and an increase of C_{TE} (emitter transition capacitance) due to the relatively low overall resistivity of the area on the emitter side of the base region. These effects, however, are normally not significantly detrimental to switching applications.

The application of electrochemical etching and plating techniques to the diffused-base alloy process have yielded excellent switching transistors, that is, the micro-alloy diffused-base transistor. Such devices, however, are relatively fragile and for extremely high speed designs. Therefore, a compromise between voltage rating and rise-time is required.

A common disadvantage of all types of transistors is the lack of control over the shape of the emitter area. With these processes, the emitter is circular, resulting in a decreasing ratio of perimeter to area as the radius is increased. Since high-current transistors require a large emitter perimeter, they necessarily have a large area that yields large capacitance values and reduces switching speeds. Accordingly, switching speed is reduced as current-handling capability is increased.

17.2.3 Mesa and Planar Transistors–General

Mesa transistors (Figure 17.2) and planar transistors use construction techniques that are better suited to many applications than the grown-junction or alloy methods. These transistors involve two basic processes:

1. The use of diffusion-masking materials and photolithographic techniques to obtain a planar structure in which all the *p–n* junctions are buried under a protective passivating layer.
2. The use of a separate collector-contact diffusion or an epitaxial growth to reduce the electrical series resistance in the collector.

In these types the original semiconductor wafer serves as the collector. The base region is diffused into the

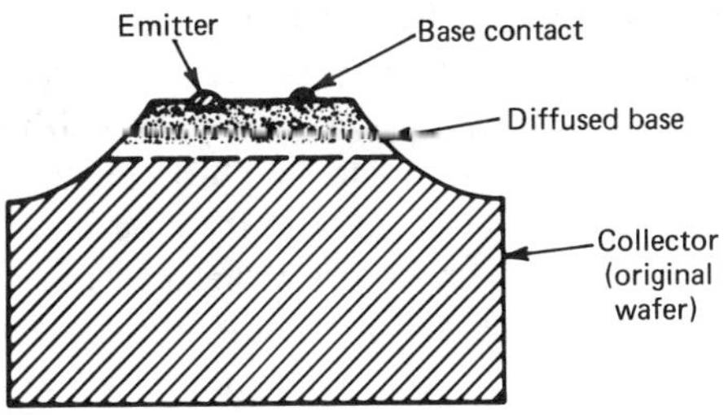

Figure 17.2 Structure of a mesa transistor.[101]

wafer, and the emitter "dot" or region is then alloyed or diffused into the base region. A "mesa" or flat-topped peak may then be etched to reduce the collector area at the base-collector junction.

The mesa structure is inherently rugged, has large power-dissipation capability and can operate at very high frequencies. The grading of the impurity concentration in the base region results in a drift field and in reduced base-lead resistance. The use of a diffused emitter region permits tight geometry control. The use of relatively light impurity concentration in the collector region results in high collector-breakdown voltages and low collector-junction capacitance.

17.2.4 Mesa Transistors

Mesa transistors are characterized by higher switching speeds, greater power-handling capability, and far more rugged construction than their alloyed counterparts. The distinctive physical fact about the mesa transistor is that the junction areas can be precisely controlled.

The bulk collector of the mesa transistor permits a low thermal impedance between the junction and the heat sink that gives mesa transistors far greater power-handling capabilities. The mesa transistor has longer storage times, but compensates for this somewhat by having faster rise times. Moreover, the mesa geometry provides satisfactory performance at higher currents and can withstand more severe environmental conditions. Mesa transistors also have switching speeds approximately equal to those of the microalloy diffused units.

The mesa transistor has some limitations. There is a limitation on the amount of current that can be drawn from the device without exceeding the power rating. A more serious limitation is the storage time resulting from the presence of this bulky collector region. The only solution to this problem is to reduce the resistivity in order to reduce the voltage rating, or to employ, in the case of silicon, some method of special doping to reduce the lifetime, which also has the undesirable effect of reducing β.

17.2.5 Epitaxial Mesa Transistors

The construction of transistors using the epitaxial process is very similar to that of the basic mesa transistor, except that an epitaxial layer of material is grown onto the basic wafer to form the effective collector region. This produces three important advantages:

A. The quality of the active semiconductor material is greatly improved.

B. The interrelation and compromises between physical parameters and electrical characteristics are minimized.

C. Compromises between interrelating electrical characteristics are minimized.

The epitaxial process permits the use of the finest-quality material to form the active region of the transistor, while supporting this active region on a far less expensive substrate, which adds the physical strength and bulk to the transistor, but contributes nothing to the performance.

17.2.6 Planar Transistors

The silicon planar process is used to produce both *n-p-n* and *p-n-p* devices featuring high-frequency response and exceptional reliability. The cross section of an *n-p-n* planar structure is shown in Figure 17.3. The principal features of the planar process are:

A. High frequency (20–150 MHz).

B. Fast switching speeds.

C. Low reverse-leakage currents.

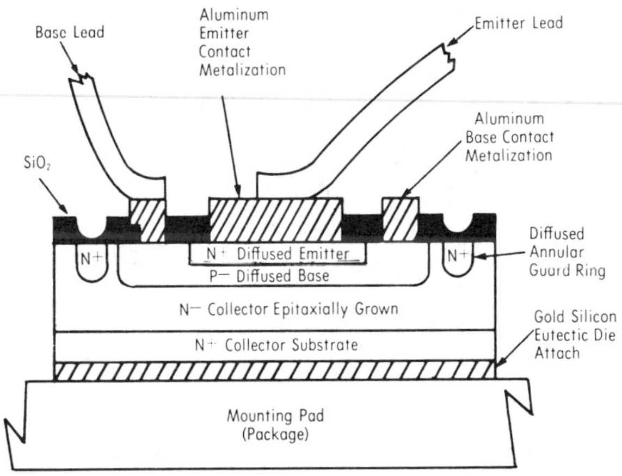

Figure 17.3 Cross section of an *n-p-n* planar die. (Reproduced from Ref. 113.)

D. Excellent beta (β) linearity.

E. High reliability.

17.2.7 Power Transistors[106,115–118]

Power transistors are unique in their varied methods of construction. Small-signal transistors and integrated circuits (ICs) are generally constructed using many steps and rather-narrow diffusions into an epitaxial layer on a low-resistivity substrate. They are usually passivated as well. On the other hand, power transistors are usually not passivated and are made by at least five basically different techniques, as shown in Table 17.2.

Power transistors are supplied in a wide variety of mechanical configurations, as shown in Figure 17.4, and used in a wide range of applications, as shown in Table 17.3.

A. *Double-Diffused Epitaxial* The double epitaxial process produces many popular high-voltage and fast-switching medium-voltage transistors. The basic double-epitaxial structure is shown in Figure 17.5. This process offers the following features:

1 True *npn*/*pnp* complementary pairs.
2 Good forward- and reverse-biased safe operating areas (SOA).
3 High power-handling capability from 4 to 350 W.
4 High-voltage capability of from 50 to 800 V.
5 High current-carrying capacity from 1 to 100 A.
6 Low saturation voltages.
7 Excellent beta (β) linearity.
8 Excellent switching speeds.
9 High-frequency usage at from 5 to 50 MHz (f_T).

The advantages of the double-epitaxial process is the degree of control inherent in the epitaxial-growth process, that is, both resistivity and base width can be extremely well controlled and a uniform impurity distribution can be maintained in both the base and collector regions. This creates transistors with excellent saturated-switching performance (low saturation voltages, fast switching speeds, and so on) and improved forward-biased SOA. Since the base width is so well controlled, devices with ratings of 1.0–100 A are readily available.

B. *Double-Layer Epitaxial* These transistors, which are not common, are used mostly for high-reliability applications. They are manufactured by diffusing an emitter into an epitaxial base region that has been deposited on an epitaxial collector region. The collector is supported by the substrate and the voltage depletes into the base and collector.

Devices of either polarity can be made with this technique. The f_T is moderately high at about 20 MHz. The performance is a compromise between that of epitaxial-base devices and double-diffused epitaxial devices, with reasonable safe area along with medium to high gain and low saturation voltage.

C. *Epitaxial Base* These transistors are made by diffusing an emitter into an epitaxial base region deposited on the collector substrate. The collector voltage depletes into the base region. The basic epitaxial-base structure is shown in Figure 17.6.

Its performance is between that of a single-diffused transistor and a double-layer epitaxial transistor. The key advantages of the epitaxial-base process is the degree of control inherent in the epitaxial-growth process, that is, both resistivity and width can be extremely well controlled, and a uniform impurity distribution can be

Table 17.2 Common Transistor Diffusions[106]

Process	Epi-Base	Single-Diffusion	Triple-Diffusion (Annular)	Triple-Diffusion (Etch-Cut)	Double-Diffusion (Epitaxial)
Characteristics	I_c = to 50 A BV_{CEO} = to 140 V SOA = Excellent f_T = 4–8 MHz SW = Slow	I_c = to 30 A V = to 140 V SOA = Excellent f_T = 1–2 MHz SW = Slow	I_c = to 10 A V = to 400 V SOA = Fair f_T = 20–30 MHz SW = Medium	I_c = to 10 A V = to 800 V SOA = Good f_T = 4–10 MHz SW = Medium	I_c = to 30 A V = to 200 V SOA = Low f_T = 40–150 MHz SW = Very fast
Complements	*N-P-N/P-N-P*	*N-P-N*	*N-P-N*	*N-P-N*	*N-P-N/P-N-P*
Outstanding features	Versatility	Ruggedness	High frequency response	High voltage performance	High speed
Typical application	General purpose NPN/PNP complementary	Power supply regulators	Inverters–Converters	TV deflection circuits	High-speed switching

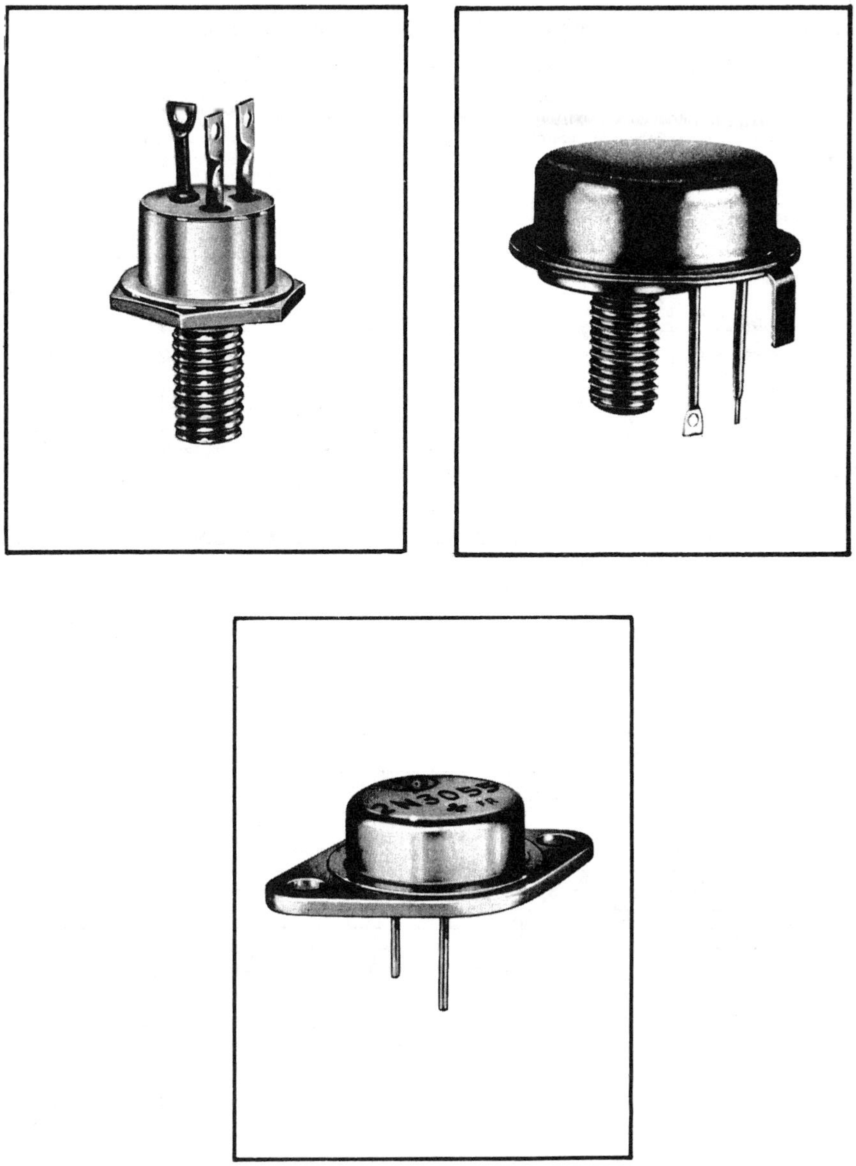

Figure 17.4 Power transistors. (Reproduced from Ref. 118.)

maintained in the base region. This creates transistors with excellent saturated switching performance. Another significant advantage of this process is that it allows the fabrication of large area devices. Therefore, high-current devices are readily available.

Many popular epitaxial-base power transistors offer the following features:

1 True *n-p-n*/*p-n-p* complementary pairs.
2 Good forward- and reverse-biased SOA.
3 High power-handling capability of 6–300 W.
4 High-voltage capacity of 40–100 V.
5 High-current handling ability of 1–50 A.
6 Low saturation voltages.
7 Excellent beta linearity.
8 Good switching speeds.

D. ***Single-Diffused*** These transistors are made by a single simultaneous diffusion into both sides of a silicon wafer. This produces two deep *n*-type diffusions into a *p*-type wafer. The collector voltage depletes entirely into the base. The basic single-diffused structure is shown in Figure 17.7.

Single-diffused-transistors are extremely rugged with

Table 17.3 Applications for Power Semiconductors (Reproduced from Ref. 118)

	RECTIFIERS		THYRISTORS/SCR's			TRANSISTORS			ASSEMBLIES				
	General Purpose	Fast Recovery	Phase Control	Fast Switching	RBDT	General Purpose	High SOA	High Pwr. Switching	Modular	Stud Mt.	Air Disc	Liquid Disc	H.V. Stacks
POWER DISTRIBUTION EQUIPMENT													
Fusion Energy Research	•	•	•	•	•						•	•	•
Generator/Exciter	•	•					•	•					
H.V.D.C. Transmission		•	•								•		
Nuclear Fission	•	•	•	•						•	•		
Solar	•	•	•	•				•					
Uninterruptible Power Supplies (UPS)	•	•	•	•		•		•		•	•	•	
Wind Generators	•	•	•	•						•	•		
TRANSPORTATION SYSTEMS													
Electric Vehicle	•	•	•	•		•		•			•		
Elevator/Electric Stairway	•	•	•							•	•		
Locomotives	•	•	•	•						•	•		
Mass Transit	•	•	•	•						•	•		
People Movers	•	•	•	•						•	•		
INDUSTRIAL CONTROLS													
AC Motor Controls	•	•	•	•		•	•	•	•	•	•	•	
AC Motor Starters			•								•		
Air Conditioning	•	•	•	•						•	•	•	
Conveyor	•	•	•							•	•		
Crane & Hoist	•	•	•								•		
DC Motor Control	•	•	•	•		•	•	•	•	•	•	•	
Electric Discharge Machining (EDM)	•		•	•		•		•	•	•			
Electric Space Heating	•		•										
Furnace	•		•							•	•	•	
Heating & Melting	•	•	•	•				•		•	•	•	•
Lift Trucks	•	•	•	•									
Lighting	•	•	•	•					•	•	•		
Machine Tool	•	•	•	•		•		•	•	•			
Off Road Vehicles	•	•	•					•			•		
Off Shore Oil Well Drilling	•	•	•							•	•		
Power Factor Correction			•								•	•	
Pump & Pipeline	•	•	•	•					•	•	•	•	
Solid State Contactors	•		•							•	•	•	
Steel Mill Drives	•	•	•	•						•	•	•	
Welding, Capacitor Discharge	•		•	•						•	•	•	•
Welding, Resistance			•							•	•	•	
POWER SUPPLIES													
Accelerators	•		•								•	•	•
Aircraft	•	•		•		•	•	•					
Aluminum Reduction	•		•								•	•	
Anodizing	•		•							•	•	•	
Arc Heaters	•	•	•								•	•	
Battery Chargers	•		•						•	•	•		
Computer	•	•	•	•		•	•	•	•		•		
Consumer	•	•	•			•			•				
Dielectric Heating	•												•
Electrochemical	•		•								•	•	
Electroplating	•		•							•	•	•	
Electron Beam Welding			•								•		•
Electrostatic Precipitators			•								•		•
Induction Heating	•	•	•	•				•		•	•	•	•
Laser	•	•	•	•	•					•	•		•
Military	•	•	•	•	•		•	•		•	•		•
Mining Power Centers	•		•							•			
Radar/Sonar		•		•	•		•	•					•
Test Equipment	•	•	•	•		•	•	•	•	•	•	•	•
Transmitters, Radio & TV													•
Water Treatment	•		•							•	•	•	
Welding, Arc	•		•							•	•		

very-low f_T and fair saturation voltage. The single-diffused process is not suitable for making p–n–p transistors.

The single-diffused mesa process is used to make many popular silicon power transistors that offer the following features:

1 Rugged construction.

2 Excellent forward- and reverse-biased SOA.

3 High power-handling capability of 5–250 W.

4 High voltage-handling ability of 35–160 V.

5 High current-handling ability of 1–30 A.

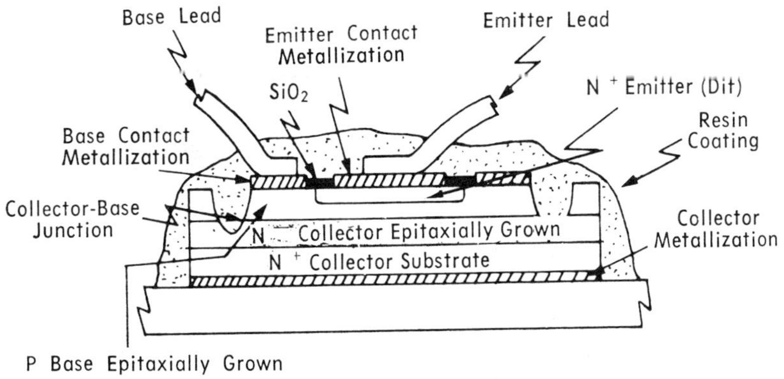

Figure 17.5 Cross section of an *n-p-n* silicon double epitaxial die. (Reproduced from Ref. 113.)

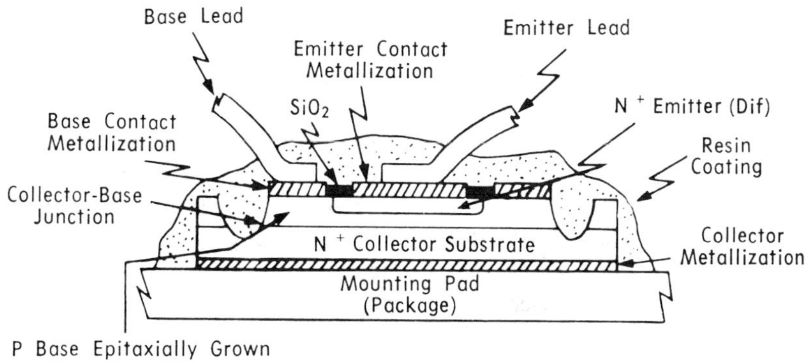

Figure 17.6 Cross section of an *n-p-n* epitaxial base die. (Reproduced from Ref. 113.)

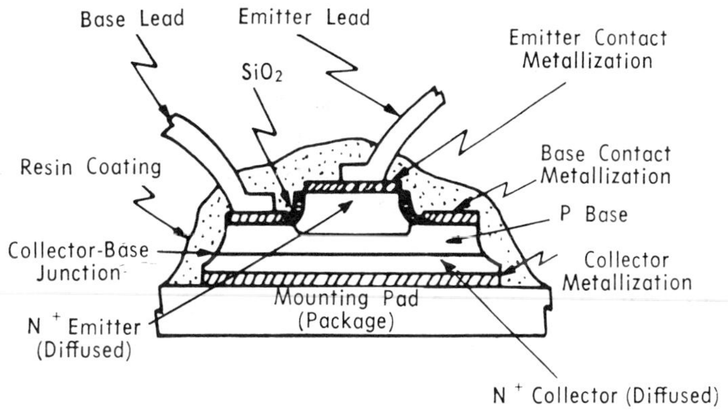

Figure 17.7 Cross section of an *n-p-n* silicon single-diffused die. (Reproduced from Ref. 113.)

6 Low saturation voltages.

7 Good high-current gain.

E. ***Triple-Diffused*** These transistors are made by three diffusions into a silicon wafer. A base region is diffused into a high-resistivity epitaxial layer, and the opposite side of the wafer is lapped to remove this diffusion. An emitter is then diffused into the base with a simultaneous diffusion into the collector from the opposite side.

A triple-diffused transistor can be made with either a shallow base or a deep base. The shallow-base device can be passivated and has medium to high f_T, fair SOA, and a medium to high collector breakdown voltage.

The deep-base device has a medium to low f_T and is

C. *Direct Coupling* Direct coupling is used primarily when cost is an important factor. Because so few circuit parts are required in the direct-coupled amplifier, maximum economy can be achieved. However, the number of stages that can be directly coupled is limited. Temperature variation of the bias current in one stage is amplified by all the stages, and severe temperature instability may result.

17.3.3 Bias Stability

Because transistor currents tend to increase with temperature, it is necessary to design transistor circuits to include a "stability factor" (SF) to keep the collector current variation within tolerable values under the expected high-temperature operating conditions.

For a given set of operating voltages the stability factor can be calculated for a maximum permissible rise in DC collector current from the room-temperature value. The calculated values of SF can then be used, together with the appropriate values of beta and base-connection resistance, in order to determine suitable resistance values for the transistor circuits.

17.3.4 Amplification

The amplifying action of a transistor can be used in various ways in electronic circuits, depending on the results desired. The four recognized classes of amplifier service for transistors are:

Class A Amplifier An amplifier in which the base bias and alternating signal are such that collector current in a specific transistor flows continuously during the complete electrical cycle of the signal, even when no signal is present.

Class AB Amplifier An amplifier in which the base bias and alternating signal permit collector current in a specific transistor to flow for appreciably more than half, but less than the entire, electrical cycle.

Class B Amplifier An amplifier in which the base is biased to approximately collector-cutoff current, so that collector current is approximately zero when no signal is applied, and so that collector current in a specific transistor flows for approximately one-half of each cycle when an alternating signal is applied.

Class C Amplifier An amplifier in which the base is biased to such a degree that collector current in each transistor is zero when no signal is applied, and so that collector current in a specific transistor flows for appreciably less than one-half each cycle when an alternating signal is applied.

A. *Audio Amplifiers* For radio frequency (RF) amplifiers operating into selective tuned circuits, or for other amplifiers in which distortion is not a prime factor, any of the above classes of amplification may be used with either a single transistor or a push-pull stage. For audiofrequency (AF) amplifiers in which distortion is an important factor, single transistors can be used only in class A amplifiers. For class AB of class B audioamplifier service, a balanced stage using two transistors is required.

A push-pull stage can also be used in class A audio amplifiers to obtain reduced distortion and greater power output. Class C amplifiers cannot be used for audio applications.

Audio-amplifier circuits are used in radio and television receivers, public address systems, sound recorders and reproducers, and similar applications to amplify signals in the frequency range 10–20,000 Hz. Each transistor in an audio amplifier can be considered as being either a current amplifier or a power amplifier.

Simple class A amplifier circuits are normally used in low-level audio stages such as preamplifiers and drivers. Preamplifiers usually follow low-level output transducers such as microphones, hearing-aid and phonograph pickup devices, and recorder–reproducer heads.

B. *Tuned Amplifiers* In transistor radio frequency (RF) and intermediate-frequency (IF) amplifiers, the bandwidth of frequencies to be amplified is usually only a small percentage of the center frequency. Tuned amplifiers are used in these applications to select the desired bandwidth of frequencies and to suppress unwanted frequencies. The selectivity of the amplifier is obtained by means of tuned interstage coupling networks.

C. *Direct-Coupled Amplifiers* Direct-coupled amplifiers are normally used in transistor circuits to amplify small DC or very-low-frequency AC signals. Typical applications of such amplifiers include the output stages of series-type and shunt-type regulating circuits, chopper-type circuits, differential amplifiers, and pulse amplifiers.

17.3.5 Oscillation

Transistor oscillator circuits are similar in many respects to the tuned amplifiers, except that a portion of the output power is returned to the input network in phase with the starting power (regenerative or positive feedback) to sustain oscillation.

DC bias-voltage requirements for oscillators are similar to those for amplifiers. Stabilization of the operating point is important because this point affects the output amplitude and waveform and the frequency stability. Operation is normally maintained within the linear portion of the transistor characteristic via constant supply voltage. Because the collector-to-emitter capacitance of the transistor affects frequency stability more than other parameters, a relatively large stabilizing capacitor is often used between the collector and emitter terminals to reduce the sensitivity of the circuit to voltage variations as well as to capacitance variations between transistors.

The maximum operating frequency of an oscillator circuit is limited by the frequency capability of the transistor used. The maximum frequency of a transistor is defined as the frequency at which the power gain is unity, that is, an input signal appears in the output circuit at the same level with no loss or gain. Because some power gain is required in an oscillator circuit in order to overcome losses in the feedback network, the operating frequency must be some value below the transistor maximum frequency.

17.3.6 Switching

Transistors are often used in pulse and switching circuits. These circuits act as generators, amplifiers, inverters, frequency dividers, and wave-shapers to provide limiting, triggering, gating, and signal-routing functions. These applications are normally characterized by large-signal or nonlinear operation of the transistor. In large-signal operation, the transistor acts as an overdriven amplifier that is driven from the cutoff region to the saturation region.

Regions of operation are similar for all transistor configurations used as switches. When both junctions of the transistor are reverse-biased (cutoff condition) the output current is very small and the output voltage is high. When both junctions are forward-biased (saturation condition) the output current is high and the output voltage is small. For most practical purposes, the output current in the cutoff condition and the small output voltage in the saturated condition may be neglected.

A. *Switching Modes*[114] Transistor switching circuits generally fall into three basic categories depending on their operating mode. These are broadly classified as (1) saturated mode, (2) current mode, and (3) avalanche mode, and are determined by the portion of the transistor output characteristic curve utilized.

The operating regions for various transistor switching modes are shown in Figure 17.11. For all modes the switch-off condition is characterized by an excursion of the load line into the cutoff region of the transistor. The operating mode, therefore, is determined principally by the DC circuit state in the switch-on condition, and by the location of the operating points.

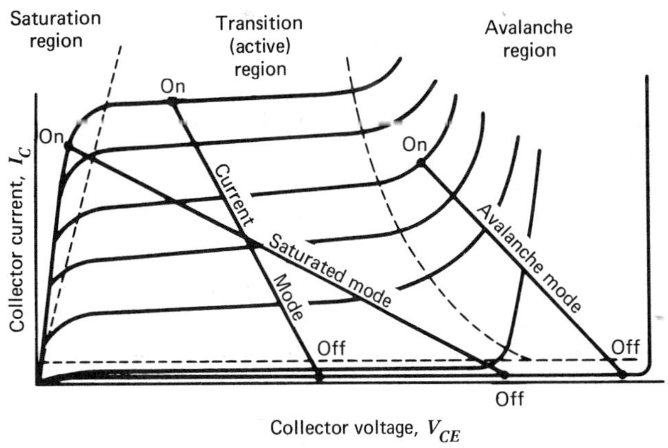

Figure 17.11 Operation regions for switching modes.[114]

B. *Triggered Circuits* When an externally applied signal is used to cause an instantaneous change in the operating state of a transistor circuit, the circuit is said to be triggered. Such circuits may be astable, monostable, or bistable.

Astable triggered circuits have no stable state; they operate in the active linear region and produce relaxation-type oscillations.

A monostable circuit has one stable state in either of the stable regions, but the circuit then switches back to its original stable state after a period determined by the time constants of the circuit elements.

A bistable, flip-flop, circuit has two stable states in the two stable regions. The transistor is triggered from one stable state to the other by an external pulse, and a second trigger pulse is required to switch the circuit back to its original stable state.

C. *Gating Circuits* A transistor switching circuit in which the transistor operates as an effective open or short circuit is called a "gate." These circuits are used extensively in computer applications to provide a variety of functions such as circuit triggering at prescribed intervals, and level and waveshape control. Because these circuits are designed to evaluate input conditions to provide a predetermined output, they are used primarily as logic circuits, namely, OR, AND, NOR (NOT-OR), NAND (NOT-AND), series (clamping), and shunt or inhibitor circuits.

An OR gate has more than one input, but only one output. It provides a prescribed output condition when one or another prescribed input condition exists. If a common-emitter configuration is used instead of the common-base configuration phase, inversion of the signal results, and the OR gate becomes a NOR gate.

An AND gate also has more than one input, but only one output. However, it provides an output only when all the inputs are applied simultaneously. As in the case of the OR gate, the use of a common-emitter configuration provides a phase inversion and provides a NAND gate.

Gating circuits are also used as amplitude discriminators (limiters), clippers, and clamping circuits and as signal-shunting or transmission gates.

17.4 TRANSISTOR TERMS AND DEFINITIONS[120]

Base A region that lies between an emitter and collector of a transistor and into which minority carriers are injected.

Beta (β) Current gain.

Characteristic Impedance The impedance seen looking into an infinite length of that transmission line.

Collector A region through which a primary flow of charge carriers leaves the base.

Current Gain The ratio of the output current to the input current. Current gain is also known as h_{fe} or beta for common-emitter transistors, and as h_{fb} for common-base transistors.

Cutoff Current The measured value of DC current when a transistor is reverse biased by a voltage less than the breakdown voltage.

Delay Time The length of time that the transistor remains cut off after the input pulse is applied.

Depletion-Mode Operation The operation of a field-effect transistor (FET) such that changing the gate-to-source voltage from zero to a finite value decreases the magnitude of the drain current.

Depletion-Type Field-Effect Transistor A field-effect transistor (FET) having appreciable channel conductivity for zero gate-to-source voltage. The channel conductivity may be increased or decreased according to the polarity of the applied gate-to-source voltage.

Drain A region into which majority carriers flow from the channel.

Emitter A region from which charge carriers that are minority carriers in the base are injected into the base.

Enhancement-Mode Operation The operation of a field-effect transistor (FET) such that changing the gate-to-source voltage from zero to a finite value increases the magnitude of the drain current.

Enhancement-Mode Field-Effect Transistor A field-effect transistor (FET) having substantial zero-channel conductivity for zero gate-to-source voltage. The channel conductivity may be increased by the application of a gate-to-source voltage of appropriate polarity.

f_{max} The maximum frequency of oscillation.

f_t The transition frequency.

Field-Effect Transistor (FET) A transistor in which the conduction is due entirely to the flow of majority carriers through a conduction channel controlled by an electric field arising from a voltage applied between the gate and source.

Frequency of Oscillation, Maximum The theoretical frequency at which the maximum-available gain equals unity (0 dB).

Gate The electrode associated with the region in which the electric field, due to the control voltage, is effective.

I_c The absolute maximum collector current.

Insulated-Gate Field-Effect Transistor A field-effect transistor (FET) having one or more gate electrodes that are electrically insulated from the channel.

Interdigitated When the base and emitter areas appear as interleaved fingers, separated from each other but aligned very close together.

Junction, Collector A semiconductor junction, normally biased in the reverse direction, the current through which can be controlled by the introduction of minority carriers into the base.

Junction-Gate Field-Effect Transistor A field-effect transistor (FET) that uses one or more gate regions to form *p–n* junctions with the channel.

Load Impedance (Z_L) The impedance seen by the transistor output terminals (ports).

Metal Oxide-Silicon Field-Effect Transistor (MOSFET) An insulated-gate field-effect transistor in which the insulating layer between each gate electrode and the channel is oxide material.

***n*-Channel Field-Effect Transistor** A field-effect transistor (FET) that has an *n*-type conduction channel.

Noise Figure (NF) The electrical-noise characteristic of a transistor, usually expressed as the ratio of device noise to the thermal noise at +25°C. This ratio is generally expressed in terms of decibels.

p-Channel Field-Effect Transistor A field-effect transistor (FET) that has a p-type conduction channel.

Peak Point The point on the emitter current-voltage characteristic corresponding to the lowest current at which the change in emitter base voltage with respect to emitter current equals zero.

Power Transistors Transistors with a power dissipation capability of greater than 1 W.

Saturation A base current and a collector current condition resulting in a forward-biased collector junction.

Source A region from which majority carriers flow into the channel.

Source Impedance The impedance seen by the transistor input terminals (ports).

Stability Factor (SF) A figure of merit as to whether or not a transistor can oscillate for some combination of source and load impedance.

Standing Wave Waves produced from the combination of reflected power with the incident power when a transmission line is terminated in some impedance other than Z_o.

Transition Frequency ("Gain Bandwidth Product") The theoretical frequency at which the common-emitter current is unity (0 dB).

Voltage Gain The ratio of the output voltage to the input voltage.

Voltage Standing-Wave Ratio (VSWR) The ratio of the peak voltage to the valley voltage of a standing wave.

CHAPTER EIGHTEEN

Thyristors

18.1 GENERAL INFORMATION

Thyristors[115,116,121–123] both SCRs and triacs, are widely used in power-control applications. With the emphasis in such applications placed on low cost, small package size, and circuit simplicity, thyristors satisfy these requirements with reliability exceeding that of electromechanical counterparts.

18.1.1 Types of Thyristors[115]

Thyristors are semiconductor devices that have characteristics similar to those of thyratron tubes; more specifically, they are semiconductor switches whose bistable state depends on the regenerative feedback associated with a *p–n p–n* structure. Basically, this group includes any bistable semiconductor device that has three or more junctions (i.e., four or more semiconductor layers) and can be switched from a high-impedance (OFF) state to a conducting (ON) state, and from the conducting (ON) state to the high-impedance (OFF) state, within at least one quadrant of the principal voltage characteristics.

There are several types of thyristors, which differ primarily in number of electrode terminals and operating characteristics associated with the third quadrant (negative) of the voltage-current characteristics. Reverse-blocking triode thyristors, commonly called silicon-controlled rectifiers (SCRs), and bidirectional triode thyristors, referred to as triacs, are the most popular types. (See Table 18.1.)

As the use of SCRs in power applications increased, the need for complete AC control became apparent. The new family of thyristor devices generated to provide bidirectional current properties is referred to as triacs. A triac can be considered as two parallel SCRs (*p–n–p–n*) oriented in opposite directions to provide symmetric bidirectional characteristics.

Thyristors are supplied in a wide variety of mechanical configurations, as shown in Figure 18.1, and are used in a wide variety of applications. (See Table 17.3.)

18.1.2 Selection[106]

To select and apply any thyristor successfully, none of its published ratings should be exceeded. Also, it is uneconomical to apply the device too conservatively. To make a proper device selection, first check all the limiting conditions of the application. Since thyristor ratings are usually specified as maximum or minimum values (worst case), determine which device best fits the application. The following steps should be taken or considered.

A. *Determine Circuit Requirements on Thyristor*

1 Voltage across and current through the thyristor in terms of circuit input voltage and output power requirements.

Table 18.1 Thyristor and Rectifier VI Characteristics[115]

Name of Device	Circuit Symbol	Electrical Characteristics	Applications
Silicon-controlled rectifier (SCR)	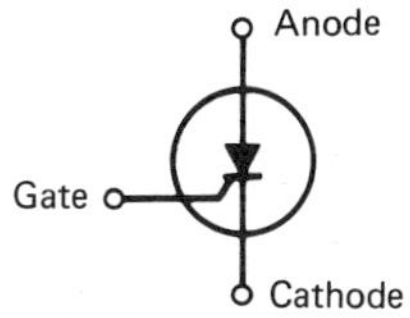		Unidirectional switching Requires commutating circuits Fast devices—4-10-μsec switching Average devices—20-35-μsec switching
Asymmetrical silicon-controlled rectifier (ASCR)	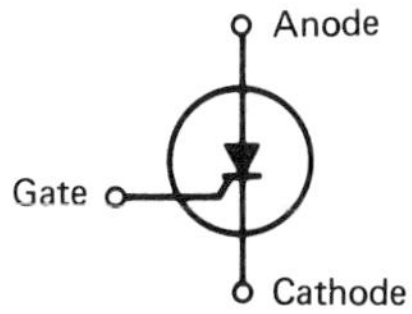	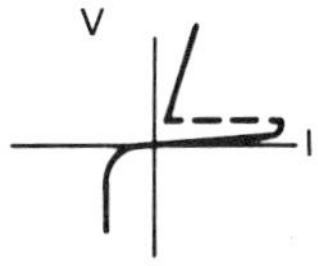	Unidirectional switching Low reverse blocking Typical turnoff time—2-6 μsec Typical turnon time—200 nsec Useful for 20-50-kHz switching
Integral thyristor rectifier (ITR)	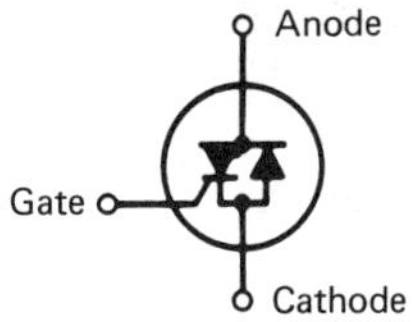	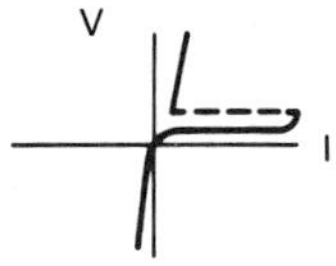	Unidirectional switching Reverse current conduction Typically faster than conventional SCRs Typical turnoff time—2-6 μsec High-volume usage in TV deflection
Gate turnoff silicon-controlled rectifier (GTO-SCR)	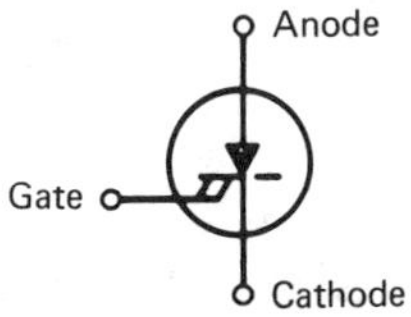	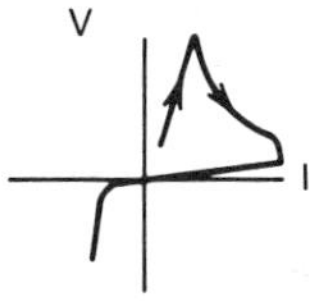	Unidirectional switching Controlled turnoff from the gate with negative-gate pulse Ideal for high-voltage transistor-type circuits Inrush to steady-state ratios of 10 typical
Triac	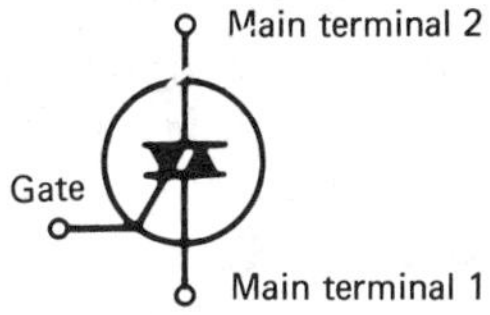	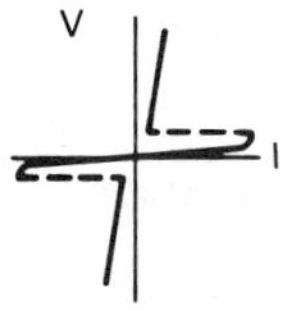	Bidirectional switching Ideal for 60-cycle AC controls Selections suitable for 400-Hz operation Surge to average current ratios of 10 typical

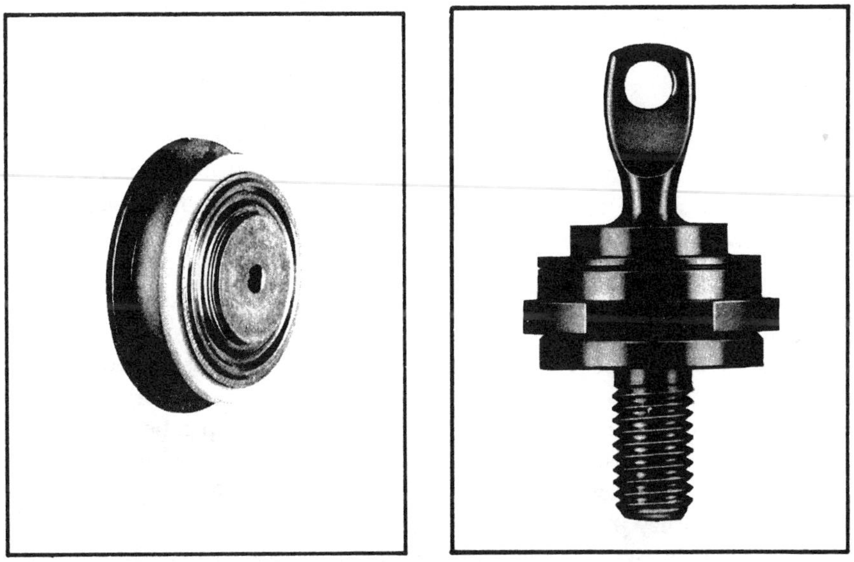

Figure 18.1 Thyristor configurations. (Reproduced from Ref. 118.)

2 Current carrying capability required of the thyristor if the current waveform is irregular, or has a high starting component.

3 Temperature range over which the circuit must operate.

4 Is a high reliability device desirable or mandatory?

B. *Select Proper Thyristor*

1 Examine device specification sheets for detailed information.

C. *Determine Proper Heatsink*

1 Check maximum allowable ambient temperature if a mounted device was selected.

2 Select proper size heatsink from curves on specification sheets.

D. *Design Triggering Circuit*

1 Know what are the thyristor triggering requirements and design criteria. (Commercially packaged triggering circuits are available using magnetic, light-sensitive or semiconductor components.)

E. *Supply Suitable Overload Protection, If Required*

1 Protect the thyristors and associated semiconductors against short circuit and other fault conditions.

2 Do not overlook normal overloads, such as cold inrush to incandescent light bulbs or starting current of induction motors.

F. *Consider series or parallel operation of individual Thyristors; Radiointereference Suppression; Frequency Response*

One of the decisions facing power semiconductor users is whether to use a power transistor or a thyristor. While in many applications the choice will be obvious, there are applications when it is not clear, and this decision is becoming increasingly difficult because of the improvements with both the power transistors and Darlingtons. (See Table 18.2.)

Power transistors have switching characteristics that are easy to control. A transistor can be turned off simply by removing its base current, while silicon control rectifiers require more complex circuitry to turn off the device. Furthermore, transistors can operate at higher frequencies than thyristors, provide quieter operation, low output ripple and good efficiency. Also, transistors are not likely to be turned on accidently by a high *dv/dt*. Power transistors are being used for precise control of dc circuits, for inverters and in switching regulator type power supplies. There are new high voltage devices which are well suited for automobile ignition systems, TV deflection circuits, inverter circuits, and choppers, to cite a few examples.

Table 18.2 Practical Power Semiconductor Maximum Performance Capabilities[a]

Type of Semiconductor	Voltage (volts)	Current (amperes)	Speed (μs)
Transistors	3000	500	0.2
Darlington transistors	1500	50	1
Rectifiers	5000	4000	10
SCRs	2500	3000	1
Triacs	1000	2000	1

[a] Maximum voltage and maximum current are mutually exclusive.

18.2 SILICON-CONTROLLED RECTIFIERS

The silicon-controlled rectifier (SCR) or reverse-blocking triode thyristor, is the most common of the four-layer *pn–pn* devices. An SCR (Figure 18.2) has three electrodes: anode, cathode, and a gate.

The SCR is the best known of all thyristors because it is a unidirectional device (current flows in only one direction) and has three terminals (anode, cathode, and control gate). The SCR is classified as a reverse blocking triode thyristor. It can control both alternating and direct current power more economically than most other semiconductor devices.

18.2.1 Construction[101]

The basic construction of the silicon-controlled rectifier consists of alternate layers of diffused silicon material that serve as the cathode, gate, base, and anode. These layers are enclosed in a special metal container, which is then hermetically sealed to maintain an ultradry atmosphere. This entire unit is mounted in a rugged case that provides protection against severe thermal environmental stresses. The pedestal below the semiconductor layers acts as a heat sink to help dissipate the heat developed internally during operation.

18.2.2 Characteristics

The SCR under reversed-bias conditions operates like a conventional rectifier. In the third quadrant of the V-1 curve (Table 18.1) reverse leakage current increases

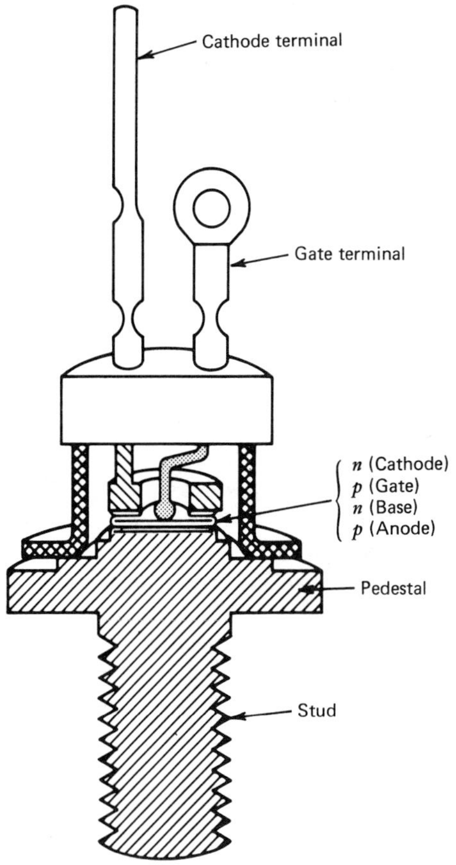

Figure 18.2 Construction details of typical silicon-controlled rectifier.[101]

slowly with increase of reverse voltage. When the increasing reverse voltage reaches the reverse blocking voltage, reverse leakage current increases very rapidly. Beyond this point, the rapid increase in reverse current creates excessive reverse power and eventual failure.

In the forward direction an SCR is electrically bistable and can be a very high impedance up to a point or can exhibit a very low impedance. In the forward blocking state, a small forward current flows through the SCR. The magnitude of this current is approximately that of the reverse current, and at some point the leakage through the device causes loss of blocking characteristics, and the device switches on.

When an SCR is forward biased and an appropriate current pulse is put into the gate, the anode-to-cathode impedance approaches zero and power is immediately transferred to the load. When the SCR is in the low-impedance state, gate current is no longer required because there is sufficient current flowing to the device to maintain conduction. However, once the voltage reverses, the device is shut off or naturally commutated, and the blocking characteristic is once again regained. This ability to initiate conduction at some time later in the cycle makes this device more versatile than the conventional rectifier. By using these functions, the amount of power to a load can be controlled from zero to full.

18.2.3 Critical Parameters

The peak reverse voltage is the maximum repetitive voltage that an SCR can withstand in the reverse direction. In addition, a transient peak reverse voltage is included in the specifications of a conventional SCR. This rating indicates maximum permissible inverse transient voltage which often appears during fuse clearing, voltage regulation, and short circuits. The rating should not be met repetitively nor should it ever be exceeded.

Another parameter of importance is transient forward blocking rating—the voltage level that can be blocked for short time periods. Intermittent conduction can be dangerous and unwanted.

Care should be exercised to make certain that the rate of anode-current rise in conjunction with this two-terminal operation does not cause destruction of the device.

18.2.4 Selection

A. *Temperature* The common denominator in all SCR ratings is a junction-temperature limitation. Most of the ratings are based on the 125°C junction temperature, which governs the blocking-voltage capability. The difference between SCR thyristor junction-temperature ratings and conventional rectifier temperature ratings is the additional SCR requirement of forward-blocking capability. This lower limitation is based on the temperature-leakage characteristics and the positive feedback inherent in SCR manufacturing. Therefore, as temperature increases, a point is reached at which the device is triggered into conduction by internal leakage currents. Above 125°C, typical devices suffer from this characteristic.

During surge and other high-current transients that reduce blocking capability, the junction temperatures go well beyond those specified or implied in technical data sheets. These temperatures are short time excursions. The maximum junction temperatures encountered in worst-case current transients may go as high as 500°C. This temperature occurs only at the junction; the other parts of the device are cooler. Modern coatings are not damaged until they exceed 250°C, and the braze joints,

300°C. Breakdown of these parts requires temperatures and durations that are not encountered in the rated conditions, so that reliability is not adversely affected within the conditions of the surge ratings.

B. ***Current and Device Losses*** Junction temperature rise, for the most part, accompanies device losses arising from controlling power in a circuit. Therefore, the device forward-drop and thermal-impedance characteristics usually establish current-carrying capability. The forward voltage determines the losses, and the thermal impedance controls the temperature rise in the package.

Most SCR thyristor applications involve converting AC power into variable DC levels. For this reason, manufacturers list ratings for these common current waveforms, including case temperature plotted against forward current. This curve enables making a meaningful temperature analysis of a thermal system using standard thermocouple practices. The device manufacturer has already accounted for the peak-to-average rating differential to facilitate this analysis. Outside the device, thermal masses are large enough to permit the conventional average rating procedure to be used. The only departure from this generality is conditions such as high-current surging, in which the transient capabilities of the thermal system must be considered.

18.2.5 Overload Protection

In any silicon-controlled rectifier circuit, precautions should be taken to protect the device from overcurrent and overvoltage surge conditions. Protection against overcurrent surges can be achieved by either preventing or interrupting the current surge, or by limiting the magnitude of the current flow by means of the circuit impedance.

The effects of voltage transients in silicon-controlled rectifier circuits can be minimized by reducing the rate at which the energy is dissipated in the devices. This "slowdown" of energy release can be achieved by relocation of the switching elements in the circuit or by a change in the sequence of switching. Other preventive methods include the change of speed of current interruption by the switching elements or the use of an additional energy source or dissipation means in the circuit.

18.2.6 Series Arrangement

Two or more silicon-controlled rectifiers can be used in a series arrangement when the total forward (or reverse) voltage is higher than the maximum voltage rating for a single device. In series arrangements, precautions must be taken to assure equal division of the applied voltage among the devices. If one rectifier carries a larger share of voltage because of leakage differences or other variations between units, it may fire inadvertently when the peak voltage across the string is large, and thereby disrupt the entire series string. Under steady-state blocking conditions, this problem can be minimized by shunting individual rectifiers with resistors of the same value to equalize the voltage drop.

Transient effects also present a problem in series arrangements. Under high-frequency voltage transient conditions, voltage division across the silicon controlled rectifiers becomes inversely proportional to the junction capacitance of the individual units. In this case, proper voltage division can be achieved by placing a small capacitor in parallel with each voltage-equalizing resistor.

In extremely critical applications, voltage division for a series arrangement can best be attained by replacing each voltage-equalizing resistor with a silicon voltage-reference diode.

Although silicon-controlled rectifiers can also be used in parallel arrangements, the circuit requirements in such applications are quite complicated and require a discussion too detailed for the purposes of this book.

18.2.7 Power Control

Silicon-controlled rectifiers are used in a large number of commercial and industrial power-control applications.

A silicon-controlled rectifier can be used to conduct during any desired portion of the positive cycle of anode voltage by applying the gate signal at the proper value of the anode voltage. This flexibility of control distinguishes the silicon-controlled rectifier from all other types of semiconductor devices.

18.2.8 DC Circuit Operation

An additional use of the SCR is in DC circuits. However, when it is forward biased, the anode current cannot drop below the holding current, and control is lost. In these circuits the SCR is switched off by diverting the current around the device so that anode current drops below the holding current, or reverse-biasing is applied for sufficient periods to allow the gate to regain control.

These applications operate from a DC source requiring commutation to control SCR conduction for circuit operation and load-waveform shaping. Regaining gate control (i.e., commutation) generally involves a capacitive bypass network or a load filter whose leading power factor allows the device to drop the load current, then be reverse-biased to form the junctions, achieve recommutation, and then block the forward voltage. Successful SCR operation in these DC circuits requires that the circuit turn-off times never become less than the device turn-off time.

18.3 TRIACS

The bidirectional switch or bidirectional triode thyristor (Triac) differs from the conventional SCR thyristor in that triggered conduction may occur in both directions. This affords economy in certain converter designs where a DC component is not necessary. Typical usage of these devices is in resistive heating of lighting controls. In these applications the Triac replaces a pair of SCRs in an antiparallel connection. The economy of a one-gate firing package, simpler heat-sinking, and ease of wiring have caused this device to display the SCR in lamp-control service and related usage. However, care must be exercised with inductive circuits. The back-to-back SCR pair has up to a half-cycle of the power frequency to recover when the current passes through zero. This is not the case with the Triac, which has a few microseconds in which to recover before the power circuit reaches holding current.

Triacs do not equal the power capability of SCRs. Present capabilities of Triacs extend to 200-A RMS ratings, and voltage capability for service on 440 Vac mains. However, within the power range of their ratings they represent an economical bulk-power control source.

18.4 THYRISTOR TERMS AND DEFINITIONS

Bidirectional Diode Thyristor A two-terminal thyristor having substantially the same switching behavior in the first and third quadrants of the principal voltage-current characteristic.

Bidirectional Triode Thyristor An *n*-gate or *p*-gate thyristor having substantially the same switching behavior in the first and third quadrants of the principal voltage-current characteristic.

Gate An electrode connected to one of the semiconductor regions for introducing control current.

Main Terminals The two terminals through which the principal current flows.

***n* Gate Thyristor** A three-terminal thyristor in which the gate terminal is connected to the *n* region adjacent to the region to which the anode terminal is connected and that is normally switched to the on-state by applying a negative signal between gate and anode terminals.

***p*-Gate Thyristor** A three-terminal thyristor in which the gate terminal is connected to the *p* region adjacent to the region to which the cathode terminal is connected and that is normally switched to the on-state by applying a positive signal between gate and cathode terminals.

Reverse-Blocking Diode Thyristor A two-terminal thyristor that switches only for positive anode to cathode voltages and exhibits a reverse blocking state for negative anode to cathode voltages.

Reverse-Blocking Triode Thyristor An *n*-gate or *p*-gate thyristor that switches only for positive anode to cathode voltages and exhibits a reverse blocking state for negative anode to cathode voltages.

Reverse-Conducting Diode Thyristor A two-terminal thyristor that switches only for positive anode to cathode voltages and conducts large currents at negative anode to cathode voltages comparable in magnitude to the on state voltage.

Reverse-Conducting Triode Thyristor An *n*-gate or *p*-gate thyristor that switches only for positive anode to cathode voltages and conducts large currents at negative anode to cathode voltages comparable in magnitude to the on state voltage.

Thyristor A bistable semiconductor device that comprises three or more junctions that can be switched between conducting and nonconducting status.

Turnoff Thyristor A thyristor that can be switched between conducting and nonconducting states by applying control signals of appropriate polarities to the gate terminals, with the ratio of triggering power to triggered power appreciably less than 1.

CHAPTER NINETEEN

Optoelectronic Devices

Semiconductors are significant in the generation and detection of light. Called optoelectronics,[114,124,125] this technology is providing light sources for readouts and indicators, as well as light-sensitive devices for light detection and control applications.

Optoelectronics is the branch of electronics dealing with devices that emit light when properly excited by an electrical signal, and devices whose electrical-conduction characteristics are influenced by the impingement of light. The development of semiconductor devices with these characteristics has led to significant changes in techniques for information display, light detection, and control.

Light-emission phenomenon can be explained in terms of semiconductor energy-band theory. An external voltage applied to forward-bias a *p–n* junction excites the majority carriers (electrons), causing them to move from the *n*-side conduction band to the *p*-side valence band. In making this transition, the electrons cross the energy gap (E_g) that separates the two bands and have to give up energy in the form of heat (phonons) and light (photons).

Each semiconductor material types has an E_g characteristic, and the wavelength (λ) of emitted light depends on the magnitude of E_g. For example, gallium arsenide material (GaAs) has an E_g = 1.35 eV and a λ_{peak} = 9000 Å. The wavelengths (i.e., color) emitted by some other materials made from gallium compounds are listed in Table 19.1.

19.1 LIGHT-EMITTING DIODES (LEDs)[126]

A light-emitting diode is a *p–n* junction diode fabricated of materials that cause light emission when the diode is forward biased. The materials used are either gallium arsenide (GaAs), gallium phosphide (GaP), or a combination of the two, namely, gallium arsenide phosphide (GaAsP). (See Figure 19.1)

When the LED is forward biased, electrons in the atoms of the material receive energy and are elevated from their normal equilibrium energy state to a higher energy level, that is, from the valence band to the conduction band. Because of the electrical potential on the LED, electrons from the *n*-side drift toward the positive terminal. As they reach the *p* side, most recombine with atoms of the *p*-type material that originlly were lacking an electron. Conversely, many electrons leaving the *n* material leave an atom of *n* material with a net positive charge (an atom lacking an electron, thus possessing a net positive charge, is called a "hole") that combines with a free electron. These recombinations cause a release of energy in the form of heat and light.

The energy of a photon of light released by recombination is equal to the difference in energy between the equilibrium state and the excited state of the electron, and is characteristic of the material. Since the wavelength of the photon is a function of the released energy, the wavelength of the emitted light is also characteristic of the diode material.

Table 19.1 Some Wavelengths and Colors Emitted by Gallium Compounds

Material	Wavelength	Color
GaAs:Zn	9000A	infrared
$GaAsP_{.4}$	6600A	red
$GaAsP_{.5}$	6100A	amber
$GaAsP_{.85}$:N	5900A	yellow
GaP:N	5600A	green

LEDs are available in colors ranging from infrared to green. Red is the most widely used color for LEDs as it provides the best combination of efficiency of light emission of the device and sensitivity of the human eye.

LEDs have found applications virtually anywhere that indicators, status displays, or warning lights are used. Because of their long operating life, small size, low power dissipation, low cost, and high reliability, they are superior to most other light sources, except where large-area illumination is required.

Circuit symbols for LEDs and other optoelectronic diodes are shown in Figure 19.2.

Most incandescent lights are rated in terms of voltage. LEDs, by contrast, are current-dependent devices since they are basically diodes. When operating from constant-voltage sources protection should be provided by incorporating a current-limiting resistor with each LED.

Lenses act to diffuse light into a broader apparent emitting area. LED lenses that produce broad, evenly diffused, light are generally assumed to be more pleasing to the eye than lenses that produce a highly intense point of light.

Visible-emitting LEDs, the link in the man–machine interface, are characterized in terms of photometric quantities. On the other hand, infrared-emitting LEDs (whose invisible light is of wavelengths longer than 750 nm) are characterized in terms of radiometric quantities.

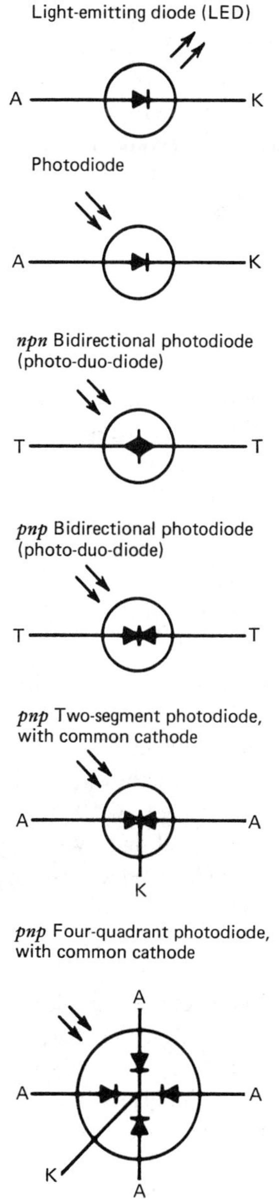

Figure 19.2 Optoelectronic diode symbols.

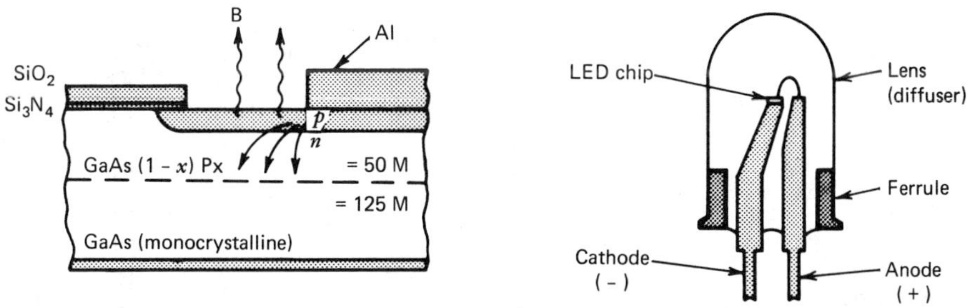

Figure 19.1 Typical GaAsP LED Cross section and package.[125]

Application requirements for IR LED sources are different from those of visible-emitting LEDs. Whereas for visible-emitting diodes a wide-viewing angle is normally important, for infrared sources a narrow beam width and high on-axis intensity are normally important.

Light output produced by infrared sources is defined by one or more of the following radiometric parameters. (See Table 19.2.)

A. ***Radiated Output Power*** **(*P* or *ROP*)** Total output of the device in all directions, measured in watts.

B. ***Radiant Intensity (J)*** Radiant flux per unit solid angle in a given direction, measured in watts/steradian.

C. ***Irradiance (H)*** The density of radiant flux incident on a surface, measured in watts/area.

19.2 LIGHT DETECTORS

Common photoconductors used with LEDs include the photodiode, phototransistor, light-sensitive Darlington amplifier, light-activated silicon-controlled rectifier (LASCR), planar silicon photoswitch (PSPS), and light-activated silicon-controlled switch (LASCS). These devices generate an electrical signal when light strikes them and provide a wide range of electrical characteristics that may be usefully exploited in applications requiring light detection or where control may be exercised through light coupling.

Light detection by optoelectronic devices is based on the response of a properly-fabricated, reverse-biased, *p–n* junction to stimulation by light. When light in the range of wavelengths to which the junction is sensitive impinges on the junction the energy released by a photon generates a hole-electron pair. Each becomes a charge carrier and, under the influence of the field caused by the reverse bias, drift across the junction. The result is current in the circuit proportional to the degree of radiation impinging on the junction.

19.2.1 Photodiode

A photodiode is simply a *p–n* junction semiconductor diode that responds to light. It is reverse-biased, whereas the light-source diode is forward-biased. The photodiode *p–n* junction creates an electric current when exposed to light. The current flows through an external load resistance, giving an output voltage approximately proportional to light intensity, see Figure 19.3.

In general, photodiode specifications are not confusing, but to select an optimum device, especially for high performance, the designer must understand the tradeoffs. There are basic parameters, in addition to signal-to-noise ratio and gain, that should be considered.

A. ***Capacitance of the p–n Junction*** Because of the reverse bias, the capacitance falls as the operating voltage increases until a point is reached when the device is fully depleted. Then the capacitance is independent of voltage and there is a truly *p–i–n* device. Thus for lowest capacitance, a high-enough operating voltage is needed. The tradeoff is that as the operating voltage increases, so do noise and dark current.

B. ***Dark Current*** This is a function of the area of a device; the larger the area the more dark current. It is measured in amperes and increases with applied voltage. The noise resulting from dark current is shot noise.

C. ***Quantum Efficiency*** Of all the solid-state photodetector devices, photodiodes have the highest quantum

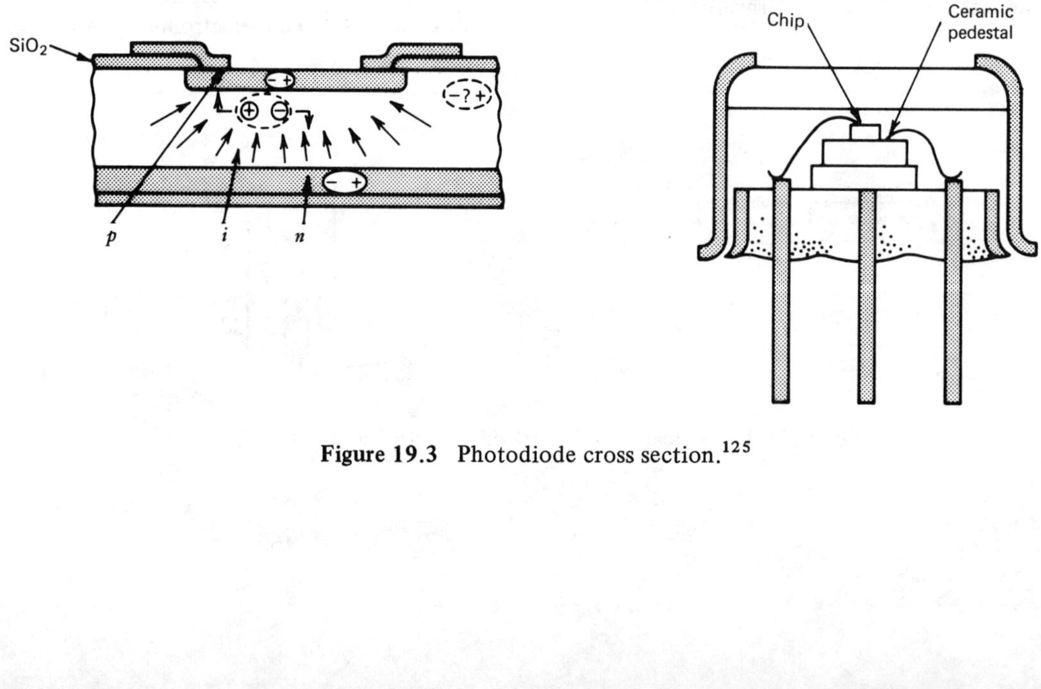

Figure 19.3 Photodiode cross section.[125]

Table 19.2 Radiometric and Photometric Parameter

Parameter	Symbol	Definition	Units	Abbreviation
		Radiometric		
Radiant Energy	Q_e		erg joule calorie kilowatt-hour	 J cal kWh
Radiant Flux	P	$P = \frac{dQ_e}{dt}$	erg per second watt	erg s^{-1} W
Radiant Emittance[b]	W	$W = \frac{dP}{dA}$	watt per sq. cm, watt per sq. m, etc.	W cm^{-2} W m^{-2}
Irradiance[b]	H	$H = \frac{dP}{dA}$	watt per sq. cm, watt per sq. m, etc.	W cm^{-2} W m^{-2}
Radiant Intensity[a]	J	$J = \frac{dP}{d\omega}$	watt per steradian	W sr^{-1}
Radiance[a]	N	$N = \frac{d^2P}{d\omega(dA \cos \Theta)}$ $N = \frac{dJ}{(dA \cos \Theta)}$	watt per steradian and sq. cm watt per steradian and sq. m	W sr^{-1} cm^{-2} W sr^{-1} m^{-2}
		Photometric		
Luminous Efficacy	K	$K = \frac{F}{W}$	lumen per watt	lm W^{-1}
Luminous Efficiency	V	$V = \frac{K}{K_{\text{maximum}}}$		
Luminous Energy (quantity of light)	Q_v	$Q_v = \int_{380}^{760} K(\lambda)\, Q_e \lambda d\lambda$	lumen-hour lumen-second (talbot)	lm h lm s
Luminous Flux	F	$F = \frac{dQ_v}{dt}$	lumen	lm
Luminous Emittance[b]	L	$L = \frac{dF}{dA}$	lumen per sq. ft	lm ft^{-2}
Illumination[b] (illuminence)	E	$E = \frac{dF}{dA}$	footcandle (lumen per sq. ft.) lux (lumen per sq. m) phot (lumen per sq. cm)	fc lx ph
Luminous Intensity (candlepower)	I	$I = \frac{dF}{d\omega}$	candela (lumen per steradian)	cd
Luminance (brightness)	B	$B = \frac{d^2F}{d\omega(dA \cos \Theta)}$ $B = \frac{dl}{(dA \cos \Theta)}$	candela per unit area stilb (candela per sq. cm) nit (candela per sq. m) foot-Lambert (cd per πft^2) apostilb (cd per πm^2) Lambert (cd per πcm^2)	cd $in.^{-2}$ sb nt ft-L asb L

[a] ω is a solid angle through which flux from point source is radiated. Θ is angle between line of sight and normal to surface considered. λ is wavelength.
[b] W and L refer to "emitted from" and H and E refer to "incident on."

efficiency, the ratio of electrical current to incident light, approaching 95%.

D. *Noise Equivalent Power (NEP)* NEP simultaneously accounts for quantum efficiency or responsitivity and noise. NEP is the amount of light that has to shine on a detector in order to produce a signal equal to the noise level. The lower the NEP, the more sensitive the device. If NEP is used as a criterion for selecting a photodiode also specify the wavelength, the test frequency, and the bandpass around the test frequency, as vendors often omit this important information.

E. *Responsivity* It is more practical to convert electrons to amperes and photons to watts than deal with quantum efficiency. However, the value of responsivity (A/W) is a function of wavelength, applied-bias voltage, frequency, and temperature. Therefore, all these parameters should be specified when a value of responsitivty is given.

F. *Detectivity* Since there is more noise as the area of a detector increases, NEP naturally tends to be worse with a large-area device. Detectivity provides a figure of merit that is independent of the area. It is defined as NEP/active area (NEP/cm^2). The higher the detectivity, the more sensitive the detector.

G. *System Requirement* (See Table 19.3.)

H. *Electrical Characteristics* (See Table 19.4.)

19.2.2 Phototransistor

The phototransistor converts light to an electrical signal in essentially the same manner as the photodiode. However, adding another junction to the device results in a transistor that amplifies the current generated at the radiated junction by a factor equal to the current gain of the transistor. Circuit symbols for phototransistors are shown in Figure 19.4.

Phototransistors, along with their close cousins the photo-Darlingtons, are by far the most popular of all photodetectors. Since they are essentially transistors, the photon-generated current is multiplied by the current gain of the device. Photosensitivity for a given light level is thus 100–500 times greater than for a photodiode.

Overall performance characteristics are fairly standard. Voltage should be 50 V or less; current capability is 1–50 mA; and typical speed, rise time, and fall time, is from 2 to 100 μsec.

Some basic tradeoffs involve active-area size, the use of a lens, speed, and gain. Generally, large-area devices are used for high sensitivity at low light levels or for high output-current applications. Small-area devices are used when high speed is a requirement. Also, sometimes the mechanical structure can become a variable parameter that is difficult to deal with in calculating device sensitivity.

When coupling speed is important, the photodiode is better than the phototransistor. The transistorized optocoupler, however, provides signal amplification. A phototransistor may be operated as a diode by taking the out-

Table 19.3 Photodiode Selection by System Requirements[127]

Choose	When
A Single-Element Photodiode (Standard Configuration)	Extremes in temperature and humidity are expected. Fast time response is required
A Single-Element Photodiode (Special Wide-Field-of-View Configuration)	The detector is close to the emitting source. Large f/number optics are used. Fiber-optics are employed.
A Quadrant-Type Photodiode (Standard Configuration)	A position sensitive device is needed. Fast time response is required.
An Avalanche Photodiode	Small-signal detection is needed. Very fast time response is required. High responsivity is required.
A Hybrid Photodiode-Preamplifier Module	Small-signal detection is needed. High responsivity is required. Very low noise is needed. Small NEP levels independent of frequency up to 10 MHz are required.

Table 19.4 Photodiode Selection by Electrical Characteristics[127]

Characteristic	*p–i–n* Photodiode	Avalanche Photodiode	Hybrid Photodiode-Preamplifier Assembly
Area (mm^2)	0.2 to 100	0.2 to 7	0.8 to 4.5
Operating Voltage (V)	0 to 225	180 to 425	Detector–45 to 335 Amplifier–±6 to ±12
Dark Current (nA)	10 to 500	15 to 100	Not Applicable
Capacitance (pF)	2 to 70	1.6 to 10	Not Applicable
Responsivity (A/W)	0.6 @ 900	8 to 20 A/W @ 1060 nm	2.5×10^3 to 3×10^5 V/W @ 900 nm
Rise Time (ns)	3 to 22	0.5 to 2	10 to 35

put between collector and base, instead of from between collector and emitter.

The base lead of the phototransistor can be used or not. When it is connected to an external voltage or resistance network the base lead provides a means to adjust gain and stabilize the phototransistor against temperature and other factors. The disadvantage of the base-lead connection is that there is a decrease in sensitivity because the base circuitry shunts some current around the base-emitter junction.

In addition to the gain offered by phototransistors, they differ from photodiodes in another important way. Because the phototransistor depends on its current gain for overall sensitivity, there is a built-in nonlinearity, as gain increases with increased light current. However,

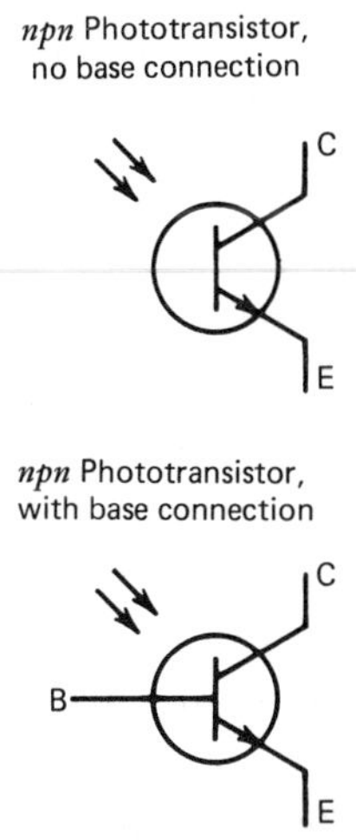

Figure 19.4 Phototransistor symbols.

with photodiodes no nonlinear mechanism is involved, so that photodiodes are usually followed by a good linear amplifier to preserve overall linearity.

19.2.3 Photo-Darlington

The photo-Darlington consists of two phototransistors in cascade on a single chip. It is basically the same as the phototransistor, except that the amplification, which is the product of the amplifications of the individual transistors, is much greater.

19.2.4 Light-Activated Silicon-Controlled Rectifier (LASCR)

The LASCR is triggered into conduction when the light falling on the base collector photodiode junction exceeds a given threshold level. Operation of the LASCR is similar to that of the silicon-controlled rectifier (SCR), with the major difference being that an external resistance between gate and cathode (in addition to the bias voltage and current) determines light sensitivity. A positive electrical signal applied to the gate can be used to trigger the LASCR, as well as to modify the light sensitivity. (See Figure 19.5.)

19.2.5 Planar Silicon Photoswitch (PSPS)

The PSPS is essentially a complementary SCR. Like the LASCR, it can be triggered by light. In addition, a negative signal, with reference to the anode, at the anode gate terminal can trigger the device.

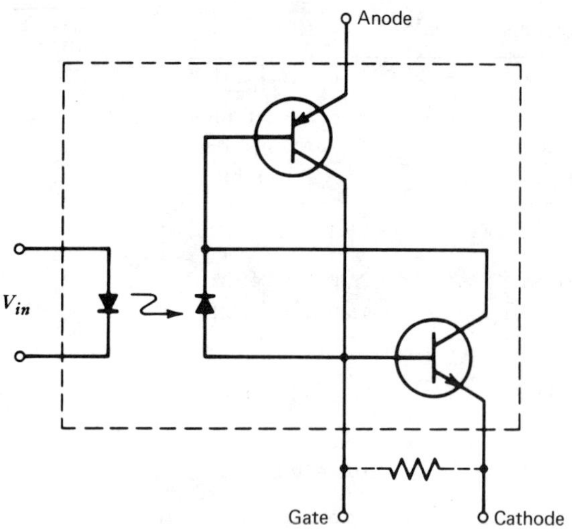

Figure 19.5 LASCR.[121]

19.2.6 Light-Activated Silicon-Controlled Switch (LASCS)

The LASCS combines the LASCR and PSPS. Having four terminals, the LASCS can be triggered by light, positive signals at the gate terminal, and negative signals at the anode gate terminal.

19.3 OPTOELECTRONIC COMPONENTS

Component characteristics will be affected by both device and package parameters. The basic optoelectronic devices can be packaged to provide:

A. Discrete detectors and emitters that will emit or detect light.

B. Interrupter/reflector modules that detect objects modifying the light path.

C. Isolators/couplers that transmit electrical signals without electrical connection.

19.3.1 Detectors and Emitters

Optoelectronic components require packaging that not only protects the chip, but also permits light to pass through the package to the chip, that is, a semiconductor package with a window. The window can be modified to provide lens action that gives higher response on the optical axis of the lens, greater directional sensitivity, and a larger aperture with less resolution.

In most commercial components the lens is also an integral part of the package for economic reasons, so that tight control of optical tolerances are compromised somewhat to optimize chip protection via the hermetic seal. This causes lensed components to exhibit wider variations, unit to unit, than simple window components, as the optical gain variations and the basic device response variations are multiplied. Due to these factors, when high-gain, highly-directional, optical systems are required it is normal procedure to recommend components without integral lenses be used in conjunction with external optics of the required quality. (See Figure 19.6.)

The other major factor in detector/emitter packaging is the choice of a plastic or hermetic package. Either is available with or without lens, although the plastic devices have the optical axis perpendicular to the leads, while the hermetic package optical axis is parallel to the leads. The hermetic package will operate at higher power over a wider temperature range and is more tolerant of severe environments. However, the hermetic package is also relatively more expensive than the plastic package. Although some components are limited to a single package type, in most applications the user must weight the technical and economic constraints of the application in order to optimize both the device and package of the optoelectronic component used.

19.3.2 Interrupter/Reflector Modules

The use of interrupter or reflector modules eliminates most of the optical calculations, geometric problems, and conversion problems in mechanical position-sensing applications. These modules are specified electrically at the input and output simultaneously, that is, as a coupled pair, and have defined constraints on the mechanical input. All the user need do is provide the input current and mechanical input, that is, pass an IR opaque object through the interrupter gap and monitor the electrical output.

Other than standard tolerance, resolution, and power constraints, the only new knowledge required is the ability of the sensed object to block and reflect IR light and an estimate of the effects of ambient light conditions providing false signals. (See Figures 19.7 and 19.8.) Once the module is characterized for minimum and maximum characteristics, it is a defined electrical and mechanical component and does not require optical design work for each new application. This puts these sensor modules in the same design category as mechanical precision limit switches, except that with the activating mechanism blocking or reflecting light instead of applying a force. Thus, mechanical wear and deformation effects are eliminated.

Most commercially available interrupter modules are built around plastic-packaged emitters and detectors.

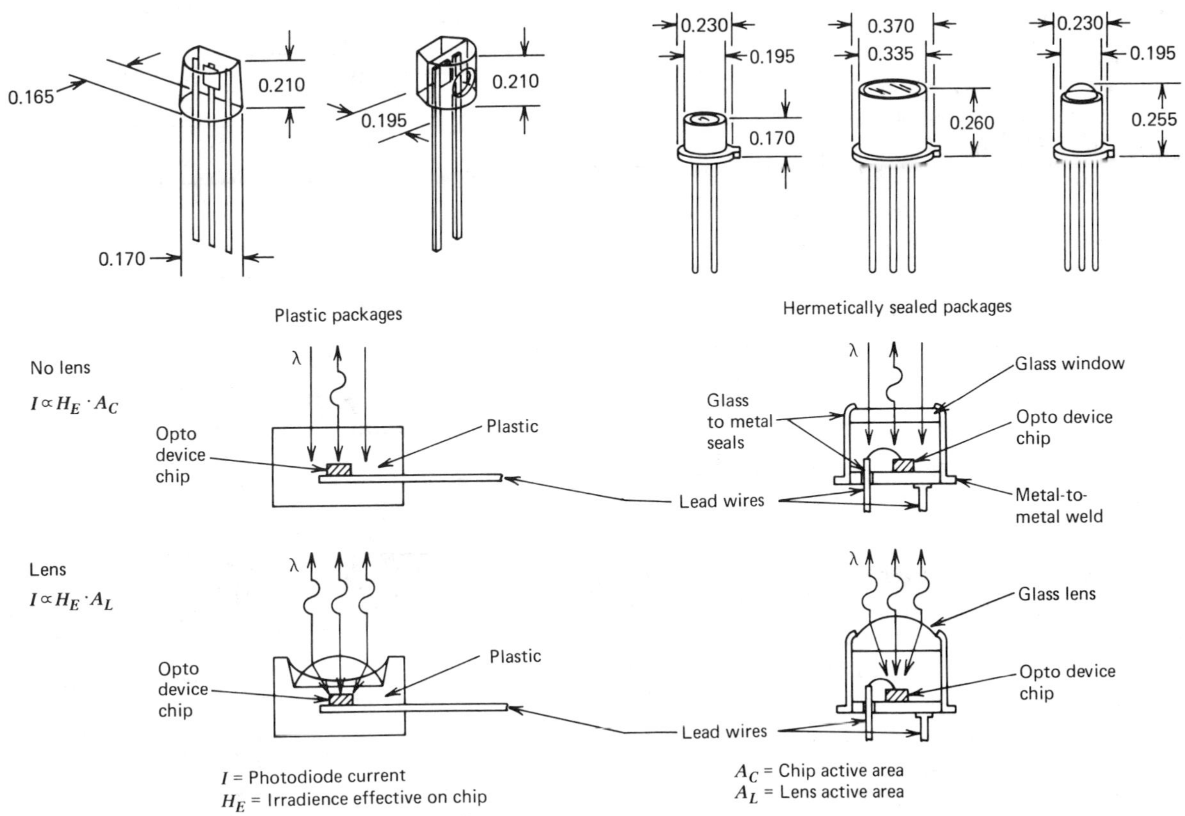

Figure 19.6 Discrete optoelectronic component package concepts. (Reproduced from Ref. 124.)

Reflective modules and other custom modules are built around both plastic and hermetic parts, depending on the cost/performance tradeoffs required. It should be noted that due to the longer, angle-critical, and generally less efficient, light transmission path in a reflector module, devices with lenses are dominant in these applications.

19.3.3 Optocouplers

An optocoupler is a solid-state switching device consisting of a light source (input) and a light detector (output).

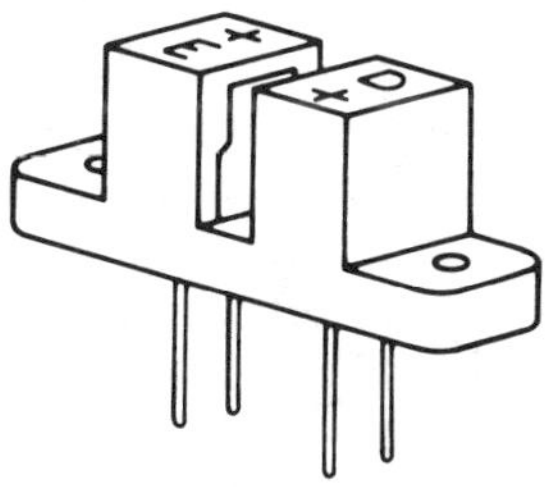

Figure 19.7 Interrupter module. (Reproduced from Ref. 124.)

In the basic optocoupler both the light source and the detector are housed in a single package that is sealed against outside light.

An electrical signal applied to the light source changes the amount of light emitted. The emitted signal falls upon, and is collected by, the detector. These input electrical signals are thus "coupled" to the output. From the output the signals perform normal electronic functions, such as driving amplifiers, triggering a thyristor power supply, or switching logic levels.

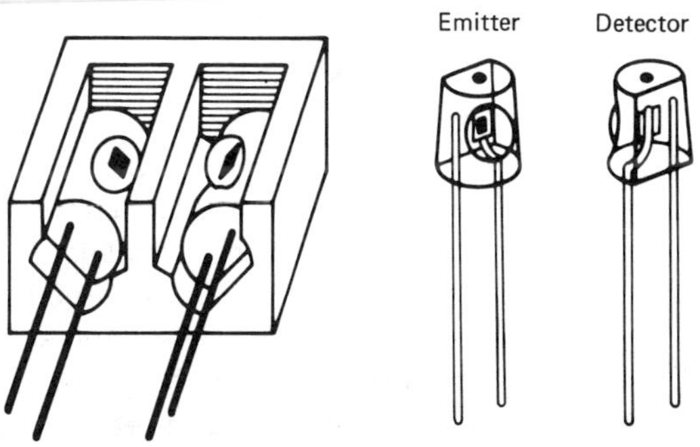

Figure 19.8 Reflector module. (Reproduced from Ref. 124.)

Circuit symbols for optically coupled isolators are shown in Figure 19.9.

The light path to the photodetector is totally enclosed in the component and cannot be modified externally. This provides one-way transfer of electrical signals to the photodetector without electrical connection between the circuitry containing the devices. The degree of electrical isolation between the two devices is controlled by the material in the light path and by the physical distance between the emitter and detector, that is, the greater the distance the better the isolation. Unfortunately, the current transfer ratio (CTR), the ratio of detector current to emitter current, that is, the effectiveness of electrical signal transfer, is inversely proportional to this separation. Thus some compromise has to be made to achieve the optimum result.

A. ***Packaging*** With the dual-inline package (DIP) the use of optical glass has proven to be a most efficient dielectric. Glass allows maximum CTR and a minimum separation distance for a given isolation withstanding voltage capability. (See Figure 19.10.) Minimum CTRs of 100% in combination with isolation voltages of 5000 V in phototransistor couplers result. Also, because of the glass dielectric design, yields are much more predictable due to easier alignment of the LED and detector and common-side wire bonding as compared with other methods of manufacture.

Although the DIP is the most common coupler package, other packages are commercially available to provide higher isolation voltage and other requirements. For very high isolation voltage requirements, 10-50,000 V, the interrupter module can be modified by the user at low cost by putting a suitable dielectric (glass, acrylic, silicone, and so on) in the air gap and by insulating and encapsulating the lead wires. For higher isolation voltages the use of the matched pair with glass or infrared transmitting fiber optics can provide a low-cost isolator. Both of these approaches utilize coupler systems that are already characterized and, therefore, easily handled from a design standpoint.

B. ***Advantages*** Optocouplers offer several advantages for switching, transferring, and isolating signals. Solid-state reliability assures long, trouble-free performance. There are no moving parts to wear or jam and no mechanical contacts to pit or stick. In addition, solid-state construction minimizes shock and vibration problems.

Operating power is lower than that for conventional relays. Also, optocouplers switch on and off in the low microsecond range, much faster than conventional relays. They transfer data from DC to hundreds of kilohertz. Although a pulse transformer couples higher frequen-

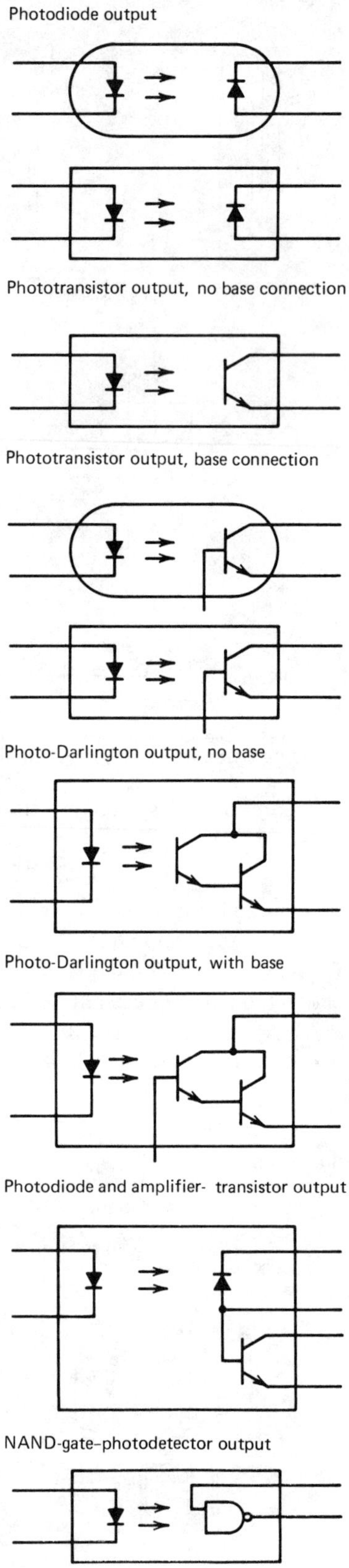

Figure 19.9 Optically coupled isolators. (Reproduced from Ref. 124.)

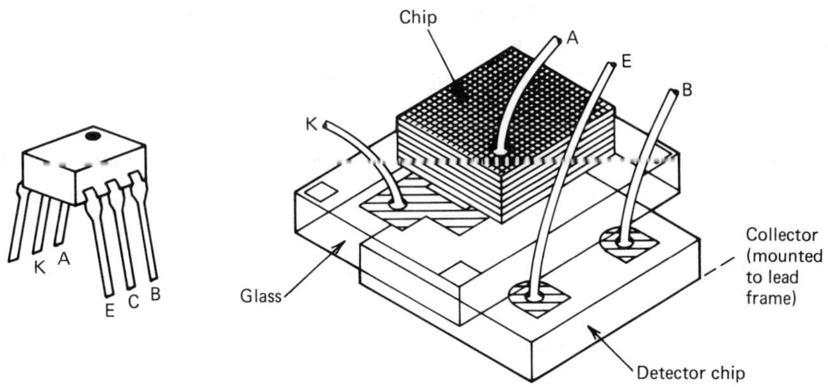

Figure 19.10 Glass-Isolation (DIP) optocoupler. (Reproduced from Ref. 124.)

cies, a pulse transformer needs a latch to restore DC information.

Not only is the optocoupler economical, but it does not require coil-suppression diodes or resistor/capacitor arc-suppression diodes, resistors, and capacitors.

C. *External Coupling* External coupling of light from the emitter to the detector provides two more forms of optocoupler devices, the encoder assembly and the reflective read-head assembly. In these units the coupling light path is external to the assembly. With the outside light path the amount of coupled light can be influenced by mechanical means. For example, an object placed in the encoder slot will attenuate or cut off light to the detector. With the read-head assembly, any change in the reflective material changes the coupled light. (See Figure 19.11.)

D. *Applications* In an automotive ignition the device is packaged so that the light path from LED to photodetector is external to the package. A slotted wheel is placed between the light source and the detector. By alternately passing and interrupting the light beam at the proper intervals this device, in conjunction with the proper power circuitry, is the equivalent of a set of ignition points that do not bounce, arc, or pit.

There are numerous other applications of optical isolators in data transmission, digital electronics, telephone switching, and so on. A very important and potentially life-saving application is in the field of medical electronics. A high degree of electrical isolation is imperative between the patients and the medical electronics linked to them. Without isolation a malfunction in an instrument, particularly one connected to the AC line, could kill a patient. Optical isolators are an excellent means of providing the required isolation.

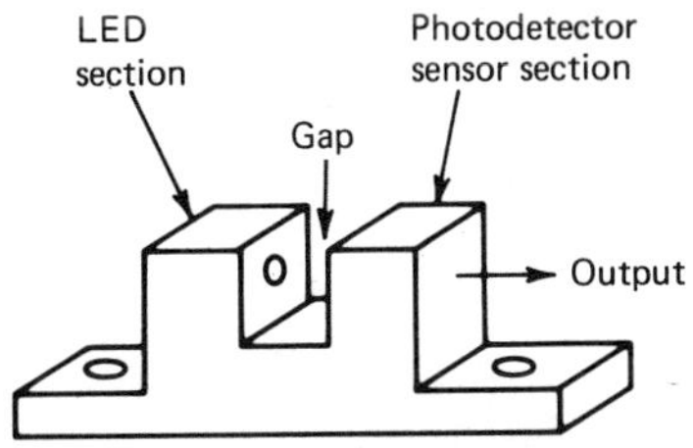

Figure 19.11 Optical coupler with external light path.[125]

E. *Solid-State Relay (SSR)* Adding a silicon-controlled rectifier (SCR) to the output of the optocoupler results in a complete package that is capable of controlling AC line power with low-level control signals. This is a solid-state relay.

The solid-state relay (SSR) is, as its name implies, a solid-state equivalent of an electromechanical relay (EMR). Instead of using a control signal to create a magnetic field to activate a switch and, thereby, connecting a power source to a load as the EMR, the SSR uses a control signal to generate light that turns on a phototransistor that fires an SCR, thereby providing power to the load. SSRs are generally more expensive than EMRs, but SSRs possess significant advantages that may be worth the additional cost.

19.4 OPTOELECTRONIC TERMS AND DEFINITIONS

Avalanche Luminescent Diode A light-emitting diode that emits luminous energy when a controlled reverse current in the breakdown region is applied.

Conversion Efficiency The ratio of maximum available power output resulting from a photovoltaic operation to the total incident radiant flux.

Dark Current The current that flows through a photosensitive device in the dark condition.

Infrared-Emitting Diode A diode capable of emitting radiant energy in the infrared region of the spectrum resulting from the recombination of electrons and holes.

Light Current The current that flows through a photosensitive device when it is exposed to radiant energy.

Light-Emitting Diode (LED) A diode capable of emitting luminous energy resulting from the recombination of electrons and holes.

Optoelectronic Device A device that is responsive to, or that emits or modifies, electromagnetic radiation in the visible, IR, or UV spectral regions; or a device that utilizes such electromagnetic radiation for its internal operation.

Photoconductive Diode A photodiode intended to be used as a photoconductive transducer.

Photocurrent The difference in magnitude between light and dark current.

Photo-Darlington Coupler An optocoupler in which the photosensitive element is a Darlington-connected phototransistor.

Photodiode A diode that is intended to be responsive to radiant energy.

Photodiode, Avalanche A photodiode that is intended to take advantage of avalanche multiplication of photocurrent.

Photodiode Coupler An optocoupler in which the photosensitive element is a photodiode.

Photoemitter A device that emits electromagnetic radiation in the visible, IR, or UV spectral regions.

Photosensitive Device A device that is responsive to electromagnetic radiation in the visible, IR, or UV spectral regions.

Photothyristor A thyristor that is intended to be responsive to radiant energy by controlling its operation as a thyristor.

Photothyristor Coupler An optocoupler in which the photosensitive element is a photothyristor.

Phototransistor A transistor that is intended to be responsive to radiant energy.

Phototransistor Coupler An optocoupler in which the photosensitive element is a phototransistor.

Photovoltaic Diode A diode that is intended to generate a terminal voltage in response to radiant energy.

Radiant Efficiency The ratio of the total radiant flux emitted to the total input power.

CHAPTER TWENTY

Special-Function Solid-State Devices

20.1 THERMISTORS[128]

The effect of change in temperature of an electronic device, particularly a semiconductor, is a problem that must always be considered. Ignoring the effects of temperature change may result in a device that operates poorly, doesn't operate at all, or, in the worst case, that destroys itself.

There are devices that put these effects of temperature change to good use, because they use temperature change to perform some useful function. One such device is the thermistor. A thermistor is a semiconductor that behaves as a thermally sensitive resistor, that is, its electrical resistance changes as its body temperature changes.

Virtually all thermistors have a negative temperature coefficient, as their resistance decreases as temperature increases. This can be contrasted with a wire-wound resistor, that has a positive temperature coefficient, that is, its resistance increases as temperature increases.

Another major characteristic of a thermistor is that it has a very high negative temperature coefficient. This means that a thermistor exhibits a large resistance change for a relatively small change in temperature. It is this large resistance change that makes the thermistor useful in a variety of applications.

20.1.1 Material

Thermistors are semiconductors of ceramic material made by sintering mixtures of metallic oxides such as manganese, nickel, cobalt, copper iron, and uranium.

20.1.2 Configurations[129]

Various mixtures of these metallic oxides can be formed into useful shapes. Their electrical characteristics may be controlled by varying the type of oxide used and the physical size and configuration of the thermistor. The most common forms available are:

A. ***Beads*** Beads are made by forming small ellipsoids of thermistor material on two fine wires tight and parallel approximately 0.010 in. apart. The material is then sintered at high temperature and the leads become embedded tightly in the beads, thus making good electrical contact inside the thermistor. The beads may be coated with glass for protection or they may be mounted in evacuated or glass-fulled bulbs, Figure 20.1. Resistance values of 300 Ω to more than 100 MΩ can be obtained in beads ranging from 0.006 to 0.100 in. in diameter.

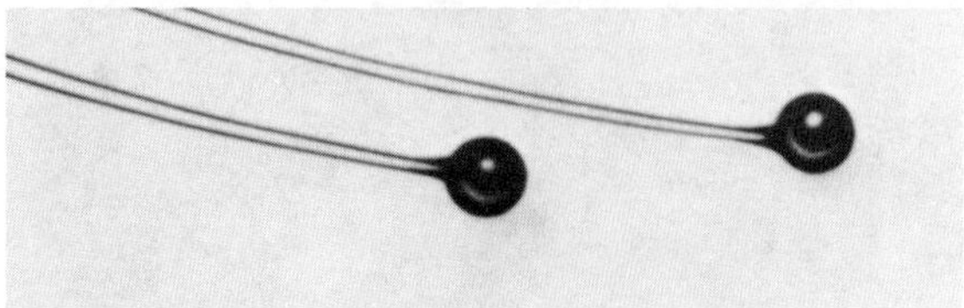

Figure 20.1 Bead thermistors. (Courtesy of Fenwal Electronics Division of Walter Kidde and Co., Inc.)

B. ***Probes*** Probes are beads sealed into the tips of solid-glass rods of up to 0.100 in. in diameter and 0.25-2 in. in length. (See Figure 20.2.)

C. ***Discs*** Discs are made by pressing thermistor material under several-tons pressure in a round die to produce flat pieces. These pieces are sintered and then silvered on the two flat surfaces. Normal disks are 0.100–1 in. in diameter and 0.200–0.500 in. thick. (See Figure 20.3.) Resistance values of 5–10,000 Ω can be produced this way.

D. ***Washers*** Washers are made like disks, except that a hole is formed in the center so that the unit can be mounted on a bolt. Several washers may be mounted together on a bolt with terminals between them, so that they may be connected either in series or parallel. Normal washers are 0.75 in. in diameter, with other sizes made as required. (See Figure 20.4.)

E. ***Rods*** Rods are extruded through dies to make long cylindrical units that are normally 0.053, 0.112, or 0.173 in. in diameter and 0.25-2 in. long. (See Figure 20.5.) Leads are attached to the end of the rods, and resistance values can be made from 1000 to 150,000 Ω. The major advantage of rods over other types of thermistors is the possibility of producing high-resistance units with moderate power-handling capacity.

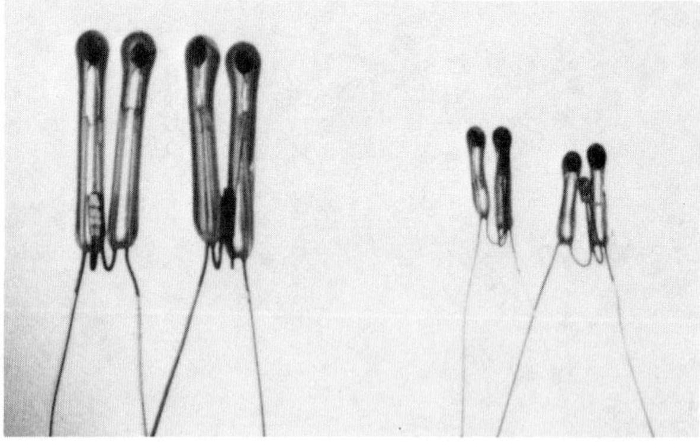

Figure 20.2 Probe thermistors. (Courtesy of Fenwal Electronics Division of Walter Kidde and Co., Inc.)

20.1.2 Thermistor Ratings

Thermistor ratings, such as stability, maximum operating temperature, and tolerance, should be considered. In addition, the following should be taken into account:

A. ***Zero-Power Resistance*** The resistance value of a particular thermistor at a specified reference temperature with zero electrical power dissipation. This temperature is usually +25°C (+77°F), unless otherwise specified.

B. ***Resistance Ratio*** The ratio of the zero-power resistance of a thermistor measured at two specified reference temperatures. The resistance ratio provides a basis for comparing the sensitivities of various thermistors with regard to temperature variations. If the ratio of one thermistor is higher than that of another, the first is more sensitive to temperature variations.

C. ***Resistance-Temperature Ratio*** The ratio between a thermistor's zero-power resistance and its resistance at

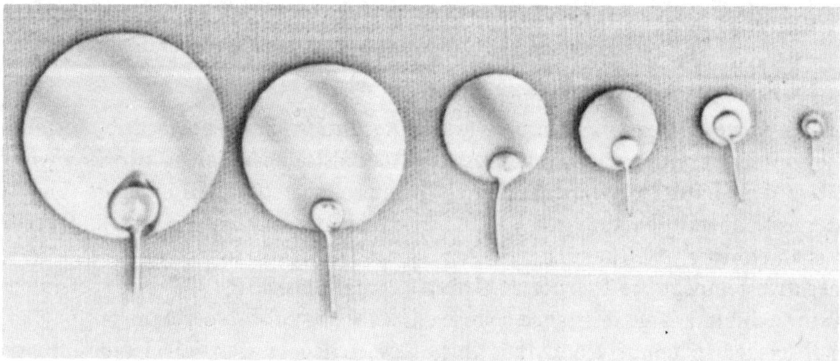

Figure 20.3 Disk thermistor. (Courtesy of Fenwal Electronics Division of Walter Kidde and Co., Inc.)

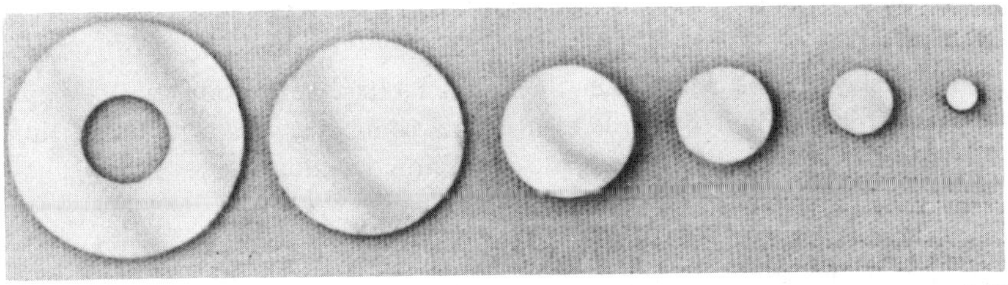

Figure 20.4 Washer thermistor. (Courtesy of Fenwal Electronics Division of Walter Kidde and Co., Inc.)

any temperature. The resistance-temperature ratio provides a quick method of determining a specific thermistor's resistance at any given temperature.

D. ***Dissipation Constant*** The ratio, at a specified temperature, of the power dissipation in a thermistor to the resultant change in its body temperature. It is specified in milliwatts per degree centigrade (mW/°C). Practically speaking, dissipation constant for a thermistor is the number of milliwatts that, when dissipated by the thermistor, will raise its temperature 1°C above its surroundings.

E. ***Sensitivity of a Thermistor*** A function of the temperature coefficient of resistance, which is the change in resistance resulting from a unit change in temperature. It is with respect to Sensitivity that the thermistor excels. Thermistors have a very high coefficient of change of resistance with temperature. Consequently, it is an ideal temperature sensor, that is sensitive to small temperature changes and requiring less amplification than other sensing devices.

20.1.3 Applications

There are many circuits in which thermistors are used to advantage. The number of successful applications has been growing rapidly and has few limitations. A few common applications of thermistors are as follows.

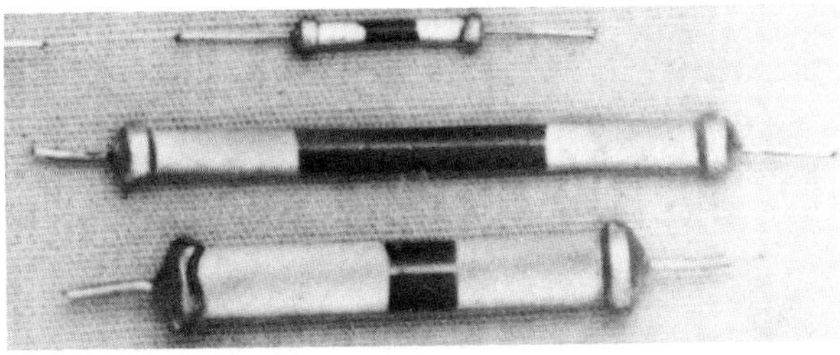

Figure 20.5 Rod thermistor. (Courtesy of Fenwal Electronics Division of Walter Kidde and Co., Inc.)

A. ***Temperature Measurement*** A simple circuit for temperature measurement consists of a battery, a thermistor, and a micrameter. As the resistance changes, the resistance of the thermistor changes and the current flow through the meter can be calibrated in terms of temperature. A more sensitive temperature-measuring circuit consists of a thermistor in one leg of a bridge circuit.

B. ***Temperature Compensation*** Most circuit elements have a positive temperature coefficient of resistance, such as copper in a meter coil. A thermistor may be used to compensate such a coil so that the meter resistance is substantially constant over a wide temperature range.

C. ***Temperature Control*** A simple temperature control can be made by placing a thermistor in series with a relay, battery, and a variable-value resistor. By adjusting the variable-value resistor, it is possible to make the relay operate at any desired temperature, as the relay will close when the thermistor gets warm and open when it gets cold.

A more-sensitive temperature control may be made by placing a thermistor in one leg of a bridge circuit, a variable-value resistor in another leg, and a polarized relay across the output. When the thermistor becomes warm the relay will operate in one direction and when the thermistor becomes cold, the relay will operate in the opposite direction. The point of operation may be adjusted by changing the value of the variable-value resistor.

D. ***Liquid Measurement*** If a thermistor is placed in series with a lightbulb and a battery, the light will operate if the thermistor is suspended in air. The thermistor heats up and the resistance drops permitting enough current to flow to light the bulb. If the thermistor is submerged in a liquid it will be cooled because of the greater thermal conductivity of the liquid; the thermistor resistance will increase, significantly reducing the current through the bulb; this may be used as a liquid-level indicator.

E. *Time Delay* By placing a thermistor and a variable-value resistor in series with a battery, a switch, and a relay a variable time-delay relay may be made. When the switch is closed, the current flow is limited by the high resistance of the thermistor, which then heats up and permits sufficient current flow to close the relay. By increasing the series resistance, the delay time may be increased, and by reducing the series resistance, the delay time may be decreased. Such delay circuits are used in many applications for which variable or fixed delay is required.

F. *Remote Control* Some bead thermistors are made with attached heaters and mounted in a vacuum bulb. If power is applied to the heater, the thermistor resistance is reduced. Such a unit placed in the input of a vacuum-tube amplifier makes a good remote gain control.

20.2 TRANSIENT PROTECTION DEVICES[130]

Overvoltage protection is necessary because signal lines are commonly subjected to a number of potentially destructive situations, as follows:

A. Analog signals may be present while the power supply is off.

B. The signal lines may receive induced voltage spikes from nearby sources.

C. Static electricity may be introduced on the signal lines by personnel or equipment.

D. Grounding problems are frequent, as AC power-line voltages at high impedance can appear on the signal lines.

E. Signal lines can be accidentally shorted to voltage sources.

Table 20.1 Transient Protection Devices

Name	VRD Polarity and Bipolarity Zener Diode	Metal Oxide Varistor	Silicon Carbide Varistor	Gas Tube Arrester
Materials	Silicon	Metal Oxide	Silicon Carbide	Gas Gap
Polarity	Polar Bipolar	Bipolar	Bipolar	Bipolar
Voltage Clamping Ratio α	$\gg 100$	$30 \sim 100$	$2 \sim 7$	–
Lightning Current Rating	8 x 20 μsec. 6 ~ 100 A	8 x 20 μsec. 250 ~ 10,000 A	8 x 20 μsec. 1 ~ 1,000 A	8 x 20 μsec. 800 ~ 6,000 A
Joule Rating	0.1 ~ 0.2 J	2 ~ 100 J	0.3 ~ 1,150 J	–
Power Rating	0.5 ~ 1.0 W	0.1 ~ 3.0 W	0.5 ~ 11.0 W	–
Dimension	$4.4^{\phi} \times 6^{t}$	$7 \sim 33^{\phi}$	$4 \sim 75^{\phi}$	$6.2^{\phi} \times 26^{t}$ $8.0^{\phi} \times 6.6^{t}$
Capacitance	25 ~ 400 pF	100 ~ 5,000 pF	30 ~ 4,000 pF	$\leqq$ 5 pF

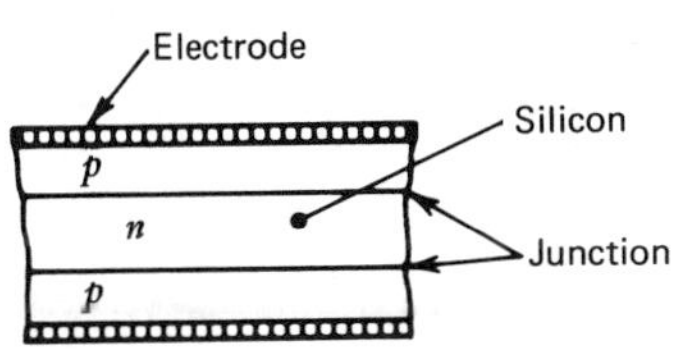

Figure 20.6 Cross section of a voltage regulator diode.

Various types of transient protection devices are shown in Table 20.1.

20.2.1 Voltage-Regulator Diode

Voltage-regulator diodes (VRDs), Figure 20.6, were developed for pulse clipping to protect contacts and semiconductors. One is the Z type, which has bipolarity-zener characteristics, and the other is the ZD-type, which has polarity-zener characteristics. The Z-type VRD can be used for both AC and DC applications, while the ZD-type of VRD is for DC use only. The ZD-type is also ideal for controlling coil surges below power-source voltages.

20.2.2 Metal Oxide Varistor

Ceramic with zener characteristics is a metal oxide varistor (Figure 20.7), which is primarily used for surge absorption. Features of these devices are:

A. Great endurance against surge voltages.
B. Excellent voltage control characteristic.
C. No follow-on current.
D. Symmetrical voltage-current characteristics.

Common applications of metal oxide varistors are lightning surge absorption, switching surge absorption, semiconductor protection, and quenching contact sparks.

20.2.3 Silicon Carbide Varistor

A silicon-carbide varistor is a bipolar nonlinear resistor which is found in various shapes, such as a washer, rod,

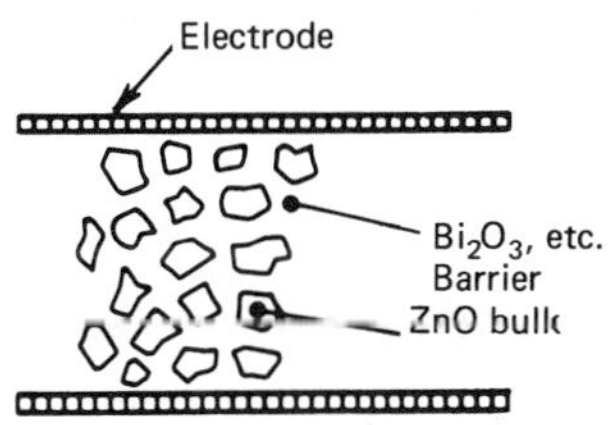

Figure 20.7 Metal oxide varistor cross section.

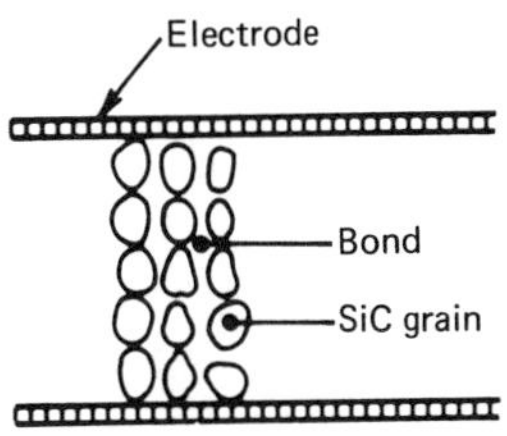

Figure 20.8 Silicon carbide varistor cross section.

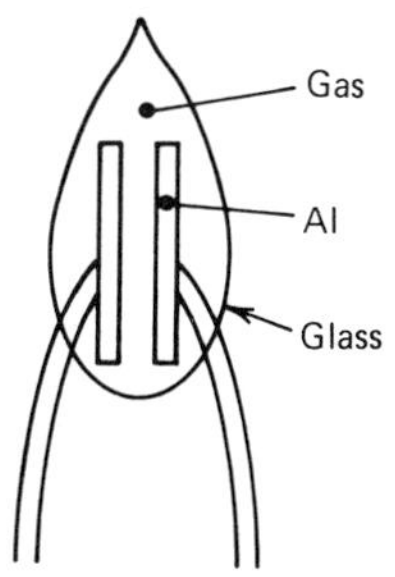

Figure 20.9 Gas tube arrestor cross section.

disc, etc., which is selected according to the application, Figure 20.8. The silicon carbide varistor is used for spark killing and protecting small semiconductor devices from an overvoltage.

20.2.4 Spark-Gap (Gas-Tube) Arrestor

A spark-gap surge arrestor (Figure 20.9) is used to protect circuits against sudden voltage overloads, in addition to acting as a high-energy voltage-sensitive switch.

REFERENCES

1 Sy Levine, "Basic Electronics for the Buyer, Part II," *Electronic Purchasing*, September 26, 1975.

2 "Military Standard—Resistors, Selection and Use of," MIL-STD-199, Revision B, Notice 5, May 22, 1978.

3 Mepco/Electra, Inc., "Film Resistors," Bulletin MB 0478 FR.

4 Charles V. Anderson, "The Metal Film Resistor, Characteristics and Technology," Mepco/Electra, Inc., 1977.

5 William Mulligan, "Need a High-Voltage Resistor?" *Electronic Design*, September 27, 1977.

6 TRW/IRC Resistors, "Metal Film Resistors—A Design Refresher," Handbook 770, 1977.

7 Cal Lockert, Victoreen, Inc., "Thick Film Resistors for Kilovolt Circuits," *Machine Design*, December 8, 1977.

8 Allen Bradley, Inc., "Military Numbering System, Fixed Resistors," Publication EC50, December 1976.

9 Herbert Levy, Ohmite Manufacturing Co., "Wirewound Power Resistors," *Electronics World*, April 1965.

10 Ohmite Manufacturing Co., "Engineering Reference Handbook," Bulletin 1100B, 1977.

11 Roy Kampmeyer, "Wirewound Precision and Power Resistors," *Electronic Design*, April 16, 1978.

12 Dale Electronics, "Standard Products Catalog," July 1, 1977.

13 Stackpole Electronic Components Division, "Fixed Composition Resistors," Bulletin 80-102.

14 Warren Yates, "Resistor Networks," *Electronic Products*, September 1977.

15 Robert C. Drew, "Precision Resistors/Networks," *Electronic Products*, May 1969.

16 Micro Networks Corp., "Designer's Guide to Thin Film Resistor Networks," January 1978.

17 Donald J. Levinthal, "Resistor Networks: A Packaging Design Decision," *Electronic Packaging and Production*, July 1976.

18 W. J. Evanzia, "Variable Resistors: Often Misused in Applications," *Electronic Products*, July 1975.

19 Davy Bursky, "Focus on Pots and Trimmers," *Electronic Design*, August 16, 1975.

20 Centralab Electronics Division, Globe-Union Inc., "Series 300 $\frac{5}{8}$″ 16mm Potentiometers," Form EO42PTO, 1978.

21 Beckman Instruments, Inc. "Cermet Trimming Potentiometers," Technical Bulletin, April 1974.

22 Carl Todd, Wilbur Hardison, and William Galvan, Bourns, Inc., "How to Kill a Potentiometer—It's Much Easier than You Might Think," *EDN*, May 20, 1976.

23 Clarostat Manufacturing Co., "Resistive Components for Industry," Catalog 9-77, September 1977.

24 George Boyd, Amphenol Connector Division, "A User's Guide to Trimmer Selection," *EE/Systems Engineering Today*.

25 Donald Epand and Ken Liddane, Siemens Corp., "Selecting Capacitors Properly," *Electronic Design*, June 21, 1977.

26 MIL-STD-198, "Capacitors, Selection and Use Of," Revision D, November 8, 1976.

27 Johanson Dielectrics, Inc., "Understanding Chip Capacitors."

28 Jack Hickey, "The World of Capacitors," *Electronic Component News*, October 1971.

29 Hy Natkin, "Capacitors: Versatile Job Performers," *Electronic Component News*.

30 Aerovox Corp., "Product Specifier: Electrical/Electronic Components," 1971.

31 Frank William Wood, Jr., Advanced Designs Inc., "Fixed Capacitors: Styles-Shapes-Mountings," *Machine Design*, June 22, 1961.

32 Illinois Tool Works, Inc., "Emcon Ceramic Capacitor Catalog," February 1974.

33 Edward L. Bowling and George S. Papadopoulos, Cornell-Dubilier Electric Corp., "Electrolytic Capacitors: Aluminum vs Tantalum," *Electronic Products*, October, 1975.

34 Mepco/Electra, Inc., "Capacitors: Aluminum . . . Tantalum . . . Film. . . ."

35 Hy Natkin, "Proper Capacitor Selection Avoids Aftergrief," *Electronic Component News*, July 1977.

36 Union Carbide Corp., "Kemet Film Capacitors," F2871, June 1977.

37 Ted Vonkampen and Paul Reeves, TRW Inc., "How to Improve the Efficiency of Your Switching Power Supply Design," 1976.

38 Gulton Industries, Inc., "Ceramic Capacitors," Bulletin H4D.

39 American Technical Ceramics, "UHF/Microwave Porcelain Chip Capacitors," 1977.

40 "Chip Capacitors Ease Circuit Packaging Problems," *Electronic Packaging and Production*, July 1976.

41 Erwin Vodovoz, "High Potency Chip Capacitors," *Electronic Products*, March 1976.

42 American Components Inc., "Monolithic Ceramic Capacitors," Bulletin 712, April 1975.

43 Mepco/Electra, Inc., "Design Data for Selecting and Applying Variable Capacitors."

44 Warren Yates, "Selecting Power Transformers," *Electronic Products*, September 1977.

45 Staco Inc., "Custom Transformers: Technical Products Guide," 1976.

46 Basler Electric, "Magnetic Components," A-1, January 1974.

47 TRW UTC Transformers, "Power Transformers and Inductors."

48 Jules H. Gilder, "Focus on Pulse Transformers," *Electronic Design*, June 21, 1975.

49 Cambridge Thermionic Corp., "Inductive Components," Catalog 501, 1973.

50 W. R. Courtney, J. W. Miller Co., "Guide to Better Coil Selection/RF Chokes and Coils."

51 Walter V. Manka, Delevan Division, American Precision Industries, "Design Power Inductors Step by Step," *Electronic Design*, December 20, 1977.

52 Magnetics, Inc., "Permalloy Powder Cores," Catalog PC-303R.

53 Arnold Engineering Co., "Molybdenum Permalloy Powder Cores," Bulletin PC-104C.

54 Ferroxcube Division, Amperex Electronic Corp., "Linear Ferrite Materials and Components."

55 Michael I. Distefano, Indiana General Electronic Products, "Adjust Ferrite-Core Constants," *Electronic Design*, November 22, 1977.

56 Magnetics Inc., "Ferrite Cores," Catalog 208S.

57 Warren A. Martin, Fair-Rite Products Corp., "Materials and Design Factors for Transformers Used in Switching Type Power Supplies," *Insulation/Circuits*, March 1978.

58 Warren A. Martin, Fair-Rite Products Corp., "Bead, Balun and Broadband Kit."

59 Fred U. Wetzler, "Power Supplies for R and D," *Research/Development*, August 1977.

60 "Modular Power Supplies," *Electronic Products*, September 1968.

61 Warren Yates, "Selecting an Electronic Power Supply," *Design News*, August 22, 1977.

62 Semiconductor Circuits, Inc., "Engineering Guide: AC/DC and DC/DC Power Sources."

63 Paul Birman, Kepco, Inc., "Key Factors in Buying Power Supplies," *Electronic Purchasing*, February 21, 1975.

64 ACDC Electronics Division, Emerson Electric Co., "Power Supply Selection Guide," *Electronic Component News*, 1974.

65 AMP Capitron Division, "Power Supply Specifying Guide," 1977.

66 William Miller, Semiconductor Circuits, Inc., "Mini-Glossary on DC Power Supplies," *Electronic Products*, February 1978.

67 Warren Yates, "Batteries: Cordless Power," *Electronic Products*, June 1978.

68 Morris Grossman, "Focus on Batteries," *Electronic Design*, June 7, 1978.

69 James Gould and Patrick Gormally, General Electric Co., "Power Up with Batteries," *Design News*, March 21, 1977.

70 "Despite Their Vulnerability, Batteries are Still Reaching New Market Highs," *EDN* April 5, 1977.

71 David J. Bak, "Long-Life Battery Provides Onboard Protection of Volatile Memories," *Design News*, April 3, 1978.

72 Union Carbide Corp., "Nickel–Cadmium Rechargeable Batteries," Data Sheet RD-1088.

73 Robert F. Wentink and Irwin Weinstock, General Electric Co., "Understanding Sealed Lead Acids Lets You Apply Them Effectively," *EDN* March 20, 1978.

74 Globe Battery Division, Globe-Union Inc., "Gel/Cell Rechargeable Batteries and Chargers," 41-2635.

75 Analog Devices, Inc., "Data Acquisition and Products Catalog," 1978.

76 Burr-Brown, Inc., "1979 General Catalog," 1979.

77 Hy Natikin, "Operational Amplifiers Proliferate," *Electronic Component News.*

78 Don Jones, Harris Semiconductor Products Division, Harris Corp., "Digital-to-Analog Converter Applications," Application Note 511.

79 "Digital-to-Analog Converter," *Design News Electrical/Electronics Design Guide*, 1980.

80 Donald B. Bruck, Hybrid Systems Corp., "Data Conversion Handbook," 1974.

81 Micro Networks Corp., "Data Conversion Products," 1977.

82 John Seaman, "New Life in an Old Work Horse," *Electronic Products*, April 1976.

83 RCL Electronics Division, AMF Inc., "Delay Lines–Lumped Constant, Distributed Constant and Variable."

84 CTS Knight, Inc., "Crystal Filters–Technical Guide Book," March 1966 (revised).

85 Electronic Crystals Corp., "Quartz Crystals."

86 Vectron Laboratories, Inc., "How to Specify Crystal Controlled Oscillators."

87 Rockland Systems Corp., "The Application of Filters to Analog and Digital Signal Processing," February 1978.

88 Electronic Precision Components Corp., "Active and Passive Filters."

89 Dennis Van Dalsen, Precision Monolithics Inc., "Need an Active Filter? Try these Design Aids," *EDN*, November 5, 1978.

90 Kenyon Magnetics, Inc., "Electrical Wave Band Filters," Catalog F-627.

91 Centralab Electronics, Globe-Union Inc., "Ceramic EMI/ RFI Filters," Catalog U-001-F1-0, October 1974.

92 Centre Engineering, "Broadband and High Frequency Low Pass Ceramic Filters," September 1979.

93 Walter Bein, Kenyon Magnetics, Inc., "How to Economize when Specifying Filters," *Electronic Products.*

94 TRW UTC Transformers, "Electric Wave Filters," Catalog 791, 1979.

95 Electronic Devices Division, Rockwell International, "New Mechanical Minifilter," Bulletin MMF-3001, April 1968.

96 Telonic/Berkley, Inc., "Rf and Microwave Filters," 1978.

97 Aladdin Electronics Co., "Compendium."

98 Werner Neuman, Corcom, Inc., "Filters Quiet Noisy Power Lines," *Electronic Products*, November 18, 1974.

99 Gary Elinoff, "EMI/RFI Filters–One Answer to Electrical Pollution," *Electronic Products*, October 1978.

100 Corcom, Inc., "RFI Power Line Filters," Catalog No. 763.

101 RCA Solid State Division, "Solid-State Devices Manual," SC-16, March 1975.

102 RCA Solid State Division, "High Reliability Devices," August 1978.

103 "Semiconductors," General Electric Co., Semiconductor Products Department, Auburn, NY, 451.27, 1976.

104 Courtesy of Westinghouse Electric Corporation.

105 "Diodes and Rectifiers," *EDN*, Semiconductor Annual, June 15, 1969.

106 Sukhbir Virk, "The State of Tunnel Diode Technology," *Electronic Products*, November 1969.

107 Courtesy of Motorola Semiconductor Products.

108 "Annual Reference Guide for Power Semiconductors," *Electronic Component News*, January 1975.

109 Hewlett-Packard Components, "Applications of PIN Diodes," Application Note 922.

110 Hewlett Packard Components Division, "Selection and Use of Microwave Diode Switches and Limiters," Application Note AN 932, July 1974.

111 Varo, Inc., Semiconductor Division, "Controlled Avalanche Rectifiers," Application Bulletin VSP 103, August 1967.

112 Varo, Inc., Semiconductor Division, "Fast Recovery Rectifiers," Application Bulletin VSP 102.

113 Silicon Transistor Corporation, "Power Transistors."

114 Courtesy of Motorola Semiconductor Products.

115 Dale Baugher and Jim Miller, "Today's Power Semiconductors Provide Prodigious Performance," *EDN* June 5, 1975.

116 Victor W. Wigotsky, "Electronic Packaging–More Opportunities for Creativity," *Design News*, October 13, 1969.

117 Leo Lehner, "Silicon Power Transistors–Design and Applications," *Electro-Technology*, February 1970.

118 Woodson J. Savage III, "Power Semiconductor User's Manual and Data Book," Westinghouse Electric Corp., DB 54-000, Table 1.1, Page 13, January 1978.

119 American Microsystems, Inc., "Guide to Standard MOS Products," May 1975.

120 Microwave Associates, "Transistor Designer's Guide," Bulletin 5210, 1978.

121 "Electrical and Electronics Reference Issue," *Machine Design*, May 18, 1978.

122 T. C. McNulty, "A Review of Thyristor Characteristics and Applications," RCA Solid State Division, Application Note AN/4242.

123 John Kessler, "Focus on Photodetectors," *Electronic Design*, March 15, 1976.

124 W. H. Sahm, "Optoelectronics Manual," General Electric Company, Semiconductor Products Department, Auburn, N.Y., 451.135, 1976.

125 "Optoelectronics," *Design News, Electrical/Electronic Reference Edition*, November 1977.

126 Monsanto Electronics Division, "Discrete LED Selecting Made Easier," Application Note AN301.

127 RCA Solid State Division, "Solid State Silicon Photodiodes," OPT-112C, May 1977.

128 Warren W. Yates, "Thermistors," *Design News*, November 22, 1976.

129 Fenwal Electronics Division, Walter Kidde and Co., "Thermistor Manual," 20H-5-877-MBP, August 1977.

130 Don Jones, "CMOS Analog Multiplexers and Switches," Harris Semiconductor Corp., Application Note 520.

Index